POSTCARDS FROM THE WESTERN FRONT

Human Dimensions in Foreign Policy, Military Studies, and Security Studies

Series editors: Stéphanie A.H. Bélanger, Pierre Jolicoeur, and Stéfanie von Hlatky

Books in this series illuminate thorny issues in national and international security, analyzing both military and foreign policy. They highlight the human dimensions of war, such as the health and well-being of military members, the factors that influence military cooperation and operational effectiveness, civil-military relations and decisions regarding the use of force, and the challenges of violence and terrorism, as well as human security and conflict resolution. Some authors focus on the ethical, moral, and legal ramifications of ongoing conflicts and wars, while others, through the lens of policy analysis, explore the impact of military and political strife on human rights and the role the public plays in shaping international policy.

Published in collaboration with Queen's University and the Royal Military College of Canada, with the Centre for International and Defence Policy, the Canadian Institute for Military and Veteran Health Research, and the Centre for Security, Armed Forces, and Society, the series plays a pivotal role in reconceptualizing contemporary security challenges – both in the academic realm and for broader publics.

15 Women, Peace, and Security
   Feminist Perspectives on International Security
   Edited by Caroline Leprince and Cassandra Steer

16 The Ones We Let Down
   Toxic Leadership Culture and Gender Integration in the Canadian Forces
   Charlotte Duval-Lantoine

17 Postcards from the Western Front
   Pilgrims, Veterans, and Tourists after the Great War
   Mark Connelly

Postcards from the Western Front

Pilgrims, Veterans, and Tourists after the Great War

MARK CONNELLY

McGill-Queen's University Press
Montreal & Kingston • London • Chicago

© McGill-Queen's University Press 2022

ISBN 978-0-2280-1189-7 (cloth)
ISBN 978-0-2280-1190-3 (paper)
ISBN 978-0-2280-1264-1 (ePDF)
ISBN 978-0-2280-1265-8 (ePUB)

Legal deposit third quarter 2022
Bibliothèque nationale du Québec

Printed in Canada on acid-free paper that is 100% ancient forest free (100% post-consumer recycled), processed chlorine free

---

Library and Archives Canada Cataloguing in Publication

Title: Postcards from the Western Front : pilgrims, veterans, and tourists after the Great War / Mark Connelly.
Names: Connelly, Mark, author.
Series: Human dimensions in foreign policy, military studies, and security studies ; 17.
Description: Series statement: Human dimensions in foreign policy, military studies, and security studies ; 17 | Includes bibliographical references and index.
Identifiers: Canadiana (print) 20220280517 | Canadiana (ebook) 20220280584 | ISBN 9780228011903 (paper) | ISBN 9780228011897 (cloth) | ISBN 9780228012641 (ePDF) | ISBN 9780228012658 (ePUB)
Subjects: LCSH: World War, 1914–1918—Battlefields—Social aspects—Great Britain. | LCSH: World War, 1914–1918—Battlefields—France. | LCSH: World War, 1914–1918—Battlefields—Belgium. | LCSH: World War, 1914–1918—Veterans—Travel. | LCSH: Tourism—France—History—20th century. | LCSH: Tourism—Belgium—History—20th century. | LCSH: Memorialization—Great Britain—History—20th century. | LCSH: War and society—Great Britain—History—20th century. | LCSH: Pilgrims and pilgrimages—France—History—20th century. | LCSH: Pilgrims and pilgrimages—Belgium—History—20th century.
Classification: LCC D524.7.G7 C66 2022 | DDC 940.3/41—dc23

This book was typeset in 10.5/13 New Baskerville ITC Pro.
Copy-editing and composition by T&T Productions Ltd, London.

Guide me, O thou great redeemer,
Pilgrim through this barren land;
I am weak, but thou art mighty,
Hold me with thy powerful hand.
> 'Guide Me O Thou Great Redeemer (Cwm Rhondda)'

Still stands His cross from that dread hour to this,
Like some bright star above the dark abyss;
Still, through the veil, the Victor's pitying eyes
Look down to bless our lesser Calvaries.
> Sir John Stanhope Arkwright, 'O Valiant Hearts' (1919)

In a sense our pilgrimage was over. But mine was not over. I was beginning to realise more and more how much the past was with me; is with me. Maybe it's true about a piece of me being in France and Flanders, in the earth or in the winds, part of me a ghost along with other ghosts walking Cambrin, High Wood, Passchendaele, talking with Raymond, Cedric, Neil, little Bert, Fletcher and the rest, looking at the world, pondering how it will go. I didn't feel ready to go back to England …
> James Lansdale Hodson, *Return to the Wood* (1955)

Contents

Figures ix

Preface xi

Acknowledgements xv

Introduction 3

1 Fragments from France and Belgium: Visiting the Battlefields, 1914–18 18

2 Postcards from the Hotel 37

3 Postcards from the Road 71

4 Postcard Scenes: Devastation 107

5 Postcards from Veterans 143

6 Postcards from Pilgrims 178

7 Postcards from Tourists 223

8 Postcards from Ypres (and Its Salient) 263

9 Postcards from Arras 290

10 Postcards from Thiepval 318

11 Postcards from Behind the Lines: Armentières, Bailleul, Béthune, Poperinghe … and Around 351

L'Envoi 374

Notes 383

Bibliography 421

Index 445

Figures

Unless otherwise attributed, all images are from the author's own collection.

0.1 Map showing the region of the former Western Front (taken from a 1936 battlefield guide). xviii

1.1 Labour delegation at Albert, February 1918. 24

2.1 The Excelsior Hotel, Ypres, c. 1920. 67

3.1 Plank road near Ypres, summer 1919. Miss Butcher, 'Curly' Allen (note the shell-case souvenir), and Miss Ellis. Image taken by Ernest Turner (of the Canadian Expeditionary Force), August 1919. (Image courtesy of the Turner family.) 79

3.2 Menin road at Hooghe, c. 1921. 81

4.1 Temporary houses, Lens, 1920. 111

4.2 Temporary buildings, Bapaume, c. 1920. 134

5.1 Veterans and pilgrims at LRB Cemetery, Ploegsteert, 1927. 157

5.2 Veterans at Menin Gate, c. 1929. 167

6.1 St Barnabas pilgrim at a grave in Lijssenthoek Cemetery, 1924. 189

6.2   St Barnabas pilgrim at a cemetery awaiting its permanent features, c. 1920. 193

7.1   Tourist excesses, Thiepval, c. 1920. 232

7.2   Servicing the needs of visitors: The Café de la Grande Mine, La Boisselle, c. 1920. 251

7.3   Playing at soldiers. Hill 60, August 1935. (Image from the album of Ivy Winifred Bradshaw née Woolston, courtesy of her great niece, Jennifer Bostock.) 258

8.1   The Cloth Hall, Ypres, 1923. 268

8.2   The British Tavern, Grand'Place, Ypres, c. 1920. 269

8.3   The Menin Gate unveiling ceremony, 24 July 1927. 280

9.1   Arras station, 1919. 293

9.2   Preserved trenches at Vimy, c. 1926. 311

10.1 William McMaster, custodian of the Ulster Tower, c. 1925. 334

10.2 The Heroine of the Ruins, c. 1920. 339

11.1 Skindles Hotel, Poperinghe, c. 1920. 353

11.2 Temporary railway platform, Bailleul, c. 1920. 367

Preface

Bob: Have you got a job yet?
Frank: Yes – I had a bit of luck – a chap called Baxter in my regiment, he was drafted out to Arras in February nineteen-seventeen and before the war he was running a sort of travel agency in Oxford Street – well, he got a Blighty one and was invalided 'ome, and believe it or not, 'e was the first one I run into when I got back last April. He'd started his business again, and things were beginning to pick up, so he gave me a job.
Bob: Travel Agency – whew!
Frank: Tours of the battlefields, I'll thank you!
Bob (laughing): That's a good one.
Frank: Some people certainly do have queer ways of enjoying themselves.

            Noël Coward, *This Happy Breed* (1939)

This history is driven by the words of British visitors to the battlefields and the visual images they took and collected. Some of those words were recorded in personal, private testimony; others were produced corporately for public consumption. Pain and anguish, pride and nostalgia, wonder and surprise were all expressed. Encountering the former fighting zones, especially in the first few years after the war before widespread and large-scale reconstruction had transformed the landscape, caused shock, amazement, awe, depression, and inspiration. None left the battlefields behind feeling indifferent about the experience, and regardless of their motivation for going, none could escape the fact that it was regarded as sacred ground, to be traversed with care and attention to its special qualities.

 The bereaved came to see the graves of lost loved ones, or their names inscribed on the memorials to the missing. Some, unwilling to accept that a loved one was, indeed, missing, came out to search for the grave hoping they would stumble upon it somewhere, or perhaps uncover an error or omission in the recording of graves.

When that hope faded and the memorial to the missing was accepted as the final place of commemoration, often there was the desire to visit precisely where the person was last seen, the place where they became one of the missing. As pilgrim testimony reveals, arrival at this place was regarded as a journey's end. It was the moment of release after a progression fraught with emotional, as well as physical, difficulty.

On the battlefields these pilgrims mingled with those who had fought over the ground, lived in the trenches, and spent their spare time in the towns behind the lines. Although Noël Coward's beautifully crafted ex-servicemen, Bob Mitchell and Frank Gibbons, from his play *This Happy Breed* may have deemed tours of the battlefields a 'queer' form of entertainment, veterans were among the most committed and enthusiastic of battlefield visitors. Tramping over the old ground allowed them to indulge in the worst of times; slipping into well-remembered estaminets allowed them to indulge, once again, in the best of times. Fallen comrades could be saluted in the cemeteries and toasted over beers in the bar. Far from burying their old identities, veterans seemed desperate to disinter them once back in Flanders Fields.

As sites of great drama and endeavour made famous through constant repetition in newspapers and newsreels, the battlefields were also a land of interest, intrigue, and curiosity. Tourists flocked to see the wonders of devastation and destruction: the trenches, remnants of trenches, rusting tanks, and miles of barbed wire being painstakingly gathered in by locals and military salvage units. But whether people should have taken a near ghoulish delight in exploring such places was an issue fiercely debated at the time. Precisely why someone was visiting the old battlefields, what they sought out, and how they behaved were matters of public interest.

A major consequence of this influx of visitors, regardless of their precise motivation, was the emergence of a whole new industry. Maps, guidebooks, reflections on the battlefields, and postcards were produced and published; a souvenir industry grew up allowing people to buy shell cases, shrapnel, helmets, bayonets and even deactivated grenades, pistols, and rifles. This remarkable range of specialist materials was sold in the hotels, bars, restaurants, and museums that rapidly grew up along the old front lines. It allowed all, regardless of their status – bereaved, veteran or tourist – to collect mementoes of their visit if they so wished. Beer, tea, chips, souvenirs, pride, grief, tears, and, for the adventurous, very muddy shoes were all part of the battlefield visiting experience between 1919 and 1939.

Despite reconstruction, which included the flourishing hospitality infrastructure inspired by the number of visitors, the return of the civilian communities, and the revival of the normal patterns of life, the British believed they stepped into a world within a world when they reached towns and cities such as Albert, Arras, and Ypres. They trod holy ground; they imagined what it had once been like; they felt the presence of ghostly armies. It was an experience few forgot or ever wished to forget.

The structure of this book is based on interpreting visions, sights, sites, and the thoughts and emotions they inspired. It is an exploration of what people saw, what people thought they saw, what people did, and what people thought they did. As such, it has three lines of sight, three positions.

The first position is primarily about the main backdrop, the landscape scenery framing the specific. It is about practicalities, how people got to the battlefields, how they got around, how they navigated a devastated world, where they stayed, what they ate, the postcards they bought, sent, and collected. During the course of my research, the reality and idea of the postcard became more and more apparent. Postcard images and snapshots of places appeared again and again. Whether pasted into notebooks and albums or bundled-up with papers, the photographic image loomed large. In the written record, people tried to describe what they had seen almost as if trying to provide a postcard image in words. At the same time, the profusion of scraps of information recorded in all sorts of different ways added to the sense of sorting through a vast archive of postcards. When combined, these seemingly mundane, quotidian aspects of life revealed that they were, in fact, crucial to the experience. This investigation of minutiae, the examination of the way people experienced the battlefields, has never been studied in such detail.

The second position privileges particular kinds of visitors: pilgrims, tourists, veterans. It examines not only what judgements and conclusions they reached on what they saw and did, but also how they perceived themselves, and others, and how others perceived them. Unlike other studies, this approach privileges and emphasizes the responses of the individual. Visitors are seen as influenced by the forces around them, whether they be the landscape, their guidebooks, their party leaders, or their fellows, but, ultimately, they engaged in their own way and on their own terms. Agency and autonomy are returned. This approach was achieved by accessing a greater range and depth of material than any previous contributor to this field (to the best of my knowledge).

The third position is dominated by specific places: some famous today, others now less well known or off the modern visitor map altogether. And this marks another departure from other studies where such spaces have been considered. Here it is argued that the precise space, the precise landscape was crucial to perceptions. The very individuality of the spaces interacted with the individuality of the visitor, creating a mutually reflective relationship. The progression towards this specificity is the obvious conclusion to a work about the meaning of place. In the interweaving of people, place, space, activity, thoughts and emotions, personal and public discourse, this work, inspired by the studies of so many scholars and using a host of material, much of it newly discovered or archived, much of it never before researched, creates a new history of a remarkable twentieth-century phenomenon whose legacy many still encounter today.

Note

Throughout the text, the spelling of place names in West Flanders will be as they were known to the British between 1914 and 1939. Thus, Ieper (as it is known and spelt today) will be Ypres; Wytschaete rather than Wijtschate, Passchendaele not Passendale, etc.

In keeping with the discourse common at the time, Chapter 10 ('Postcards from Thiepval') uses the term Ulster as synonymous with Northern Ireland. However, the ancient province of Ulster consists of nine counties, only six of which were included in Northern Ireland following the partition of the island of Ireland.

On footnoting: to save the reader from having to check multiple reference markers, where one source dominates the evidence used in any particular paragraph, the material has been cited in one note flagged at the end of the paragraph, rather than at the conclusion of each sentence or quotation.

Acknowledgements

This project was completed thanks to the help and encouragement provided by many people and institutions. My wonderful colleagues in the 'Gateways to the First World War' centre, which is funded by the Arts and Humanities Research Council (AHRC) have been inspirational and supportive throughout. Brad Beaven, Helen Brooks, Will Butler, Alison Fell, and Emma Hanna are owed particular thanks for accompanying me on a fourteen-mile walk across the Ypres battlefields on a freezing January afternoon. Enormous thanks are also due to Lucy Noakes and Sam Carroll. Particular thanks go to Zoë Denness, Qian Lu, and Kirsty Corrigan for helping to arrange the many research trips and archival visits required to complete the research. Sophie De Schaepdrijver and Tim Bowman were selflessly generous with their expertise and time, as was Stefan Goebel, who read the manuscript and made many valuable observations. For advice on specific points, I am grateful to Dominiek Dendooven and Serge Durflinger. Thanks are also due to Mario Draper for kindly including me in a massive open online course (MOOC) he developed for the School of History at the University of Kent, which provided another opportunity to walk the battlefields and think through their commemorative frameworks. I also owe a huge debt to my wonderful PhD students. The themes of this book were developed in discussions with Natasha Silk, Oli Parken, Amy Harrison, Megan Kelleher, Jennifer Turner, and Tim Godden, who provided so many fascinating reflections on grief, commemoration, beliefs and superstitions, landscapes, and the details of Imperial War Graves Commission (IWGC) design and architecture.

I would also like to thank the wonderful team of volunteers who joined me in two projects funded by Gateways to the First World War to explore newspaper coverage of battlefield tourism: Peter Alhadeff, Mark Allen, Hazel Basford, James and Susan Brazier, Steve Dale, Charles Davis (Australia), Malcolm Doolin, Valerie Ellis, Simon Gregor, David Hearn, Jan and Richard Johnson, Gill and Roger Joye, Pat O'Brien, Stephen Miles, David McGregor, Rod Saunders, Julie Sears, Jonathan Vernon, and Eve Wilson. Working together, we produced three booklets: 'Visiting and Revisiting the Western Front 1919–29: The Experience of Australians' and 'Visiting and Revisiting the Western Front, 1919–1939' (which explore the different Australian and British experiences, respectively), and 'Zeebrugge: the Making of a Legend'. These can be accessed via the project website (www.gatewaysfww.org.uk).

In addition, I would like to record my appreciation to the staff of the archives, libraries, and museums I visited in the course of research: Piet Chielens and his colleagues at the In Flanders Fields Museum, and the regular attenders of our joint seminar programme, especially James and Sue Brazier. At the Commonwealth War Graves Commission, Gareth Hardware, David Richardson, Sanna Joutsijoki, Geert Bekaert, Glyn Prysor, George Hay, Andrew Fetherston, and the archives team were all immensely helpful. Debbie Manhaeve at the Passchendaele Memorial Museum proved to be an enthusiastic supporter of the project throughout. Derek Gallagher and Sir Edward Crofton of the Friends of St George's Memorial Church, Ieper, also provided great encouragement, as did the Western Front Association team, especially Colin Wagstaff, David Tattersfield, and Simon Phillips. In Ieper, I would like to thank Ruth and the team at the Albion Hotel, who were such warm hosts on my innumerable visits to the city. I am also indebted to George Godden and Mary Setchfield, the children of two IWGC gardeners, for sharing their memories and photographs of Ypres.

Richard Baggaley at McGill-Queen's University Press provided expert guidance, and I am extremely grateful for his enthusiasm for the project from the start, and for the valuable comments supplied by the early readers of the complete draft.

Every effort has been made to secure the necessary permissions to reproduce copyrighted material in this book, though in some cases it has proved impossible to trace holders. If any omissions are brought to my attention, I shall be happy to include appropriate acknowledgements on reprinting.

Finally, my heartfelt thanks go to my wife, Jacqui, and our children, Fabian (who didn't get his name by accident!) and Tilly, for their patience, support, inspiration, and love. They allow me to disappear, often for days at a time, to tramp and cycle the battlefields and allow me to live with the battlefields, an obsession and addiction that never lessens.

This book is dedicated to the memory of my father, Brian Stanley Connelly, London cab driver, lover of London, lover of history. Thanks, Dad.

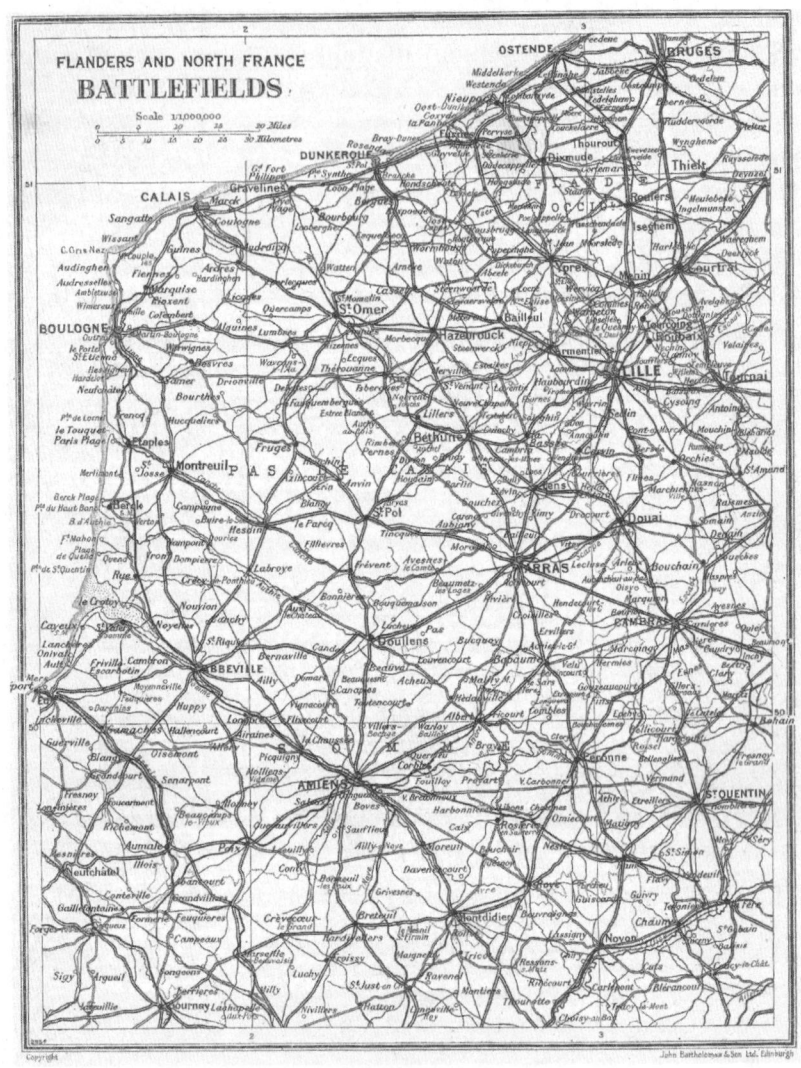

Figure 0.1 Map showing the region of the former Western Front (taken from a 1936 battlefield guide).

POSTCARDS FROM THE WESTERN FRONT

Introduction

HISTORIOGRAPHY, METHODS, IDEAS, AND SOURCES

The great privilege of being a historian engaging in a project such as this is the enormously interesting breadth of materials available. Of course, this is a piece of history, using the methods of the historian, but the historian acts in an incredibly eclectic manner, drawing upon a range of different disciplines. By doing this, it has been possible to apply an equally broad, and fascinating, range of approaches and ways of thinking about the sources. But, being a work of history, it also has to start by acknowledging a huge debt to David Lloyd's 1998 work *Battlefield Tourism: Pilgrimage and the Commemoration of the Great War in Britain, Australia and Canada, 1919–1939*, which established the benchmark for such studies.[1] Lloyd dealt with the perceived differences between tourist and pilgrim and the tensions these created, the particular approaches of veterans, the needs of the bereaved to see the precise place where their loved one was commemorated, how the Unknown Warrior created a universal alternative, and how patriotic and imperial values were woven into a culture of remembrance, making commemoration both an individual and a communal process. Bruce Scates has also contributed an extremely valuable study in his examination of Australian responses to the battlefields, *Return to Gallipoli: Walking the Battlefields of the Great War*. Australian experiences were also explored by Bart Ziino in his fascinating study, *A Distant Grief: Australians, War Graves and the Great War*. Taking up many of the themes of Lloyd, Scates, and Ziino, this book focuses even more closely on the importance of precise place and how it was explored and imagined by British visitors between 1919 and 1939. It is also about gender, power, and identity, how people recorded their memories, and the audiences with whom they chose to share their experiences and feelings. Thanks to the huge advantages offered by digitization, it has been possible to explore these themes through a much greater range of material than Lloyd could access when he undertook his study. In particular, it uses local newspaper sources on a grand scale. These sources have proved invaluable, not only for revealing in great detail the massive volume of pilgrimage and battlefield visiting, but also for highlighting the cultural constructs built on and around the way visitors behaved and the things they did. Crucially, it also argues that visitors were not solely vehicles for delivering or emphasizing the objectives of those who led them to France and Belgium. People might have been led or guided to these particular spaces, but they were not tablets onto which the messages of others were inscribed. Visitors, especially those labelled pilgrims, brought their own memories and inscribed their own memorial texts on the landscape.

This book is also about material culture: souvenirs (particularly postcards), photographs, badges, tickets, clothes, maps, booklets, certificates. It is about big bits of material culture: cars, trains, buildings (in the form of hotels, restaurants, and cafes), as well as cemeteries and memorials. All are considered for what they meant to the experience of visiting the battlefields. As Nicholas Saunders's work on the First World War has revealed, new perspectives can be achieved by examination of the conflict's material culture, landscapes, and archaeology. It is a position shared by the In Flanders Fields Museum (Ieper) with its belief that the landscape constitutes the last, and enduring, witness to the conflict. At the same time, as David Lowenthal has put it, 'no physical object or trace is a self-sufficient guide to bygone times'.[2] It is for this reason that the material culture is wrapped up in written documentary evidence, whether handwritten, typed, printed, for public or private use and dissemination, for the commercial or public sector. Because it relies on documents, it also privileges certain viewpoints: officials and civil servants working for a host of organizations, journalists writing for newspapers local and national, people who wanted to make private or public testimony (sometimes both). Often those who were making public statements did so assuming the right to sum up, judge, and interpret the actions of many others. Such records and statements were equally often created on behalf of particular groups who wanted to influence or dominate a debate, mood, or perceptions of the war and its dead. As Jay Winter has remarked, collectives 'try to come as close to the microphone of public discourse as they can, and this *prise de la parole* is part of what defines their collective character'.[3] In this instance, it is a broad range of collectives including charity groups, veterans groups, the Imperial War Graves Commission (IWGC), and tourism companies, as well as individuals who just wanted to tell others what they had experienced.

At the heart of this history are the men and women attempting to make sense of mass death and destruction on the Western Front, some doing so because they had suffered the agony of loss, and others out of sheer curiosity, lured to the landscapes and places they had lived with throughout the conflict but never had the chance to see in person. When trying to express their thoughts and feelings and describe the sights, many visitors faced precisely the same problem as wartime combatants: how was it possible to state in mere words precisely what they had seen, precisely how they felt? The crisis of written communication identified by Paul Fussell, Samuel Hynes, and many others can be detected in the post-war testimonies of battlefield visitors. It can be felt most strongly in the comments on the devastated landscape. After surveying the wasteland, people were left at a loss as to how to convey the scene. By the same token, by the end of their tours, most visitors were keen to state how far they had been enriched by the experience (intriguingly, this could, perhaps, be seen as a prefiguring of Adorno's paradox).

The acquisition of insight and understanding through personal experience allowed many to reach the same conclusions as the middlebrow writers on the war: the conflict had been appalling and had caused waste and desolation, but, they argued, it had been worth it, provided humanity learned the correct lesson. It was a lesson they had to believe, for the alternative – that it was a worthless, futile, exercise – was too awful to contemplate.

Writing was used to transfer and communicate what the eye had seen. Seeing, smelling, and touching the places where men had fought and died inspired people to create a written record, which so many wished to do even when consciously noting the difficulty involved in making that record. However, accompanying the words was the visual, in particular the postcard or the snapshot from a cheap camera. As the work of Roland Barthes, as well as many others, on photography has revealed, it is an extremely complex medium. A seeming slice of reality, the capturing of a precise moment in time and space, a photograph is a bundle of contradictions and overlapping intentions and imperatives. The photographer, the commissioner, the purchaser, or the user all apply their own meanings to the image. The photograph then has its own life, acting on others in different ways in different times and places. In the form of a postcard, the purchaser–recipient relationship comes into play, as does the nature of any message inscribed. When it came to battlefield visiting, the postcard was regarded as indexical: it was the empirical record of a historical and present reality. At the same time, it was also commemorative and mnemonic: it was the permanent evidence of that reality and a symbol invested with spiritual and emotional power. Visitors to the Western Front were obsessive purchasers of postcards and photographers of the devastated zones, which were collected, mounted in albums, or kept with other souvenirs of the visit. Seeing was believing. It allowed the unbelievable to be comprehended. Interestingly, most visitors appear to have purchased postcards to keep as souvenirs rather than to send home; they were lovingly curated in collections or carefully slipped into books at the relevant page. The postcard was a vital visual reminder to the purchaser, rather than something given to others.

Among those undertaking the journey, and leaving quantities of testimony, were veterans. There has been much research into the position of Great War veterans, but relatively little focuses on their role as battlefield visitors. Much of the historiography has focused on veterans as political activists, whether in campaigns for better pension entitlements or on a grander scale as radical elements helping to disturb the wider political scenario. This study looks at an aspect of veteran culture that has been largely overlooked: their attitudes towards the precise ground over which they fought and the places they were billeted on their return to those sites. Moreover, battlefield visiting by veterans often saw them engage intensively in the precise details of their actions and then place those experiences in the broader operational framework in which they had participated. While engaged in these activities, veterans seemed to do so without any overt reference to the growing controversies over the way the war was fought. What Brian Bond labelled the post-war 'battle of the memoirs' is barely detectable in veteran battlefield visiting. In looking at the relationship between veterans and the battlefield, the complex ways in which they understood their war experience is revealed. Drawing on a range of writings, John Pegum has provided the most extensive investigation, arguing persuasively that the restoration of the battlefields left veterans feeling increasingly dislocated as they lost their ability to orient themselves. The intensity of their emotions and recollections was intimately linked to a sense of place, and as that place disappeared, they felt robbed of the ability to connect with their memories. Such was the depth of their disorientation that veterans almost envied the dead for remaining in possession of the landscape through their graves in the war cemeteries. The evidence explored in this study reinforces Pegum's conclusion, but also challenges it. Veterans undoubtedly felt something akin to betrayal at what they found when visiting the restored world in Belgium and France. However, the evidence also shows that they found it very easy to reconnect once they penetrated the surface. As will be shown, they moaned and rejoiced in almost equal measure. There were always enough mnemonics across the battlefields, not least in the cemeteries themselves, to point them back to their old selves. Whether this experience represented a facing down of sites of trauma and, through that, catharsis, is less certain. By contrast, their enthusiasm to return and their actions once back on old ground often reflected a sense of personal and corporate pride resulting in an outpouring of nostalgia. Labelling veterans as men and women trapped by their wartime experiences needs to factor in agency: if these men and women were ensnared by their past, unable to escape it, then battlefield visiting suggests it may well have been a voluntary position.

Men other than veterans visited the battlefields, usually as fathers or brothers of those who served. Research into the home front male is relatively scarce, and the male as home front father is rarely the focus of studies. Historians of emotions have touched upon the subject, but this is often through the prism of the relationship between a father and an adult male child on the fighting front during the war. This study picks up the baton at the Armistice and takes it into the post-war world, showing what the home front male made of the battlefields and the war when contemplating them in retrospect. However, it is a study far more biased towards explorations of female reactions. There is now a rich historiography of motherhood and family life in the conflict, and female relationships with relatives at the front have been examined in many excellent works. There is also much literature on the role of woman as mourner, but very little relates to women as battlefield visitors seeing the sites of agony and endeavour for themselves. Important to this study is the work of Janet S.K. Watson who explored the class differences inherent in perceiving engagement in the war as either a form of duty or one of work. As she also highlights, war writing in the twenties and thirties increasingly came to define the experience in negative terms, making disillusionment and disgust the dominant tone within literary discourse. The women who went on battlefield pilgrimages do not conform to this model; rather, they continued the wartime mode of duty. Dutifully they went to their loved ones' graves; dutifully they carried out the rituals of commemoration before returning home having fulfilled a duty that was not so much a burden as a noble, and highly necessary, cause. As they often found, it could be a pleasure, too. Alison Fell has pioneered new approaches here in her recent exploration of the female relationship with the battlefields in her focus on women veterans, and their particular sense of what constituted important, sacred space in France and Belgium. As she has shown, female veterans represent as complex a range of opinions and reflections on their wartime service as their male counterparts, which, once again, makes it difficult to use blanket terms such as traumatized or disillusioned.

For the bereaved, a battlefield visit was central to the process of coming to terms with loss, facilitating movement through the phases of grief. Elizabeth Kühler-Ross identified five stages of grief: denial, anger, bargaining, depression, and acceptance. As she revealed, they were not necessarily distinct from each other and there was much overlap, but the different elements were perceptible. The bereaved moved towards the acceptance stage by regaining agency and personal control through visiting the sites of loss and commemoration. For the pilgrim visiting the cemetery, the physical, material object and the performance of rites in the spaces created and curated by the IWGC acted as a surrogate funeral, and thus a crucial part of the grieving process was finally completed. The archaeologist-anthropologist Michael Rowlands has stated that, within the bereavement process, the funeral is crucial:

> Death, as a threat to the continuity of this theoretically static world, must be negated and, as one might expect, the funeral is the principal means by which this negation occurs. Discontinuity itself is denied through emphasizing continuity in a life-regenerating process which often requires a separation of the time-bound, polluting aspects of primary rites [the disposal of the body] from the regenerative aspects of the secondary rites on which the reintegration of permanent order depends [reflection on the soul, spirit, and legacy of the person].[4]

The actual built setting of the memorial or cemetery and the focus on the name on the headstone or panel was crucial, for material objects act as memory triggers and transmission of the past as each person interacts with them. Beyond the commemorative marker, whether headstone or name on a panel, the bereaved were also dealing with the idea of the dead body. And the dead body was material culture in its own right. Thomas Laqueur has argued that the corpse has agency: living humans cannot escape the dead body and their presence forces the living to notice them, to engage in a relationship with them. To put it another way, the living exist not only in a continual discourse with death itself but with the specific remnant of human life, the body. As a result, human societies have forged processes and rituals focused on the dead body. And, as Laqueur maintains, the First World War, as a massive exercise in death causing bodies to be buried and lost far away from their families and home communities, forged an obsessive desire to list and categorize the dead. New forms of commemoration sprang up, but the same fundamental issue remained: the special relationship with the dead body, or the need to assuage the dread of its absence.

This history is therefore also an exploration of emotions. Humans do not have a universal, inherent, and 'natural' shared set of emotions. Instead, 'emotions are constructed through practices and ways of communicating'.[5] As Sarah Tarlow has noted, culture defines emotions. But, as she also states, within a culture, emotional responses vary according to age, gender, groupings and associations, and individual psychologies. This study concurs with her conclusion, as the influence of these different positions and states can certainly be seen in the responses of veterans, the bereaved, and tourists to the battlefield sites. Tarlow also contends that within commemorative activity the main 'emotional pivot' was one of 'regret and grief' rather than patriotism or nationalism. As such, this maintained a continuity with nineteenth-century funerary and grieving practices. However, creating a solid distinction between nationalist and patriotic reinforcement on the one side, and grieving processes on the other, ignores the subtle blending that occurred in battlefield visits. When veterans mourned and respectfully remembered their fellows, they also commemorated their military corporate identity and its relationship with God, King, and Country.

For those who defined themselves as pilgrims, or were given that label by others, the concept of being on a pilgrimage, and the emotions and mindset associated with it, ensured they were seen as very different from those off on a holiday or tourist jaunt. By the twentieth century, and particularly driven on by the First World War, the term 'pilgrimage' had emerged from its religious roots and was increasingly being applied to ostensibly secular journeys that were deemed to share similar characteristics. When 'pilgrimage' is used to describe those travelling to see war cemeteries, memorials, and other places of commemoration, the blurring between the religious and secular is obvious. However, as Peter Jan Margry has pointed out, the 'true' religious pilgrimage remains a different phenomenon, despite similarities in external appearances: 'visits to graves, shrines and special places display parallels in rituality, materiality or (religious) vocabulary, ... [but] these say little about their religious meaning'.[6]

Nonetheless, the battlefield and war cemetery visitors often perceived themselves to be pilgrims on a spiritual journey. They conformed with all the indicators and attributes of pilgrimage identified by Ian Reader and Tony Walter: pilgrimage as 'a journey out of the normal parameters of life, the entry into a different, other, world, the search for something new, the multiple motives of participants, ranging from homage and veneration' to the more quotidian and commonplace.[7]

Reader's and Walter's definition of the pilgrim as someone seeking to enhance or affirm their 'existence on one or more levels, that may make him/her whole, more complete ... [and] affirm a sense of cultural identity' by deliberately stepping outside the usual rhythms and patterns of everyday life can also be seen in the motivations, manner, and behaviour of war graves pilgrims. Through this conscious movement into a liminal space, it can also be argued that the pilgrim moved closer to the dead, they found themselves in a world connected with society but not existing in the midst of it for the duration of their pilgrimage. The war graves pilgrimage then became the process whereby the cathartic, alternative funeral was held as the person reached the grave, and by going through this ritual, the way was paved for the return to society.

Having been through the collective, and individual, experience of the pilgrimage, the war graves pilgrim gained a sense of order, community, and group consciousness. Pilgrimage thus created the link between Tarlow's identification of grief as the key driver of commemoration and the sense of group identity expressed in terms of national or local identity, patriotism, and pride. Further, war memorials and cemeteries, the rituals they inspired, and the visitors who performed them were combined in homage to the sacrificial act undertaken by the dead. Through this process identity was increasingly subsumed into a mystical, shared, ancient origin, escaping any taint of the present. At the same time the dead were (and are) made ever present among the living through the memorial. The living visitor reflected on the dead, forging a continual circle of interaction between the living and the dead, with the pilgrimage as the instrument for achieving that communion.

What linked the pilgrim and tourist was the sense of playing a particular role in a particular space. This was something they shared with soldiers' experiences in wartime. Long before they could travel in safety and at their leisure as veterans, soldiers explored as tourists. When given a few hours, or the luxury of a day or so, soldiers examined sections of the battlefield, as well as sites behind the lines and beyond. And, as nearly all were civilians temporarily serving as soldiers, they viewed the world with the eyes of tourists. Tourism studies has identified the processes of tourism and visiting different places as a performative act involving 'the inscription of roles, and the following of appropriate techniques and directions'.[8] This makes sites and landscapes the theatrical space in which a dramatic performance is acted out by the tourist. The tourist-performer arrives physically and mentally clothed for their role, carrying with them their innate interests as well as the cultural influences that have shaped them. Stage-management of behaviour comes from the impact of the site itself as it influences engagement and movement through the space, and as Paul Gough has argued, the memorial sites of the Western Front, such as Newfoundland Memorial Park on the Somme, guide the visitor according to a very definite plan. Walking becomes part of the performance, which is often directed or managed by others: the site has a set pathway, or a guidebook encourages a particular route. At the same time, walking opens up the senses to the wider space, its nature and form. A performative relationship is created between site and individual.

There can be little doubt that the battlefields, cemeteries, and memorials of the former Western Front encouraged a sense of performance in the visitor as they trod an immensely dramatic stage. And treading was the vital mode. Whether the prioritized mode of transport or one used only sparingly, the visitor had to walk, and that created a network of paths, formal and informal, across the landscape linking cemeteries, memorials, and other sites. It is this element, the very nuts and bolts of moving across and through the landscape, that is often overlooked in memory studies of commemorative sites; the impact of human communications networks, roads, railways, paths, and tracks has not been considered significant. Many studies of commemorative sites take what might be called a helicopter approach, in which they hover over one site or memorial and consider its nature and meaning, often contextualizing it within the *immediate* environment, before moving on to another without reflecting on the physical accessing of that place by visitors and how that process may influence interactions with it. And, above all, it was the process of walking that created those networks of memory and commemoration, so important for those regarded as true battlefield pilgrims. Robert Macfarlane notes that paths have an ability to transcend time-based boundaries and retain memory, 'as if time had somehow pleated back on itself, bringing continuous moments into contact, and creating historical correspondences'.[9] Paths also speak across time, forging relationships that either endure or fade away: 'Paths are the habits of a landscape. They are acts of consensual making ... Paths connect. This is their first duty and their chief reason for being. They relate places in a literal sense, and by extension they relate people ... [but] paths are consensual, too, because without common care and common practice they disappear.'[10]

Paths were both part of the spaces of the former battlefields and the links between them, which, in turn, highlights the vital importance of the way space has been conceived, understood, experienced, and imagined. Landscape studies have argued that humans have a complex relationship with the environment around them: humans both shape the environment and absorb it; landscape reflects human intervention and forms it, as echoed in UNESCO's definition of landscape as a combination of natural form and human intervention. Within the landscape, certain spaces have been labelled or given special status by humans, and through that are transformed from space to place. Selena Daly, Maria Salvante, and Vanda Wilcox adopted a human-centric emphasis in *Landscapes of the First World War*:

> [Landscape is] limited in space, and must have some kind of boundaries; we cannot speak of 'the earth's landscape' but rather of many landscapes. Consequently, landscape is an inherently anthropocentric idea, requiring human

interactions, at the very least from a viewer or viewpoint, since it is not delimited by an inherent geographical feature but instead defined by the person (or people) who is observing, describing or representing it. It is human categorisation and interpretation which distinguish one landscape from another.[11]

People live their lives in relation to the landscape, carrying it around in their minds and memories, as well as in their daily interactions. Despite adopting this human-centric vision, Daly, Salvante, and Wilcox are sympathetic to Pamela J. Stewart and Andrew Strathern's view that concepts of place, community, and landscape intersect and overlap. This means the divide between the 'natural' environment and the 'human' becomes blurred. Thus, for environmental historians, understanding humans as part of a natural environment and not as some kind of alien interloper is important. This means grappling with the crucial double helix of how humans inscribe space and how space creates its own rules guiding humans into how to use and define it. In this complex relationship the naming and categorizing of the landscape has a further effect on how place and space are conceived: 'The names of locations within areas record the forms of human experience that have occurred within them.'[12] For battlefield visitors this was of definite interest, for it meant a rich web of British names assigned by the soldiers, often made permanent in cemetery and memorial titles, combined with French and Belgian place names that had been hammered into the private and public spheres by constant repetition in wartime reportage. The real and mythic landscapes existed side by side.

Historical geographers such as Hugh Clout and Dries Claeys have led studies of the restoration process bringing together huge amounts of data in their forensic examinations. The landscape examined in this book is one devastated by the weapons of modern warfare. It was a world humans had made unnatural and then attempted to 're-naturalize' through reconstruction. Despite the relative speed with which it was restored – a development that often amazed British visitors – the landscape never fully lost the marks of war. In the intersection of the cities of the dead built by the IWGC, the restoration of the towns and villages, and the recovery of the landscape can be seen something akin to the concept of 'Thirdspace' developed by the urban theorist and political geographer

Edward Soja. He argued for a place where the real and imagined met in a continual process of narrative creation shaped by the way humans engage with space. Soja's work overlaps with that of Michel de Certeau, who embedded walking into the creation of imagined space. Walking freed the individual from the power of corporate schemes and organizations expressed through media such as maps. Although a person might be influenced by knowledge acquired in a range of ways, walking introduced spontaneity and movement according to personal experience, interest, or need.

In his examination of the South Africa national memorial at Delville Wood on the Somme, Jeremy Foster has posited the idea of the continual, triangular conversation between the 'natural' site, the inscriptions placed upon it by Herbert Baker (the memorial's architect), and the visitor. Therefore, what might be called a site-place survives, not solely because it has been re-encoded and re-reified, at least in part, by successive generations or groups, but because the process involves the visitor bringing their preconceptions to the site and entering into a dialogue with it that is driven by the combination of what the originators intended, the 'primal' effect of the landscape on human senses, and the meanings the visitor ascribes to it. For Foster, this interaction between the primal appeal, the work of the creators, and the visitor is the 'intersubjective' effect. '"The potentialities of meaning" are glimpsed ... as part of the lived, corporeal encounter with the site that sets up a dialectic between the "imagined" and the "physical", the historical and associative.'[13]

The world of the former battlefields was a web of these relationships, with the cemeteries and memorials embedded in a landscape undergoing restoration at its heart. Yi-Fu Tuan argues that the human mind interprets space according to geometric designs and has an innate concept of spatial organization. Developing his theory, Tuan believes place is an object that defines the wider space. Places surround, frame, and make space. Across the devastated, and then restored, former battlefields, the cemeteries acted as perfect expressions of geometric principles, ordering the way the mind and eye saw and perceived space and the landscape. The cemeteries and memorials then acted from the inside out: the visitor perceived the outer, former battlefield space from the 'platform' they provided. The geometric forms conceived by the architects and framed by the horticulturalists engaged with the imaginations and preconceptions of the visitors as they sought to make sense of the landscape.

It was the staff of the IWGC that created the dominant geometric forms, and the history and architectural and horticultural principles of the IWGC have been studied from many angles. Architectural history has made a significant contribution, but it has focused overwhelmingly on the contribution of the influential principal architects (including Edwin Lutyens, who has inspired a considerable bibliography). This study goes beyond that focus on creation of the cemeteries and memorials to examine their effect by exploring the interaction between the designed and curated cemetery space and the visitor, how the two developed a relationship and how it was expressed, and how these spaces were perceived against the wider landscape. A significant question regarding the cemeteries has been the extent to which they built up a cult of the fallen soldier, partly through disguising the true horror of war almost to the point of misleading people about the experience.

Tim Godden's pioneering work[14] has pushed the debate in a new direction by emphasizing the importance of the junior architects in the design of the cemeteries. Identifying their veteran status as crucial, Godden argues that their war experience shaped fundamentally their understanding of space, orientation, and proportion, resulting in designs full of mnemonics, some obvious, others beautifully subtle, relating the cemeteries back to the wartime landscape. His interpretation is highly persuasive when the reactions of veterans to the battlefields are studied. The landscape made sense to the veteran, no matter how 'disfigured' by restoration; when the veteran stood in the cemetery surrounded by comrades, alive and dead, and gazed around, he saw his old world. This ties in with Jill Bennett's argument on the difference between 'ordinary memory', which places trauma in a temporal context and framework, and 'sense memory'.[15] The former can be made accessible and intelligible to the outsider, but the latter is felt internally by the affected person and cannot be externalized easily. Here artists can step in as interpreters helping to vent the internal, incoherent, or inexpressible. Through interpreting pain, the artist can communicate with those affected while inviting others to attempt to engage with it. Moreover, the sense memory may also bodily affect the person, and artists can, wittingly or perhaps even unwittingly, recreate that in their work. In turn, this may trigger a reaction in the individual, leading to therapeutic and cathartic ends, as the artist manages to tap the flow of trauma-induced pain. The IWGC's architects were the artist-transformers of the battlefield, tapping into their own memory, their perceptions of pain, misery, and loss, and through that created a beautiful conduit for the channelling of emotions.

Within the cemetery and memorial spaces individual memories could overlap and intersect with those of other visitors. When formal collective rituals were performed, these internal processes were harnessed in external, organized expressions usually combining grief, sorrow, pride and patriotic sentiments, and national history. These attempts to create collective memory have been much discussed by academics and other commentators. The sociologist Émile Durkheim, and the group he inspired, which included Marc Bloch, Maurice Halbwachs, and Lucien Febvre, identified the socially constructed nature of all memory. For Halbwachs, an individual's memory was profoundly shaped and influenced by the wider social group. Further, the social collective memory would fade away and be replaced by others. This meant forgetting was as important as remembering. Here, collective memory was not racial or encoded in great works of art and material culture but embedded in social structures and groupings. Seeing as these groupings were forever changing, the solid became fluid.

Collective memory, it might be argued, is the end product of a series of exchanges rather than some great ethereal concept that exists over and above individuals. Pierre Nora's development of these ideas led him to the importance of places and spaces as repositories of, and antennae for, memory: *Les lieux de mémoire*. Memory sites, he argues, have a strange, reflexive power: they mark actual, fixed, verifiable points and events in human history while standing outside time. Such sites attract reverence and respect for their seemingly timeless quality, and it is this element that fuels the circular relationship: people invest them with significance; in doing so they constantly bring their own preoccupations and preconceptions to these places, thus renewing them. The memory site is both solid and fluid at precisely the same moment. In the 1920s the work of the IWGC and others created instant sites of memory, which have remained true to Nora's definition ever since. At the time, as Jay Winter so persuasively argued in his *Sites of Memory, Sites of Mourning*, the power of these sites flowed from their ability to inspire a sense of continuity with the past, allowing individuals the comfort of feeling rooted in their own past as well both the recent and distant pasts of their wider local, national, and (in the case of the British Empire) international communities.

Investigating the motivations, actions, and experiences of those who visited the battlefields between the end of the Great War and the outbreak of the next global conflict has been a fascinating experience. To understand the range of evidence those visitors have left to us has meant examining the sources through a host of lenses that have helped to tease out meanings. It has also demanded much thinking about the landscape and space, and the ways they are interpreted and conceived. The themes and threads running through this study are many, overlapping, and interlocking, just like the routes in a good battlefield guidebook of the 1920s.

1

Fragments from France and Belgium: Visiting the Battlefields, 1914–18

From this little mound, whose shapelessness and horrors the wild mustard which is springing up all round is trying to hide, you can see the whole battle ground. When the field of Waterloo is forgotten the Butte de Warlencourt will be remembered.

Birmingham Daily Press (13 June 1917)

Here and there, as we came and went along the dolorous Ridge, we chanced upon great monumental crosses standing high on well-built cairns ... This great Ridge of Vimy is truly an altar bathed in the rich red blood of thousands of high and gallant hearts.

John Oxenham (March 1918)

We came across a boot with a foot in it and shell hole with a boot showing itself. Pulling it out, the leg above the knee was there.

J. Carrigan, Labour Party official (March 1918)

Visiting the battlefields of the Western Front was by no means the preserve of peacetime. People came and went throughout the conflict. Indeed, the first visitors arrived soon after the earliest battles were fought; they were the vanguard of a steady stream who crossed the Channel for a variety of reasons. Some came as individuals or with one or two close friends or family members. Such visitors were often responding to the harrowing news that a son, husband, or brother was dangerously ill in hospital. Others were simply consumed with curiosity: they wanted to know what the battlefields were like. Some came in groups, usually as part of official tours arranged for propaganda and publicity reasons. A select few, those best connected, were given very special consideration. Whatever their reason for undertaking the journey, visitors engaged with a range of people – men and women, soldiers, nurses, local people, representatives of charitable organizations – while witnessing various aspects of a mighty war machine engaged in the manifold and quotidian operations of modern war. Much of what those visitors saw, did, reported, and recorded foretold the experiences of all who came after them when the guns were, at last, silenced for good.

Newspapers started publishing stories of actual and prospective battlefield travellers during the autumn and winter of 1914.[1] A schoolteacher, Miss Daisy Kessel, managed to arrange a trip to the Marne front – the lines having advanced – through her local travel agent in December 1914. Driven by a determination to witness history unfold before her eyes, she saw for herself the misery and devastation caused while also collecting an impressive number of souvenirs.[2] A few weeks earlier, Lady Violet Cecil had set out to find her son's grave. Although she was unsuccessful in her quest, Lord Killanin continued the search and solved the mystery by locating the graves of a number of officers and men. He oversaw their reburial and brought home many of their personal effects.[3] Others came to see relatives in hospital, often making arrangements in great haste and fearing that they might arrive too late; these trips must have been emotionally taxing, to say the least.[4] Hearing that her son had been wounded, but having no further detail, Isabella St John set out on an odyssey. Bluffing her way past suspicious French officials, she managed to get close to the actual front lines, tracking down her son to his billet, where she was finally reunited with him.[5] Such adventures were largely concentrated in the first half of the war, before the army's bureaucracy and administrative systems moved into top gear and regulated the flow of civilian travel to France and Belgium far more tightly.

THE ARRANGEMENT OF OFFICIAL VISITS

As the movement of individuals and small groups on personal missions declined, the number of official parties grew. Among the invited guests was F.S. Oliver. A highly influential thinker on British politics and imperial unity, Oliver knew many prominent people and had played a role in bringing David Lloyd George to the premiership in 1916. In the autumn of 1917, when the Third Battle of Ypres was at its height, Sir Douglas Haig wrote to Oliver in his 'own hand, asking me if I would pay him a visit of a week at least, and longer if possible'. Concerned at the slow progress of the battle, Haig probably wished to make a favourable impression on an influential person who could then report back to London in suitably positive terms. Much excited by the invitation, Oliver accepted at once and started to make preparations. On arrival in France, he discovered Sir Douglas Haig's own car, complete with field marshal's pennants, 'furled of course', and chauffeur ready to deliver him to General Sir Hubert Gough's Fifth Army headquarters. Despite his status as exalted guest, things became quite eccentric when Oliver reached Gough's headquarters, for the entire house, as well as all outhouses, were in use as offices and firmly shut up for the night. Eventually, he found an open kitchen door. On entry he stumbled across a deckchair and, exhausted by his long day and very late arrival, slumped down in it. After a while, an orderly discovered him and quickly bundled him into a guest room, where he slept for a few hours. Woken for breakfast, he was joined by General Gough, who was sporting 'a very youthful suit of pyjamas'.[6]

An earlier high-profile guest was Maurice Hankey, secretary to the War Council and the Committee for Imperial Defence, who accompanied Lloyd George's predecessor as prime minister, Herbert Asquith, on his official visit in July 1915. In Ypres, the party explored the city, which was under intermittent shellfire, causing Hankey much concern, for the prime minister seemed determined to linger. Worse still, Asquith insisted on measuring the largest shell crater, while General Plumer, commander of the Second Army, 'looked daggers at Milne, his chief of staff, who had arranged the programme'. The sense of anxiety was somewhat relieved with a dash of Pythonesque humour when the party reached the ruins of the Cloth Hall. Lord Kitchener, a keen collector of sculpture and art, noticed the few statues left in their niches high up on the remaining walls of the great building. He clearly thought an opportunity had presented itself, as was noted by Milne, who remarked that the statues were 'in greater danger than they had ever been from German shells'. After lunch, the eccentricities continued. Touring the Belgian lines at Fort Knocke, the water levels were so high that 'the trenches were alive with frogs hopping around one's feet'. The next day they toured the French lines around Arras and, from the commanding high ground at Notre Dame de Lorette, Hankey was treated to a 'splendid view over La Bassée, Lens, Loos, Ecurie, Souchez, Mont St Eloi and other scenes of heavy fighting north of Arras'.[7]

Hankey returned to Ypres in January 1918, soon after the punishing Third Battle of Ypres had run its course. He found a battlefield riddled with water-logged shell holes and punctuated by German pillboxes that were being converted for use by the British.[8] The tour then moved on to the First Army front and the smashed region around Arras and Lens. Alas, this was a mere curtain-raiser to the real scenes of destruction further south, on the old Somme front. Once on the old Somme battlefield the party was keen to see the German lines, but the host, General Byng, was extremely apprehensive and did not want to take them to any location where they might be spotted and provoke retaliation. Byng's fears were realized when an enemy observation balloon became aware of their presence and a barrage commenced soon after. 'We then moved off in as dignified a manner as we could assume (personally I felt rather in a funk!)', Hankey noted in his diary. Taken to another, safer observation post overlooking Péronne, they saw a landscape bathed in sunlight but 'desolate and shell-strewn, including, around the post, masses of burst gas-shells'.

Despite the important intelligence-gathering mission of which he was a part, Hankey's reflections on the tour made it sound a little like playing at soldiers, with the odd splash of enemy activity merely adding a thrill rather than exposing them to the full realities of life at the front:

> So ended our trip. We had travelled for hundreds of miles for five days through stricken, deserted, despoiled and utterly ruined villages. We had been several times under shellfire and once under aimed shellfire. For a part of every day we had worn steel helmets and gas masks at the 'ready' position. We had watched several air fights and much shelling of aeroplanes. We had seen many German planes but ten times as many British. We carefully studied the whole strategical and tactical situation on maps and on the ground. We have seen a part of the defences both in front and rear. We have conversed with a great number of officers, and seen the men under all sorts of conditions.[9]

On his official battlefield tour of November 1917, Austen Chamberlain also exuded a sense of skirting the edge of a cauldron without staring into its depths, advising his brother, Neville, to take plenty of warm clothing, a tweed suit, strong boots, and gaiters or knickerbockers, as well as decent suit for the evening.[10] Fine dining was guaranteed after a day of trudging between shell holes.

Far less elevated socially were the British trade union officials and representatives of trade and commerce invited out to France in a scheme that commenced in the spring of 1918. It was F.S. Oliver who suggested this formal programme: having been inspired by his own visit, he believed it would have a similar effect on others.[11] The intention of the trips, arranged by the Ministry of Munitions in conjunction with the National Chamber of Trade, was to show those involved in munitions production the vital importance of their home front work to the maintenance of military operations on the Western Front.[12] Like the higher-status visitors, these munitions workers and trade unionists were reliant on official chaperons who took them on carefully organized tours of different battlefields, which usually meant places that had been recently captured and thus were a fair way from the front lines. Among these official guides was the distinguished journalist C.E. Montague, who had managed to enlist in the army and gain a commission in 1914, despite being forty-seven. Montague was ambiguous about his role as official guide. Seeing civilians take a ghoulish delight at the battlefields left him feeling ashamed. For him it was rather like playing a role in a bad-taste charade:

> I feel a kind of grudge against the mere sightseer who comes out to see the war as a sort of show, accompanied by all sorts of luxury and petting. It seems they were rather scared at the place I had brought them to on Sunday, where the shells were falling about, and I have been rebuked, not very gravely, for imperilling the army's guests – not very gravely, because I think we all feel in our hearts that the sightseer's only chance of saving his soul alive is that he should get a taste, if only for a few minutes, of the kind of thing that our soldiers are bearing all day.

He refined his craft as a guide and prided himself on being able to estimate just how hefty a whiff of danger each guest desired and could manage.[13]

Regardless of the cynicism shown by guides like Montague, most visitors felt their time in France gave them direct, personal experience, which they believed imparted a deeper knowledge and understanding. Despite his privileged access to information about the war, F.S. Oliver felt he knew nothing until he witnessed the battlefields for himself. 'I have to confess that I had not the faintest idea of the proportion of the thing, either spiritually or materially, until I went out and saw with my own eyes and heard with my own ears,' he wrote in his diary.[14]

Austen Chamberlain described his official tour very much in terms of the Cook's tourist, breathless at the hectic schedule of impressive sights, but by its end he had learned something that only personal testimony could provide. Driven across the Somme battlefields on the Albert-Bapaume road, he saw 'a scene of awful desolation'. On the following day he was taken up to Ypres, where he was driven 'to a spot with the suggestive name of "Hell Fire Corner"'. The plan was to get as far as the Messines ridge in daylight, but this was aborted due to the shortness of the November day. The next morning he saw Indian labour units at work, and in the afternoon met the head of the air services, Hugh Trenchard. Attempting to summarize the essential nature of the battlefields in a single sentence, Chamberlain deemed them to be 'much as expected *but rather more so*' [original emphasis].[15] The reality was starker, bolder, and bigger than he had imagined. He had learned something.

THE VISITOR EXPERIENCE

Tramping over the battlefields, catching their legs on wire, and clambering in and out of old trenches gave the delegates on trade union visits the same heightened sensation, inspiring even deeper admiration for the soldiers. One was recorded as saying: 'The papers gave us an idea of what it is like out here, but we couldn't picture it as being as bad as this.'[16] These wartime visitors summed up the desire of all who were to follow them: the need for understanding, and, for the bereaved, a way of framing memories of a lost loved one.

The realization that wartime visitors gained so strongly was the slow descent into the world of ever-more intensive military activity, of the layers and layers of military infrastructure before arriving in a landscape increasingly pulped and desecrated. The transformation started at Calais, with exposure to a world so very different from home

Figure 1.1   Labour delegation at Albert, February 1918.

made more confusing by encounters with the familiar. On a mission to visit her wounded son, a Mrs Rose and her daughter saw 'British, French, Belgians, Indians everywhere'. 'How we had dreaded landing in a strange country with only a vague idea of its language,' she wrote. But then came the realization that 'surely we are mistaken. This is not France, but England, for everywhere we looked we were met with English faces and voices in their hundreds.'[17] For those visitors being taken to the battlefields, the first thing encountered was the amazing mix of personnel, services, and materials required to undertake major military operations. It was here Oliver made his first acquaintance with the Chinese Labour Corps. He saw Chinese labourers at work and described them as 'a pleasant type of animal; very smiley, and apparently contented with their lot. I understand that they are very good workers and so far, have caused little trouble.' Reading Oliver's comments in solely racial terms might just miss some of the spirit of his comment, for he called himself one of the 'various strange animals' who were guests of general headquarters. There was, however, a distinct sense of aloof superiority in his reaction to Tamil labourers 'swarming over the countryside' engaged in salvage work.

He recorded their appearance in detail noting their 'grinning white teeth, and crisp foozly beards of jet and glittering beady eyes' almost as if describing exhibits.[18] Isabella St John stumbled across a party of Indian troops camped in a wood. Witnessing their suffering at the severe cold provoked her sympathy and a sense of incongruity. Turbaned soldiers trying to cope with the grip of a French winter was not something she ever expected to witness.[19] It added to the sensation of being in a weird world with its own rules and nature.

Speeding through the darkness of an autumn night in a staff car, Oliver was struck by the almost ghost-like nature of the world outside. He was aware of people everywhere, but the blacked-out countryside compounded by driving in and out of fog banks made it an uncanny experience. Having had his imagination focused on the front line trenches, Oliver had never tried to visualize the scene behind the lines. As a result, the sight of two streams of traffic, one going up to the lines and the other coming back, was 'full of novelty'. Drawing closer to the front, the sound of the guns grew in intensity and the mud and muck increased. The combination of effects and sights left him feeling as if he had stumbled into 'one of those scenes one used to see in the Norwegian Fairy Books of Trolls working in the mountains'.

Arrival at the edge of the battlefield provoked shock. Finally reaching the Menin Road, Oliver at last glimpsed something of the fighting line, where everything had 'a certain leaden aspect'. Nonetheless, he was sensitive enough to realize that in not seeing the actual front line he had not been exposed to 'the real *heart* of the war'. The closest he came to the nitty-gritty of the battlefield was after clambering up onto the Ypres ramparts: gazing out across the landscape, he was amazed by the sight. Ironically, one of the things he noted was an issue commonly remarked upon by visitors, which was the lack of drama in their immediate vicinity and the sense that the real activity was elsewhere. Experiencing for himself the phenomenon of the empty battlefield, try as he might he could not detect the position of the guns that were roaring all around him. This left Oliver with the distinct sensation of being in it but apart from it: 'My own feeling was that I was under some kind of strange spell, so that I had to keep reminding myself that what I saw was really going on.'[20] It was a sensation made all the more intense by his guides, who seemed much more interested in swapping snatches of gossip about the progress of the fighting than actually observing it.

The intensity of the destruction, the immense scale of each battlefield, and the difficulty of interpreting what was going on created another dislocating irony. On the one hand, it was utterly awesome and almost incomprehensible, but on the other it was utterly dull and enervating in its mundane similarity. John Masefield said the old Somme battlefield was 'a difficult thing to describe without monotony, for it varies so little'.[21] After the war many echoed this point. There was a further difficulty. How was destruction to be described? D.S. Doig, editor of the *Dublin Daily Express*, drew upon a biblical allusion, calling Ypres 'the abomination of desolation spoken of by the prophet'. Scrabbling to find a comparison, he told his readers, 'You have seen the ruin made by the felling of two or three decrepit tenement houses. Ypres is like that, multiplied to the nth power.'[22] Ypres, no longer a city and with no inhabitants to be seen, sobered him by its silence, broken only by the occasional sound of heavy guns. Unaware of the city's pre-war architectural glory and unable to imagine city life as it had been, such was the intensity of the damage, Oliver felt only emptiness at the scale and intensity of the destruction. Drawing on precisely the same image as Doig, he compared it to slum property being cleared for redevelopment, and thus a sight 'much more dreary than tragic'.[23] Despite their different emotional experiences, Doig and Oliver agreed on the essential point: Ypres was the acme of destruction. This was a mantra repeated by every visitor for the next twenty years.

If the ruins shocked but ultimately numbed in their sheer nihilism, the sight of smashed woods and skeleton trees was something that caught the imagination and long left an impression. The early bombardments of the war initiated the process whereby trees were transformed into the weird, haunting skeletons that so affected visitors. In the autumn of 1915, D.S. Doig found the Ypres-Poperinghe road strewn with branches from trees hit by gunfire, but not yet the wrecks such trees were to become.[24] Two years later, the hammering had taken its toll, leaving Oliver far more moved by their condition than by the devastation of Ypres. His guide told him this was a common reaction.[25] According to a journalist who visited the battlefield in the autumn of 1916, the trees on the Somme were a collection of 'ragged stumps' looking 'like the ends of all gardeners' brooms'.[26] Gaunt and stripped of organic essence, trees were transformed into strangely mechanical structures, as if they were bits of discarded military kit. Unlike the towns, which could be compared to condemned houses awaiting clearance by a redeveloper, the state of the woods and even individual trees was a complete novelty.

The flip side of desecrated nature was its power to declare war on war by fighting back with irrepressible vitality. The same journalist, haunted by the ragged stumps of the Somme, saw sap oozing from them. Walking on to the ruins of Fricourt village in autumn 1916, he saw them knee-deep in clumps of fresh grass. A long and bitter winter followed, but when spring finally blossomed into action the Somme was transformed, stunning those who had known it during the fighting. The artist William Orpen was little short of stupefied by the contrast. He had left the battlefield 'the most gloomy, dreary abomination of desolation the mind could now imagine', but on his return he found miles of baked white chalk gleaming in the sun and covered with brightly coloured wildflowers visited by swarms of white butterflies. Nature had blended it 'into one wonderful combination of colour' turning it into 'an enchanted land'.[27] John Masefield had predicted as much on his early tour of the Somme front, noting that 'one summer with its flowers will cover most of the ruin that man can make'.[28]

THE MORAL LESSONS DRAWN FROM VISITING

Almost every visitor who saw the devastation of the Western Front drew the moral conclusion the British government desired: it was the fault of the Germans. A member of a Labour delegation from South Wales told a local newspaper of his determination to play his part in winning the war 'after seeing the wanton destruction carried out by the Germans'.[29] In January 1915, a Coventry councillor who had recently visited France returned utterly shocked by what he had seen, all of it caused by the vicious invasion of the enemy.[30] Seeing the state of Arras and Péronne confirmed Henry Oakes, a senior trade union official, in his belief that German militarism had to be stamped out. In fact, it left such an impression on him that he changed his mind about reprisals on Germany, arguing the only way to teach the Germans a lesson they would remember was by inflicting equal devastation on them.[31] The corollary of such passionate statements was sympathy for the French people forced to suffer at the hands of so savage an invader. In Albert on the Somme front, F.S. Oliver saw a town gradually coming back to life in the wake of the German retreat. Everyone seemed to be digging for possessions they had carefully buried in their gardens.[32] A visitor from Kenilworth said people in Britain 'must be thankful that we are not suffering as the French people are. We saw sad scenes of desolation but were told that we have seen very little compared with the damage that had been done in other parts of France and Belgium.'[33]

Aside from the battlefields themselves, an important aspect of the official tours was ensuring guests saw sites of significance to them, together with their localities. When D.S. Doig joined an Irish delegation visiting the battle zones in the autumn of 1915, a circuit of Irish units was included, as well as the ruins of a convent that had once housed Irish nuns.[34] A South Wales delegation was given the emotionally charged experience of seeing the graves of Cardiff City battalion men in cemeteries on the Somme battlefields.[35] Another visitor taken to see the graves of local men was Henry Oakes. Standing in the cemetery, Oakes removed his hat, remarking 'never before has the sadness of it all touched me so much'.[36] Such visits could also reassure the bereaved that the army took its responsibilities towards the dead seriously. In January 1918, the editor of the *Huddersfield Daily Examiner* included a letter from a local man describing the way the war graves were marked and tended on the Western Front. It was published in the hope that 'solace will be afforded to many by the knowledge that the graves of those who have fallen are being tended with loving care in those stricken and devastated regions'.[37] These visits thus acted as two-way propaganda and morale boosts. Where union delegates were shown troops at the front, they could express the appreciation of all at home for their commitment and determination to achieve victory. Then, on return, they could report the high morale of the British Expeditionary Force (BEF) and the impressive infrastructure dedicated to supporting soldiers both in and out of the front lines.

Admiration for the resilience and humour of British troops was universal among all visitors to France and Belgium. Isabella St John felt deep respect for the fortitude shown by the soldiers she watched doggedly going about their tasks despite the freezing weather.[38] During his time travelling among all sections of the BEF, F.S. Oliver was much struck by the sense of purpose, determination, and cheerfulness exhibited everywhere, which contrasted with the weariness and anxiety expressed on the home front.[39] Henry Oakes was brought face to face with the 'tragedy of the whole business' when he saw casualties being loaded onto the hospital ships. At the same time, he admired the efficiency and quality of the evacuation system and very obvious care devoted to the wounded, as well as the degree of humanity and humour shown by all involved in the process.[40] Seeing the soldiers go about their usual routines also allowed visitors to pick up on their culture and habits. D.S. Doig was delighted by the soldiers' slang, informing his readers that the Tommies had renamed Ypres 'Wipers', Poperinghe 'Pop', and Wytschaete 'White Sheet', and that places like Des Pierres farm had been relabelled 'Despair Farm'.[41] His amusement increased when his driver asked a French soldier in perfect Franglais, 'Savvy voo la root a la mayson le Generalissimo Anglais?'. Revealing his status as a literary man, Doig drew upon Chaucer, calling this the 'French of the school of Stratford-at-Bow, but it serves'.[42] Other visitors encountered the increasing numbers of women from Britain and the Empire that were serving in France and Belgium. Mrs Rose was deeply impressed by the nurses she met. 'I don't often go into raptures about a woman,' she wrote to her local newspaper, 'but the matron of the Australian hospital compels one's whole love, and admiration. My daughter and I often spoke of her, and she is often in my thoughts.'[43]

As well as in nursing, women were employed by the Church Army, Salvation Army, and YMCA to work in the various rest centres established to give soldiers a bit of home comfort and entertainment when out of the line. Branching out from such activities, such organizations were soon helping relatives arriving to visit the wounded and sick. It was from this infrastructure and system of arranging travel and accommodation and guiding visitors at every stage of their journey that the War Graves Visitation Services emerged after the Armistice. Mr E. Bending, the stationmaster of Wellow, near Bath, was among those extremely grateful for such assistance. After returning from seeing his son in hospital, he wrote to his local newspaper assuring 'any who may be called to France and feel timid of the journey that they need have no fear whatsoever, for they are met by representatives of the YMCA at every point both going and returning. I knew this Association was doing good work, but never dreamt of the vast organisation they have in France, and the wonderful hospitality and kindness shown by the ladies under its banner to the soldiers and their relatives, all entirely free of charge, which I for one shall never forget, nor able to repay'.[44]

CONVEYING THE MESSAGE TO THE HOME FRONT

Trying to explain the overall feel and look of this world to people at home was difficult. One much used method was to draw upon comparisons with Britain. Judging by its echoes in post-war writings, a highly influential text was C.E. Montague's introduction to the propaganda publication *The Western Front: Drawings by Muirhead Bone*. Montague divided the British Western Front into three zones, each resembling a UK region. Between the Somme and Arras was a country like Salisbury Plain, with its chalk downland. North of Arras there began 'a black country, where men of South Lancashire feel at home' due to the mills and mines of this industrial region. There was then the gradual fade into Flanders, a region of cities great in the Middle Ages, but now quiet. On this sandy plain there were echoes of Rye and Winchelsea, the medieval ports of Sussex, and the rich, well-worked agricultural land around it.[45] Wartime visitors trod this path marked out by Montague. F.S. Oliver said the area south of Ypres was reminiscent of Berkshire and Sussex. He wondered whether the run of hills had similarities with the Cheviots before deciding that the Flemish variant were longer and undulated more obviously. He found the landscape behind the Somme lines equally reminiscent of home. Here was a 'gently rolling country with winding roads and hedges and woods – very much like that ... of England'.[46] John Masefield agreed, describing the landscape of the Somme as much like the English downland and chalk regions and comparing it to the Chilterns and parts of Berkshire and Wiltshire, with the River Ancre running through its chalk course echoing the Thames at Goring and Pangbourne.[47] This categorization of the battlefields according to certain human and natural features, and their similarities with home, established the way the landscape would be viewed and interpreted throughout the twenties and thirties. In turn, this established the lingering idea that much of the Western Front was somehow, and in some mystical way, owned by Britain.

John Masefield's sensitivity to the landscape and his eye for comparisons marked him out as someone perfectly in tune with Montague's concept of the ideal visitor. Sent as an eyewitness, a sort of everyman, he felt the weight of responsibility on him to produce a work that all would find valuable. In trying to achieve that aim he took on the twin roles of observer-interpreter *and* comforter. This resulted in a short book, *The Old Front Line*, a masterpiece of elegy and guidance. He translated the abstraction of maps into vivid pen portraits of place, and in the process he provided the prototype of so many guidebooks to come. Believing his primary audience to be the bereaved, those trying to imagine 'where their dead are buried', he merged the topographical with the spiritual.[48] His starting point was the cartographic, but led the way into a world both physical and metaphysical. *The Old Front Line* was a best-seller and ran to a number of editions. Masefield satisfied the need to know the battlefields by understanding this desire to know them in different ways.

The famous music hall star Harry Lauder was another determined to penetrate the mystery of the battlefields for reasons Masefield and Montague understood and appreciated. Captain John Lauder was killed on 28 December 1916 near Bapaume on the Somme front. Heartbroken, his father travelled to France primarily on a mission to entertain the troops. On reading his powerful account of the tour, it becomes clear that the energy he poured into the effort was a way of channelling his grief and turning it into something positive by doing something for the men he admired so much. Around his busy schedule he included a visit to his son's grave. Taken across the Somme battlefields to the cemetery by highly sympathetic guides, he was shown to the very row, then the guides slipped away and left him to walk to the grave alone. Prostrated by grief, Lauder collapsed to the ground as memories flooded over and through him. Finally emerging from this reverie, he considered the many 'broken-hearted ones at home in Britain' and how they were longing for the day when they, too, 'might gaze upon a white cross, as I had done'.[49] After the war it was the mission of many to realize that desire as they guided mourners to the small patch of ground or inscription panel sacred to them. It was the mission of the Imperial War Graves Commission to ensure each place truly met the definition of sacred.

For those who, like Lauder, had immersed themselves in the battlefields and the scale of the military effort, the return home could be difficult. Isabella St John promptly fell ill, which she believed to be a 'reaction from such abnormal mental effort'. Nevertheless, she believed herself blessed 'in having been permitted to bring to accomplishment that which with the whole strength of my being I had so ardently and intensely desired'.[50] F.S. Oliver was left with the same sensation as many soldiers: reality was France and Belgium, while home was a distorted mirror image of true life. Back in London he felt the disorienting sense of being in a dream, whereas in France there was clarity through the 'confidence, cheeriness, irrepressible vigour and high spirits. I suppose as I go on thinking about it I shall get the explanation sometime, but it seemed rather like coming out of the real world into a shadow world.'[51] Many since have felt something similar.

SOLDIERS AS BATTLEFIELD VISITORS

Exploring the battlefields and meditating on their meaning was not an activity confined to civilians. Soldiers were equally as keen. A particular preoccupation was paying tribute to comrades by finding graves and erecting memorials. John Tucker, a private in the Kensingtons, wanted to visit the grave of his commander, Major Cedric Dickens (grandson of the author). In preparation for his visit they made a rail for the grave and painted an inscription on the cross, but on cycling to the site where Dickens was buried Tucker could find no clear trace of it.[52] Thwarted in this attempt to salute their leader, Tucker and his chums cycled across the Somme visiting ruins, exploring old shelters and trenches, and generally grubbing around among the detritus of the fighting.[53] In February 1918, Private A.W. Robins of the 2 South African Infantry Regiment had a similar experience. After taking part in a special commemorative service at Delville Wood, he and his chums visited the cemetery and then took the opportunity to explore the remains of the wood.[54] Soldiers could be tourists, too.

For Lieutenant Charles Carrington some free days in the spring of 1917 provided the opportunity to visit the grave of his brother. On reaching the spot, he saw it was marked by a substantial white cross clearly inscribed with the name and regimental details and decorated with a painted New Zealand fern. As the cross was dirty and a little unsteady, he cleaned and fixed it more firmly before covering it with green turf excavated nearby. Recounting the story in a letter to his mother, he told her the grave was in 'a terribly desolate place', estimating 'it will be two or three years before even the grass grows there again', but he managed to offer a glimpse of the future, for he added it 'might be quite a pleasant little dell when the grass and the trees grow again'.[55] In March 1918, Brigadier W.R. Ludlow of the 47 Division went down to the Somme battlefields to look for the grave of his son. Accompanied by an interpreter, he travelled by train, alighting at Albert, where the two men made enquiries at the corps' graves and burial office. Given advice and information, their next stop was the Directorate of Graves Registration and Enquiries (DGRE) office, where they were shown the maps marking the areas of the battlefield cleared and recorded by the unit. 'It did not include the Serre and Beaumont-Hamel areas and a good many others, but they had already located many thousand graves and made a large number of cemeteries', Ludlow wrote. Offered the assistance of a guide and a car, he and his interpreter set off for Serre, which was still difficult to pinpoint accurately due to the utter devastation. Equipment, as well as skulls and bones, was scattered everywhere, while thick clumps of grass had grown up, covering many of the white crosses dotting the landscape. Eventually they reached Serre Road Cemetery No. 1, which the DGRE officials believed the most likely burial place. Exploring it, Ludlow found some fine memorials, including a stone obelisk to a French regiment, but no sign of his son's grave. The DGRE office in Albert promised to search the surrounding area when Ludlow reported this news on his return.[56] Further searches were made, but it was not until May 1931 that Captain S.W. Ludlow's body was discovered and identified. He was then buried in Serre Road Cemetery No. 2, just a few hundred yards from the cemetery his father had searched. His parents chose 'Dulce et decorum est pro patria mori' as the inscription on the headstone.

PREDICTING POST-WAR BATTLEFIELD TOURISM

Given the number of civilians, soldiers, and others keenly engaged in viewing the battlefields, it was not long before there were predictions of a post-war tourism boom, and with it came questions over the propriety of such activity. In the run-up to Easter 1915 a travel agency was reported as running advertisements for battlefield tours over the holiday period. The *Manchester Courier* condemned such activity as 'abhorrent to every good citizen'. No one, it argued, should be allowed to see the 'battlefields as if [they] were a peepshow got up for their special delectation'.[57] Despite the clampdown on tourist adventures, the prospect of battlefield visiting remained a subject of discussion and speculation. A journalist who had returned from an official tour of the Somme battlefields could see no other destiny for them, declaring with matter-of-fact simplicity, 'The tourists will come ... in their thousands.'⁵⁸ Others took a rather arch and ironic tone. Telling his wife that he had gained the knack of battlefield guiding, C.E. Montague believed he had a lucrative post-war career mapped out: 'I do believe I shall be one of the best-equipped guides to the battlefield in existence after the war and could make quite a decent subsistence taking millionaire Americans round it for the rest of our lives.' He returned to the same point in another letter to the family: 'After the war I can make my living by conducting American tourists over the great battlefield.'⁵⁹ Of course, in associating battlefield tourism with Americans, Montague was also implying there was something rather gauche and uncouth about the concept. The British press agreed: Americans would want to see the battlefields as tourists because to them they were simply sights of wonder and not sacred altars of sacrifice made in a noble cause.⁶⁰

Bruce Bairnsfather, creator of 'Old Bill', a cartoon stereotype 'old sweat' British soldier who soon gained legendary status, was amused rather than disgusted by the idea of tourists swarming to the Western Front. In 1918 he produced a cartoon of post-war Ypres. In his vision the city was surrounded by a high wooden fence covered in posters reading: 'To the mine craters. Explosions daily at 3 and 8', 'To the ration dump. Real swearing by a ration party kindly lent by the Royal Warwickshire Regt.', 'Ypres open all the week', 'Cloth Hall entrance', 'Admission 1/-'. People of all ages and backgrounds queue to go in through the turnstile, while a commissionaire stands at the door. Above it all a balloon hovers, advertising 'Ypres'.⁶¹

Wry humour was expressed by another British officer as he claimed the landowner of a crater site was in for a bonanza after the war. 'All he has to do', he wrote, 'is to collect the scraps of Boche wire (barbed) that are lying about, stick a few posts in around the crater, erect a tea-house, and there you are – but not forgetting the ticket office. He might go further if he is industrious and enclose a series of dug-outs still in existence close by, make a huge collection of Boche mementoes, and sell them to the visiting public. [Thomas] Cook would undoubtedly include this crater in his itinerary. That Frenchman is a lucky man.'⁶²

'Patrol', a regular columnist in *Answers*, also saw an economic opportunity. In September 1918, he recounted meeting two officers who were trying to buy a hotel in Poperinghe with a view to exploiting the post-war tourism market. The columnist believed it was a wise investment, but he switched his focus from the tourist to the pilgrim, arguing the battlefields would soon be 'the Mecca of sorrowing pilgrims from the old and new worlds ... these will be the shrines to which the widows and orphans, the sorrowing parents and lovers, will repair on a great pilgrimage of proud and pious homage'. Fearful of what rampant commercialization might bring, 'Patrol' warned the battlefields 'must not be degraded into remaining through future centuries mere show places for tourists, the haunt of the guide, their pregnant silence disturbed by the raucous megaphone'.[63] The battlefields were special. Mere tourism was utterly inappropriate for such sacred sites. At the same time, the imprint of the visitor was already stamped into the landscape. In Ploegsteert Wood ('Plugstreet'), just inside the Belgian border, two trenches were quickly labelled 'Tourist Line' and 'Tourists' Peep' in recognition of Lord Balfour's visit. He was rapidly followed by large number of VIPs brought to this relatively quiet section of the front to see the actual fighting positions.[64] A few miles beyond Ploegsteert, a small cemetery on the Chemin de la Blanche was begun in June 1917. The site was also the place where official guests dismounted to make the rest of their visit on foot. The cemetery name summed up that function with beautiful simplicity: Motor Car Corner.

*

During the war people came to France and Belgium out of necessity to visit sick and injured loved ones, but they also came out of curiosity and intrigue. Although the latter type of visitor was sometimes labelled ghoulish in their interest and attitude, the fascination was, nonetheless, easily understandable. With newspapers and the newsreels full of images of the front, it was only natural that people wanted to see the reality of it, and for many the interest was far more than a simple craving for a thrill: the desire to know truly, to experience truly, and to understand truly was genuine. But the arrival of visitors brought with it the fear that one day soon unabashed commercialism would transform the battlefields from the sacred to the profane. And a commercial operation was, indeed, already emerging. Michelin published its first guide, dedicated to the Marne battlefields, in 1917. Clearly sensitive to the idea of exploiting the dead for gain, the company framed its approach with great care, transforming visiting into a moral and intellectual necessity. Witnessing the sites (and sights) would bring senses, soul, spirit, cognition, and interpretation together: 'Such a visit should be a pilgrimage, not merely a journey across the ravaged land. Seeing is not enough, one must understand; a ruin is more moving when one knows what has caused it; a stretch of country which might seem dull and uninteresting to the unenlightened eye, becomes transformed at the thought of the battles which have raged there.'[65] The outlines of the post-war debate on battlefield visiting had emerged while the guns were still hammering on the Western Front.

2

Postcards from the Hotel

At the present moment it is difficult to give very detailed information regarding travel to the battlefields.

Daily Mail Handbook to the Battlefields (1919)

L. & Y. and N.E. Railways Belgium and the Battlefields via Hull and Zeebrugge. T.S.S. Duke of Clarence will sail regularly from May 13th to October 5th, 1920. Boat trains will run to and from Hull (N.E.R. – Riverside Quay Station).

Advertisement for resumption of services to Zeebrugge (April 1920)

Hotel Skindles, Poperinghe, Best Hotel in the Salient, Homelike, Electric Light, Bathrooms, Central Heating.

Advertisement (1920)

The pent-up desire to visit the battlefields, and particularly that of the bereaved to see the places where their lost loved ones were buried or fell during the war, flooded out the moment the Armistice was announced. Although this desire was totally natural, and one looked upon with respect and understanding by the government, the difficulties thrown up by this seemingly simple issue were many and complicated. With the belligerents still a long way from a peace treaty and the British army and its allies taking station on the Rhine to ensure German co-operation, cross-Channel shipping remained focused on military needs. Communications across the devastated zone within France and Belgium were parlous and the shredded human infrastructure was straining to cope with the slow return of former inhabitants. And the battlefields were not just an appalling mess; they were a dangerous appalling mess riddled with collapsing trenches and dugouts, unexploded ammunition, barbed wire, and, of course, a myriad of corpses requiring proper burial. It is little wonder the government wanted to dampen enthusiasm for cross-Channel visiting.

GETTING TO THE BATTLEFIELDS

For the British people, however, the Armistice was popularly understood as the end of the war, and they could see little reason why there should be any delay in permitting travel. It was partly an entirely understandable failure of imagination. Despite the many images of the devastated battlefronts that circulated during the war, it was difficult to comprehend just how much of north-eastern France and Flanders had been affected. Few could compute that a region such as the Somme was inflicted with a battlefield nearly thirty miles across leading to the destruction of close on 400,000 hectares of land.[1] A *Dundee Evening Telegraph* report on Armistice Day itself revealed this very naivety and inability to grasp the realities of the situation, stating that the Thomas Cook travel agency 'already [has its] arrangements in practically complete form for visits to the various battlefields by those who have lost relatives'.[2] In making such a claim the story opened up another issue. Namely, whether the commercial travel sector would be allowed to control and dominate the highly emotive process of visiting war graves. For many, the bereaved had the right to make the journey at government expense, having suffered the agony of loss of their loved one in devoted service to God, King, and Country. Such was the intensity of the argument it became the subject of parliamentary debate a mere twenty-four hours after the Armistice came into effect. The under-secretary of state for war was asked whether visits by the bereaved to the graves of their loved ones would be covered by state funds. Very probably aware of the huge budget this would demand, but equally sensitive to the emotional climate of the moment, he largely evaded the question by promising to consider the proposal closely.[3]

If the War Office had hoped to avoid further statement, it was quickly disabused, as the Directorate of Graves Registration and Enquiries became a barometer of public interest and feeling. Within a week of the Armistice, it had received over 3,500 letters requesting burial location details from bereaved families.[4] The motive behind this surge in enquiries must have been the awakening of expectations. With the end of hostilities, the chance to visit the grave of their fallen relatives had finally arrived. Unwilling to risk being seeing as indifferent to these appeals, the War Office sought to clarify the situation through a public statement. The influx of enquiries and the desire to visit were acknowledged in an announcement making 'it known generally, to prevent disappointment, that at present, and probably for some time to come, it will be impossible, owing to military reasons, to make arrangements by which these visits could be permitted'. Wishing to avoid charges of indifference, the announcement concluded by keeping the issue open to enquiry and correspondence: 'It is desirable that any application on this subject, or as to the location of individual graves, should be made in writing.'[5] Revealing its antennae were highly tuned to public interest, a *Daily Mail* editorial took up the issue on 25 November 1918. While praising the recording of graves and the ongoing maintenance of the military cemeteries, it noted that only by visiting the actual graves would the bereaved truly find comfort. Assuming that 'the authorities are busy already on some scheme of shepherding this great pilgrimage of the relatives of hundreds of thousands of men', it hoped preparations would soon be completed and travel could start 'sooner than is officially promised'.[6]

Aware that the issue of visiting was relevant to its own work, the Imperial War Graves Commission (IWGC) decided to act quickly. Just over a week after the Armistice, the commissioners sat down to consider 'what arrangements needed to be made to enable relatives to visit the graves abroad'.[7] With the situation in France and Belgium so unclear there was a need to gather information and discuss possible solutions. Two months later the Commission returned to the issue and had to face the additional problem of imperial demobilization, which added to their difficulties. Many Dominion troops and civilians preparing to make their journey home after war service wanted to visit the battlefields and graves before their departure. The Commission and the War Office were sympathetic to this desire but rather hamstrung by the realities on the ground, with the single biggest difficulty being that of transport. With much of the French and Belgian railway network in the former battle zones straining to deal with the daily demands of military traffic (especially as the Army of Occupation was being established in Germany), there was little slack in the system. A cabinet meeting in March 1919 discussed the difficulties of rail transport in France and could not estimate at that point when normal conditions might return.[8] Unable to predict exactly when the French and Belgian authorities would allow the resumption of normal arrangements, during the first half of 1919 the British government did its best to dissuade all travellers from even attempting to visit the former fighting zones, providing regular reminders of the region's inaccessibility but with the promise that restrictions would be lifted as soon as conditions improved.[9] Also attempting to plan around the uncertainties was the commercial travel sector. The first post-war edition of Muirhead's travel guide to Belgium, with its revised title of *Belgium and the Western Front*, was candid on the transport crisis, reporting the 'suspense and uncertainty' prevailing over all travel and accommodation arrangements.[10]

However, the sheer determination to find graves and see the battlefields in person meant the region was by no means hermetically sealed, and it became clear that tourists, ex-servicemen and women and families alike were gaining access. As reports from visitors filtered into circulation during the spring of 1919, and as delegates started to gather for the Paris peace conference, expectations of normal travel conditions rose. At the end of May 1919, the newspaper *Answers* published a column in which it looked forward to the supposedly imminent end of travel restrictions, allowing for movement 'uncontrolled by military regulation and unfettered by ... passes'.[11] The statement was, however, premature, for the War Office was still unable to give a precise date on the resumption of normal cross-Channel traffic. The *Daily Mail* pointed out that technically it was possible to obtain a passport and then find a way across the Channel, 'but there is no organisation on the other side for getting about'.[12] A few weeks later, the *Daily Express* was reporting few signs of the French railways relaxing its travel restrictions, making holidays to France and destinations beyond unlikely. Quoting an interview with a Thomas Cook representative, the paper confirmed that small parties were visiting the battlefields privately, with tours costing somewhere in the region of FF800 to arrange, but a cheaper option of day trips to the French battlefields from Paris was possible.[13] Further evidence of battlefield visiting came from the Church Army, which reported in June 1919 that it was already assisting people and had 'organized a system of escort for persons visiting war graves in [France and Belgium]'.[14]

The contradiction between an official travel ban and people clearly getting to and from the battlefields fuelled impatience. Reacting to this frustration, the London *Evening News* expressed its sympathy for those desperate to see the graves of loved ones but advised staying put until conditions improved. Ballast was added to the argument by emphasizing the sheer chaos of the fighting zones: 'No one can describe the awful desolation and devastation' was noted as a prelude to the hard practicalities; there were virtually no hotels, trains were slow and infrequent, and motor transport was difficult to obtain. For these reasons, a little more forbearance was urged.[15] Harriet Hutchinson, writing in the *Nottingham Journal* in September 1919, added a different note of caution. Given the intensity of grave concentration work being undertaken, she said relatives might arrive with definite information as to the location of an isolated grave via a grave marker marked on a map, but find it had 'disappeared', as a DGRE team had exhumed and reburied the body in a war cemetery.[16]

Reinforcement of the official message came in a sobering account of a battlefield tour published in the *Aberdeen Press and Journal* in June 1919. Having experienced the actual conditions in France where 'neither food nor accommodation [was] available', the Press Association journalist wanted people to understand the sheer chaos of the devastated zone. Pitching her piece directly at a female audience, she wrote: 'There are in England thousands of mothers, sisters, and widows of those killed in the war who find it difficult to understand why, six months after the conclusion of the armistice, facilities are still refused for visits to the graves of their dear ones in France. In this country [France], where the people have to a much greater extent been brought into close contact with the horrors of war and their sequels, the situation is better appreciated.' She then went on to recount a dramatic tale with some hair-raising moments. As well as the dangers posed by unexploded ammunition and other war materials, the frontier-like nature of life in the devastated zone proved intensely scary. On returning to the 'terrible sight of Péronne' after a frightening journey in the pitch black down a ravaged road, she and her companion were traumatized by 'a party of roughs, half a dozen men who would have thought nothing of knifing us for our money'. Fortunately, some British soldiers arrived and assisted them.

Shaken by their experiences and unwilling to continue their journey, the two women required accommodation for the night. Unfortunately, the options were almost non-existent. The British officer informed them he did not think a rough male military camp was a suitable place, detailing his orderly to guide them to the French Red Cross nurses' depot. On their arrival, the superintendent said the accommodation was already overcrowded and there was no way she could squeeze in two more. However, she did know of a woman who was prepared to take in lodgers and she was willing to guide them to her house. It proved another fruitless trek, as again all the available space in this house was occupied. The situation was now looking bleak indeed. Mulling over the matter, the superintendent said there was a common hut for returning refugees, but she could not in all conscience recommend it to two foreign women. Instead, she suggested an alternative plan: their car could be garaged for the night in the yard of the nurses' hostel (the importance of this offer revealed much about security in Péronne, as the superintendent clearly believed it would not be there in the morning if left on the street); it was further agreed that they should go back to the British officer and beg for help.

Returning to the British military headquarters, the women explained their position. Highly sympathetic to their plight, the British officer they had met earlier provided them with a meal and then decided to hand over the room he shared with his orderly. Two camp beds were found, and the orderly even lent them his pillow, which had been in his pack since Mons. Despite this kindness, the two women found it very difficult to sleep, for rats scurried continually and they heard the sounds of someone dragging heavy metal across the street, while the strength of the wind increased overnight, causing the remnants of damaged buildings to collapse. The whole night-time drama was rounded-off by the sound of salvaged ammunition exploding. The next morning, being early summer, the sun was well up by four o'clock, and so the two ladies decided to get up and motor on to Paris.[17] The fearful reality of life in the devastated zone is encapsulated by this harrowing account and it should have served as a warning to any who thought navigating the devastated zone would be similar to a pre-war jaunt through northern France.

Despite such stark calls for patience and understanding, many remained determined to make the crossing, especially those bereaved whose loved ones were buried in the base hospital cemeteries and along the old military lines of communication. Having no intention of going anywhere near the fighting zone, these visitors simply wanted to cross the Channel and venture only a few miles inland. For these people, the official ban must have seemed an extremely blunt instrument. Tempers were frayed still further on 1 July 1919, when the British government restated its position that travel to the battle zone regions was not permitted.[18] However, the signing of the peace treaties in Paris from late June and the formal acknowledgement of victory and peace signalled the start of a slow shift towards pre-war travel conditions and the official position rapidly began to crumble.[19] At the seventeenth meeting of the IWGC in November 1919, Sir Fabian Ware, vice-chairman, creator, and presiding genius, reported to the governing council of commissioners that some 60,000 people had crossed the Channel to visit the cemeteries and battlefields during the course of the summer. This influx had placed pressure on the region. 'The French were getting rather disturbed about the large number of people who were going out', he told the commissioners, 'as it was throwing a great strain on the country owing to the shortage of food and accommodation.'[20]

The major drivers of this surge in visiting and the resumption of normal travel conditions were commerce and economics. For the town and port of Dover, the reopening of normal cross-Channel services was crucial for local prosperity and employment. The people of Dover had experienced much consternation when, in the early spring of 1919, the South Eastern and Chatham Railway recommenced limited passenger services from Folkestone to Boulogne while dedicating its Dover services to the transporting of troops, goods, and mails to Calais. Although the Dover–Ostend crossing reopened at the end of January 1919, it took another year for a limited Dover–Calais service to be re-established, and it wasn't until the autumn of 1920 that anything like a full service was in operation.[21]

With this gradual return to something like pre-war schedules, the railway and cross-Channel steamer companies began to re-advertise and added the new element of battlefield visiting to their copy. In 1920 Belgian State Railways declared the three-hour passage from Dover to Ostend to be 'the best route for the Western Battlefields and all parts of Belgium and the Continent'. By placing the battlefields in a package with access to other parts of the continent, it made a battlefields tour but one element in a jolly holiday.[22] That same spring the Hull–Zeebrugge service by the Lancashire and Yorkshire North Eastern Railway recommenced.[23] Many northern newspapers carried the story, clearly based on the railway company's own press releases. 'There is every sign that the service ... will be an extremely popular one again', declared the *Hull Daily*

Mail, 'especially as it gives the North of England voyager a unique opportunity for visiting the battlefields of Belgium and Northern France.'[24] The *Burnley News* said the well-planned service provided the advantage of 'allowing visitors the daylight of two full days to enable them to visit the battlefields of Belgium and other famous places in the neighbourhood of Zeebrugge'.[25] The *Manchester Evening News* added that 'across the water the railway companies have arranged a marvellously interesting and instructive week-end', which ensured a detailed look at the battlefields.[26] Such moves heightened expectations of a glut of battlefield visitors. In May 1920, the *Dundee Courier* ran a series of articles on battlefield visiting. The first was published under the headline 'Battlefield Tours. Where to go and how to get there. Continental Invasion this Summer.'[27] A year later the Harwich–Zeebrugge service was restored, leading to a similar profusion of stories about the Great Eastern Railway's ability to deliver the tourist to Belgium and the continent as well as the battlefields. In a glowing testimonial the *Chelmsford Chronicle* told its readers that 'the new thrice-weekly service is so arranged that those who wish to make a pilgrimage to Belgian battlefields or to the grave of some fallen hero, but cannot spare more than a single day from business, can do so comfortably in the allotted time'.[28] Clearly wishing to drum up trade from those interested in war graves visits, in 1922 the railway companies running cross-Channel services offered concession tickets for groups of twelve and over 'travelling under the auspices of recognized associations'.[29]

Despite this rush back to normality, the transition period made careful planning of a battlefields trip extremely important. The South Eastern and Chatham Railway, advertising in *The Pilgrim's Guide to the Ypres Salient* in 1920, provided firm instructions on preparations: 'Owing to the existing conditions affecting Continental Travel, intending passengers are advised to make their journey arrangements well in advance.'[30] According to the *Gloucester Echo*'s correspondent, 'the journey involves a certain amount of physical and mental strain if undertaken and carried through rapidly, the traveller should be in fairly good health'.[31] And the strain started with the Channel crossing: 'The steamer going and coming is packed. You must be patient in spite of tedious formalities and keep a sharp look-out you do not omit any formality.'

Opening up the Channel ports got people into France and Belgium. Progressing beyond that point was another issue. With access to, and navigation around, the devastated zone being so difficult during the immediate post-war period, there was little to stop the emergence of a lucrative trade in specialist tourism. John Frame, the founder of a large travel agency business, recognized that the war had created a new form of tourism. Soon after the Armistice he set out with his two sons on a fact-finding mission. He enlisted the help of the YMCA on arrival in France and was given every assistance to explore and assess the possibilities. This raises the question of how Frame explained his mission to the YMCA representatives. Either it was such early days that the YMCA had not yet woken up to the idea of commercial exploitation of grief, or he managed to convince them of the sincerity of his desire to assist the bereaved; he certainly maintained that his interest lay in helping the bereaved find graves, rather than seeing the battlefields as sites of touristic wonder.[32] For others the possibility of overt battlefield tourism was very much on the horizon. A columnist in *Answers* writing at the end of May 1919 predicted average costs for a battlefield tour undertaken in 'closed cars' and 'accommodated in historic chateaux' at around £5–£10 per day depending on the degree of comfort and luxury required.[33] These figures represented a sum beyond the working class and a distinct stretch for anyone below the wealthier echelons of society.

For the wealthy and influential there were many bespoke services. This was especially important given the warnings against independent travel due to the lack of available accommodation and the general conditions.[34] Writing primarily for a Canadian audience of his March 1919 battlefields tour, the journalist John W. Dafoe said it was necessary not only to hold all the required paperwork, but also to employ an expert military guide. The devastation and lack of accommodation made it a wilderness unnavigable for the 'unattached civilian'.[35] Guiding services were also strongly recommended by Sommerville Story, a British writer long domiciled in Paris. His 1920 guidebook, written with at least half an eye on a wealthy American market, advised 'the personally conducted tour … there is much to recommend this in the case of visits to the battlefields. In the first place one's route of travelling will be followed in accordance with expert guidance and after careful study, and in the circumstances of the battlefields this is a great advantage.'[36] Thomas Cook recognized this need very quickly, advertising in the summer of 1919 a thirty-five guinea guided tour travelling 'first class throughout' in 'superior hotel accommodation', but not including transfers and entry to 'the places of interest visited'.[37] Within a short space of time the battlefield tours of the famous travel agency were ubiquitous. A correspondent writing for the *Aberdeen Press and Journal* produced something akin to a free advert for its services. Determined 'that we should visit the battlefields as soon as permission was granted', the correspondent contacted Thomas Cook, confident a tour could be arranged 'under the auspices of this great tourist agency, in perfect comfort, and with a thoroughness that left nothing to be desired'.[38] Although the cost of thirty-five guineas for the trip to Belgium (or forty-five for the French tour) was deemed on the high side for some, it was pointed out that this covered first-class travel throughout, included passport services, porterage of luggage, and private motor travel accompanied by expert guides.

Ability to capitalize on the scale of its operation did not mean Thomas Cook stood alone; it faced competition from other, smaller concerns entering the market. By February 1920 the Imperial Travel Bureau, with head offices in Sloane Square, was offering battlefield tours costing fifteen guineas per person.[39] Western Front veteran Captain R.S.P. Poyntz, with offices in Regent Street, London, commenced advertising in May 1920, promising 'private visits to battlefields and war-graves' at first-class standards throughout from ten guineas per person. Efficiency, convenience, even luxury, and authenticity of experience were the selling points stressed by these firms. Authenticity came through the status of the personnel. The Franco-British Travel Bureau offered 'personally conducted tours by ex-officers'.[40] The Imperial Travel Bureau advertised itself as part of the Association of Officers, Ltd. Captain Poyntz listed his credentials, including his BA (Oxon) degree and that he was 'BEF 1914 and 1915–1918'. By delineating 1914 so strongly in his advert, Poyntz was also revealing his membership of the hallowed Old Contemptibles (nickname of the original British Expeditionary Force), and therefore deliberately trading on the enhanced status of these particular veterans.

Thomas Cook also knew the value former service personnel added to the team. The 1919 tours had been under the charge of a 'competent courier', but by 1921 the firm was stressing its battlefields offerings were assisted by 'several ex-Officers who served in various sectors of the Front, whose services have been secured for this purpose'.[41] This move was possibly in response to criticisms from authorities such as 'Old Brigade' (almost certainly the pseudonym of

Western Front veteran and battlefield-guide author, T.A. Lowe) in the *Pall Mall Gazette*. Commenting on Cook's battlefield travel plans in July 1919, he was sceptical about the quality of the guides: lacking first-hand knowledge, they would be capable of little more than sketching out general overviews.[42] A khaki tunic, some medal ribbons, military riding boots, and a cane reassured a visitor that all was well in hand.

Of course, the stress on officer status also implied the reassurance of having the right class of person in charge, and with it a sense of trustworthiness. This was especially important for those who had opted for a degree of independence, making their own way to France and Belgium but then requiring expert assistance and interpretation once on the battlefields. As with the fully inclusive tour, this option grew with great rapidity. 'Old Brigade' told his readers those 'preferring independent travel and ... less exacting' could opt for the nine-and-a-half guinea 'popular' tour available from many outlets once they had made it over the Channel.[43] A *Gloucester Echo* correspondent who visited a grave in August 1919 wrote admiringly of the service provided by two former officers based in Amiens who could supply cars, book hotels, and provide luncheon baskets 'at a reasonable price'.[44] (This was probably the outfit run by Messrs Gregson and Griffiths.[45])

Establishing credentials was extremely important for these businesses and they rooted them firmly in their officer status. Lieutenant-Colonel E.P. Cawston was the founder of the Battlefields Bureau, which, like the Imperial Travel Bureau, showed off its 'Association of Officers Ltd' affiliation. Cawston added an extra touch of class by boasting offices at the Chateau des Trois Tours in Brielen, a village just behind the old front lines at Ypres, deep in the fabric of the battlefields, where, as much travel literature also stated, the Ypres League and IWGC also had offices. Described by Cawston as 'the finest centre for Tours in the Battlefields of Belgium', the Chateau was the 'Headquarters of the First Canadian Division during the Second Battle of Ypres and the Headquarters of many a British Division and Brigade from that date to the Armistice'.[46] The location of Cawston's business therefore gave it the ring of trust and implied official recognition through sharing space with the IWGC. The London end of the operation underpinned the desire for status, being based on Piccadilly, at the heart of the West End and clubland. Cawston employed two former members of Queen Mary's

Army Auxiliary Corps and thus maintained the presence of veterans throughout the firm.[47]

Acutely aware that such services were beyond the means of many, Fabian Ware and the IWGC acknowledged that there 'would be serious trouble if the organization of this matter were not taken in hand'.[48] With the IWGC focused on getting its cemetery construction scheme underway, it had neither the time nor funds to engage directly in travel to the sites. Ware told the Commissioners that under their charter they 'had no power ... to help visitors to get to the Cemeteries or to run a touring agency'. Rudyard Kipling, literary adviser to the IWGC, supported Ware in suggesting a role for the commercial travel industry: 'Cook and Sons or other agencies could always arrange to take them [visitors] to the Cemeteries provided they could obtain correct information as to the name of the Cemetery.'[49]

Nonetheless, Ware sought the opinion of the commissioners as to whether the responsibility should be added to the IWGC's charter commitments. Most were wary of taking on further responsibilities and instead expressed the hope that the YMCA, Church Army, and Expeditionary Force Canteens, all of which had built up sophisticated infrastructures to help keep soldiers entertained when out of the line, could divert their personnel, materials, and extensive network of prefabricated buildings to helping visitors.[50] Wishing to explore the option, the IWGC convened a conference with the YMCA, Church Army, and Salvation Army. All of the charities were happy for their facilities in France and Belgium to be used by visitors, but all agreed that those unable to pay should have their costs met by the British government.[51]

Ware was right to keep the IWGC engaged with the issue, for it had a high public profile. Writing to his local paper in June 1919, 'A. Mac.' bemoaned government parsimony over the question of free travel to war graves. Accepting that it might not be possible to provide entirely free travel in all instances, he advocated subsidies: 'The rate charged might be made so that bereaved folks might get their heart's desire gratified at a low cost.' Seemingly a veteran, he finished his appeal with a touch of bitterness: 'Some ... politicians have short memories for the lads who held the gate "Ypres, 1914, and after."'[52] Introducing a talk about the work of the IWGC in the spring of 1921, Folkestone's mayor assured the audience, especially the veterans present, that the issue of gaining financial assistance for relatives to visit the cemeteries was of great concern to them all and he was interested in the possibility of launching a fund dedicated to it.[53]

FINANCING THE VISITS OF THE POOR

Determined to keep the matter before the government, a deputation from the National Federation of Discharged and Demobilised Sailors and Soldiers (one of the organizations later folded into the British Legion) met the prime minister, David Lloyd George, in February 1920 to discuss the situation. Presenting their demands, the NFDDSS members expressed their opinion 'that a refusal to arrange free facilities is entirely false to British sentiment', and drew the Prime Minister's attention to relatives who were 'denying themselves the necessities of life in order that they may save sufficient money with the view of a possible visit in the future'.[54] Lloyd George was sympathetic to the notion, telling them 'I thoroughly agree that every facility ought to be given to the relatives of deceased soldiers to visit their children's graves', and thereby privileging parents as the chief mourners.

He asked Winston Churchill, as Secretary of State for War, to look into the matter. On reporting back to the cabinet Churchill outlined a series of scenarios and implications. He started by dealing with the suggestion of a fifty per cent reduction in the cost of a third-class return fare to be offered to two near relatives of a soldier buried in France or Belgium. Working on the assumption that there were half a million graves in France and a forty per cent take-up of the offer, he believed the cost would be somewhere in the region of £500,000 for travel alone. Believing that some might wish to abuse the system, Churchill pointed out the necessity of bureaucratic machinery to oversee any such programme, leading to greater expenditure. Further, the bereaved would probably require assistance in France and Belgium, which demanded yet more administrative staff and infrastructure. Churchill estimated the cost of the complete package to be around £2 million.[55] Unwilling to commit to such a sum, the government faced incessant questioning in Parliament, to which it responded with a stream of vague responses. Finally, in August 1919 a decision appeared to have been taken when a War Office minister made an unequivocal statement in the Commons that free travel would not be made available to relatives of the bereaved.[56] However, the matter was too sensitive to disappear entirely, and the cabinet was left with no choice but to rethink.

Clearly reluctant to give up on the idea of providing financial assistance, Churchill discussed the matter with the Treasury.[57] Once again railway charges and passport costs were identified as lying at the heart of the problem.[58] The Army Council took up the question of railway expenses with the Ministry of Transport, requesting its view on the 'possibility of affording a substantial reduction in the cost of railway travel in the United Kingdom and transit by railway owned steamship across [the] Channel' for relatives wishing to visit war graves. At the same time, the Foreign Office was approached for a reduction in passport fees demanded by 'the strong public sentiment in connection with visits of these relatives to the battlefields'. While the Foreign Office proved receptive and agreed to make arrangements with the Belgian and French governments for special war graves passes in lieu of a full passport, the Ministry of Transport stated its inability to provide any help as it lacked both the powers to enforce an agreement on the railway companies and the funds to provide any direct subsidy.[59] With no one wishing to accept financial liability, but with a deep sense of public interest in the matter, the problem was batted around Whitehall during the summer of 1920. Eventually, a Treasury official suggested that a direct grant to organizations such as the YMCA was the only way of allowing poor families to visit the cemeteries. Whether it could be dumped onto the War Office's existing budget was another question. Desperately looking for solutions, the Treasury turned to the IWGC, enquiring whether it might be able to pick up the bill. Fabian Ware was very quick to quash this suggestion, citing the Commission's charter and its exclusion of 'any service for conducting relatives personally to the cemeteries or finding for them accommodation or transport'. Returning to the hunt for funds, the Treasury wondered whether the profits accrued by the Expeditionary Forces Canteens were a possible source of funds, but this, too, was eventually excluded.[60]

Lacking a solution, the problem rumbled on into the spring of 1921, when the Treasury finally agreed to make a one-off grant to the various charities running war graves visitation services. Power to distribute the funds was devolved to the Army Council, which transferred £25,000 to the YMCA in two instalments (the first in 1921–22 and the second for the financial year 1922–23), while £15,000 went to the Salvation Army and £10,000 went to the Church Army in two instalments between 1921 and 1923. This almost immediately sparked an inter-service debate, as the Admiralty and the Air Ministry enquired as to whether the grant was confined to relatives visiting army graves in France and Belgium. The War Office quickly agreed to include those visiting naval graves, on the understanding that they numbered only some 300 (presumably casualties of 63 Division were excluded from the calculations and were deemed to fall under the army). Far more difficult was the question of relatives wishing to visit the graves of naval casualties buried at the Royal Navy's main Scapa Flow base in the Orkney Islands. Admiralty officials discussed the situation among themselves in the spring of 1922, before reaching the conclusion that the War Office would be unlikely to include travel within the United Kingdom, while also noting the irony that accessing 'remote places like Scapa' was much more difficult than 'relatives going to Belgium and France'.[61]

The nineteenth-century language of poverty and class is distinctly perceptible in such debates over what the grant did – or did not – cover. The War Office used exactly the same terms as the Victorian Poor Law Commissioners and Boards, referring to the need to support 'deserving and necessitous cases', and the war graves charities were instructed to apply 'stringent verification through the existing machinery [of] the bona-fides of all applicants for assisted visits'. The conditions ran on with their own inexorable logic and thoroughness. British parents or direct relatives of a soldier who served and died with Dominion forces were eligible, but both relatives of Dominion soldiers who had migrated to Britain after their death and British-born relatives who had emigrated to the Dominions since August 1914 were not. Further, no one could apply more than once; not more than two people could apply to visit a grave, and 'each visit shall be made direct to the grave and back and shall not be regarded as a means to any other end, e.g., a tour of the battlefields, or a journey to some place beyond that where the grave is situated'.

As can be seen, such was the intensity of the desire to guard against exploitation of the system, and especially against the casual tourist intent on having a good time, that even a battlefield tour was explicitly excluded. This meant that, if their loved one was buried at any distance from the fighting zone, the exclusion of such a tour stripped the bereaved person of a context in which to frame their grief. In addition, it also firmly excluded relatives of the missing. At this point, the memorials to the missing were not yet even the briefest of sketches in an architect's notebook, and so the only place a bereaved person could go was the battlefield, perhaps to identify the location where their lost loved one was last seen. As for the organizations actually arranging relatives' visits, the YMCA, Salvation Army, and Church Army, knowing they were bound to operate under these War Office guidelines, promptly agreed a common policy on implementation, and by spring 1922 they were at last ready to expand the services they had been running since 1919.[62]

As they set to work, a hierarchy and moral economy of grief was established. The original list of eligible family members ran: widow, children, step-children, foster children; parents, step-parents, foster parents; grandparents; sister, stepmother, adopted sister; brother, stepbrother, adopted brother. This was later refined to: mother, father, widow, 'betrothed wife if still unmarried', daughter, son, sister, stepbrother, step-parents, foster parents, stepchildren, foster children, grandparents, stepsister, stepbrother, adopted sister, and adopted brother. In both instances female family members were placed higher than male: widows and mothers ranked higher than fathers, and sisters were placed above brothers. However, only an unmarried fiancée of the deceased was eligible. On one level this was part of the desire for economic stringency, and perhaps grasped at the idea that a former fiancée would have found additional financial support through her subsequent marriage and thus could be excluded. Secondly, and more prejudicially, there was the implication that, having found a spouse, the grief of a now-married fiancée was somehow of a lesser rank. Class associations and stereotyping were also perceptible and are evident in the response to the Air Ministry enquiry as to whether Royal Flying Corps and Royal Air Force graves were covered by the grant. It was noted that relatives of the air services would probably not prove a drain on resources due to the 'comparatively small number of casualties ... and of the fact that a large proportion of the casualties were those of officers'. According to this sentiment, officers were *ipso facto* gentlemen, and therefore their families were highly unlikely to require any kind of financial assistance in order to undertake their visit.[63]

Intertwined with the grant to subsidize visits for the poor was the question of free passports, which was another request the NFDDSS pushed hard with the government.[64] As the cost of passports was an acknowledged difficulty, the Foreign Office accepted the task of brokering a substitute form of documentation with the Belgian and French governments.[65] At least one Foreign Office official was highly sympathetic to this need, deeming it 'clearly right ... [that relatives visits] should be facilitated in every way'.[66] Agreements were reached with the Belgians and French, and the new system came into effect in September 1921. In order to gain a war graves pass an applicant had to go through a two-stage verification process. First, a police officer or secretary of a Local War Pensions board had to authorize a photograph of the applicant (Certificate A). Second, the IWGC had to provide full details of the grave and apply its stamp to Certificate B. The pass was then issued and valid for ten days.[67]

These decisions and the instigation of mechanisms to oversee them simply codified practices already adopted on the ground by the various charities overseeing war graves visits. By 1919 the Church Army, Salvation Army, and YMCA had become veteran operators across the battlefronts. Aimed at giving Tommy wholesome home comforts when out of the line, these organizations had huge and sophisticated infrastructures, especially on the Western Front. With stacks of vital supplies as well as staff blessed with a wealth of local knowledge, their hutted encampments became magnets for visitors. Partition walls were soon erected to create small cubicle bedrooms; common rooms were carpeted with rugs and furnished with chairs, tables, and, often, a piano and liberally splashed with vases of flowers to create cheery, homely spaces. The final addition was a chapel, answering the vital requirement of a contemplative space, a place for mourning and consolation.

Further, as Christian organizations, the Church Army, Salvation Army, and YMCA accepted as part of their mission the need to offer consolation to the bereaved and recognized a visit to the grave as an essential component of mourning. Much of the perceived value of their operation then rested on the implicit and explicit moral distinctions these organizations made between the battlefield tourist and the war graves pilgrim. All three made it clear that their services were for the bereaved only. 'We have also organized a system of escort for persons visiting war graves in the district,' stated the Church Army's annual report for 1918–19, in the process making its criteria absolutely clear: 'Only parents, widows and other relatives, or intimate friends, are permitted to join the parties.'[68] In Ypres its hostel was reserved for grave visitors travelling with the Church Army, but independent travellers would be accepted provided there was space and that they were seeking a particular war grave.[69] The YMCA worked to the same principle, devoting its chain of huts to 'the relatives of the fallen'.[70] Wishing to assist these organizations, the NFDDSS urged the government to permit the Disposal Board, the body set up to sell off unwanted British military equipment, to offer the army's accommodation facilities in France and Belgium 'to such organisations as are able to make use of hutments or temporary structures, on purchase, as hostels for necessitous "grave pilgrims"'.[71]

FINDING A PLACE TO STAY AND THE WORK OF THE WAR GRAVES VISITATION SERVICES

Demand for accommodation and cemetery guiding services increased rapidly, and as more funding became available, additional facilities were provided. St Barnabas Hostels expanded from their original base at Calais to a second hut in the city, handed over by the Canadian Red Cross, and branches were opened at Albert, Amiens, Arras, Béthune, Cambrai, Hazebrouck, Poperinghe, and Ypres.[72] In March 1920 the *Times* reported that 'the YMCA have moved up to the devastated areas a number of their huts originally engaged in war-work with the Armies, and have fitted them up for the accommodation of visitors'.[73] Like the St Barnabas Hostels, the YMCA operation spread across the battlefields, with hostels at 'Armentieres, Bethune, Arras, Albert, Cambrai, Peronne, Amiens, Calais and Boulogne' accommodating 300 people.[74] Listing its facilities in this way created a pilgrimage route across the battlefields, working its way south from Ypres, through the battlefields of the Franco-Belgian border, down to the Somme, and arcing out to Cambrai and Péronne before working its way back to the coast through Amiens. Sir Arthur Yapp visited the former Western Front on behalf of the YMCA in January 1920. He saw for himself the work being done at the organization's Ypres centre to support relatives visiting the graves of their loved ones. According to Yapp, over 500 people had already made use of the facilities, including French visitors and a British woman who had travelled from her home in Italy to find her son's grave. (This was very probably the mother of Edwin Benbow, who lived in La Mortola, Ventimiglia.[75]) Another YMCA official, Arthur Hickman, made a whistle-stop tour of his organization's hostels that saw him tick off the crucial details of their network: the hut at Albert served the Somme area and its 270 cemeteries; near the Péronne hut was a cemetery containing many Indian graves; at Cambrai an old Army Education service hut had been recycled, and there were also hostels at Arras, Béthune, and Armentières, as well as that at Ypres.[76]

The Salvation Army's network of hostels consisted of Ostend and Ypres in Belgium, and Boulogne and Arras in France. In addition, it also had a facility in London, providing pilgrims from across the United Kingdom with a staging post in their travels to and from the battlefields.[77] Experienced staff managed the hostels, and every relative was given a personal guide to the grave of their loved one. As with the other charities, the Salvation Army's intention was to 'give every facility possible and to make their journey as easy and comfortable as the circumstances will allow'. In a festival of statistics, it could demonstrate its immense success. Between 1920 and 1927 it arranged for 6,349 people to visit a grave either free or through partial payment, and for 20,785 doing so paying their own expenses. During that same time period nearly 5,000 people had stayed at the Ypres hostel, and over 5,000 at the Arras branch. In 1921, 7,000 people had been advised or assisted in some way with their war graves visit, while in 1927 its Ypres hostel had served nearly 6,000 hungry pilgrims with meals.[78] Praise for its efforts was lavish. The journalist and writer Henry Benson, a regular contributor of articles on visiting war graves that were syndicated to many newspapers, always spoke highly of the Salvation Army's commitment to help the bereaved, especially the female bereaved. He stated that he had heard women with accents from across the Empire wandering around France and Belgium desperately looking for a grave and it was the Salvation Army that took them under its wing and ensured they got to the right place. Benson advised everyone to approach the Salvation Army if they wished to visit a grave. He praised its degree of expertise, its eye for every detail, and its deep knowledge of every requirement of the pilgrim. Accommodation in its strategically placed string of hostels, each including a good supply of hot water, comforting and familiar food, and caring and insightful guides, were all part of the Salvation Army package, and it was all done with the minimum of trouble on a not-for-profit basis.[79]

Like the Salvation Army and YMCA, the Church Army turned its wartime infrastructure over to war graves visitors in 1919. In Ypres, its initial base was formed from its original huts next to the prison and close to the British reservoir cemeteries. Subsequently, the huts were replaced with a purpose-built hostel, opened by Earl Haig in

1924.[80] By 1928 nearly 8,000 people of 'all classes including a large number from the Colonies' had been accommodated at its Ypres facilities.[81] Demand was such that in 1927 it undertook an extensive overhaul of its Ypres centre to provide better accommodation, particularly in bedrooms.[82] Many of its visitors had made the journey on assisted passages, but for those who could contribute, the Church Army set fees commencing at £3 3s. And like its sister organizations the charity prided itself on its attention to individual needs: 'From the commencement the personal touch has been maintained by sending people in small groups of 2 to 15 people whenever possible, the Guides being our own officers or sisters or a clergyman who escorts the party throughout.' In addition, the Church Army was responsible for a chapel, again in an army hut, and held services for both visitors and the growing British community permanently domiciled in Ypres.[83]

The Ypres League also had plans to build a permanent hostel in the city but first established a temporary building. St Barnabas Hostels was approached as a potential collaborator, an offer it gratefully accepted, and a joint management committee was established to oversee the project. St Barnabas Hostels had already leased a building and the task was now to convert it into bedrooms and a rest room providing 'a quiet retreat with sympathetic attention' to all the requirements of a pilgrim visiting Ypres. The Parminter brothers, both veterans and established in pilgrimage-tourist services, were the Ypres League representatives on the subcommittee, with Captain P.D. Parminter undertaking to act as its agent and guide for all visitors to the area.[84] The League then took a house in the Rue Surmont de Volsberghe in Ypres, which was fitted out as a headquarters containing a well-stocked library and a large room with armchairs and writing tables – 'the two things that are apt to be lacking in Continental hotels'. It had the additional quality of being next door to Parminter's garage and business, Wipers Auto Services, and his daughter assisted with the running of the centre, which was thought to be something 'ladies will be glad to hear'.[85] During the twenties, the British Legion also built up its presence in Ypres, opening Haig House in the city centre. The veteran A.W. Keith made it his Ypres base and was extremely impressed by the efficiency with which it was run by a fellow ex-serviceman, former Sergeant Reeder, along with his Belgian wife. Keith thoroughly recommended it for its bright, cheery rooms and good food.[86]

The war graves visitation charities took great care to ensure the quality of their facilities, as was revealed by an IWGC inspection in 1928. The Salvation Army hostel was deemed to be 'very comfortable, clean and from all appearances well managed'. A similar conclusion was reached on the Church Army equivalent, which was judged 'comfortable, clean, [and] airy'.[87] 'Rover' was full of praise for the team at the Ypres hostel run by Mr Tarr, who so devotedly guided people to graves 'and in Miss Beazley the mothers bereft will find a sympathy beyond all price'. He was equally delighted by the team at Arras, as they provided a friendly welcome allowing 'the stunned soul' to relax after the harrowing sights of destruction and misery. At Cambrai two Tynesiders told him of the warm welcome they had received at the local YMCA hut. Among the staff were members with relatives buried nearby, which would surely have aided their great sense of '"fellow feeling" for other mourners [which] "makes them wondrous kind"'. At the Albert hostel the pilgrim could stay in decent accommodation and be loaned the use of a car to aid their visit.[88]

The work of the three organizations was clearly much appreciated by the IWGC; Ware wrote that 'they are really doing wonderful work and are so modest about it and have quietly helped and supported us in many of our difficulties'.[89] Maintaining these facilities placed a considerable burden on these organizations, which in turn meant frequent appeals for funds. These were often framed in gender terms, reifying the position of the mother. 'Will you send a poor mother to France to visit the grave of her loved one for the first time?' asked a St Barnabas appeal, which further asserted, in a deliberately heart-rending tone, that 'For the £4 you subscribe the Mother goes over and comes back in comfort, and the happiness you will give her will more than repay you. Fifteen hundred Mothers are waiting to go.'[90]

The provision of such services also recognized the fact that the commercial travel companies might exclude those of more modest means, or worse still, exploit them knowingly. The Reverend Matthew Mullineux, former chaplain to New Zealand forces and founder of St Barnabas Hostels, was deeply moved at finding many pilgrims 'miles from their hotels vaguely wandering about in search of Cemeteries and with no sign-posts to guide them thereto'.[91] Similar sentiments were expressed by Henry Benson, who had witnessed 'mothers from little hamlets in Scotland, wives who have crossed the seas alone from Canada, and sisters from the factory districts of the Midlands and Lancashire, wandering helplessly and hopelessly about, looking for someone who might advise them'.[92] The *Yorkshire Evening Post* recounted two pathetic stories of lost pilgrims. One told of an old man found wandering around Amiens in great distress, and with no money, desperately searching for his son's grave. The second report concerned a poor woman who carefully saved her money to visit her son's grave near Arras. Having made little preparation before she left, she then spent hours walking many miles from the town before fainting from fatigue at the doorway of the British Army's Directorate of Graves Registration and Enquiries office. 'For these people guides seem to be urgently needed,' stated the 'Gossip of the Day' column in the newspaper.[93] Together, these stories formed the Parable of the Poor Pilgrims.

The possibility that such vulnerable people might fall prey to the unscrupulous was something that haunted Mullineux and inspired his condemnations of disreputable tourist companies, hoteliers, and taxi drivers.[94] Determined to remedy the situation, he founded St Barnabas to assist the poor and needy war graves pilgrim. Run by a small committee of 'devoted ex-war workers', the team soon opened a hostel. Calais became the headquarters of the operation, complete with an office containing books, maps, and all relevant railway timetables. The first pilgrim the team assisted was a lady who had been quoted £35 for a trip to Loos. St Barnabas managed to take her for £14, even after hiring 'an expensive French motor-car' and the pilgrim insisting on paying for the guide herself. By the time of publishing its pamphlet (probably in 1924), the trip could be made for £4 due to the economies of scale St Barnabas had achieved.[95]

As Henry Benson stated in a widely syndicated article, the scale of operations meant that 'from a pecuniary stand-point the cost will be infinitely less than if it were undertaken alone and unpiloted', and as the *Hendon and Finchley Times* noted of the Salvation Army scheme, every expense was 'reduced to the lowest possible sum'.[96] The War Office grant then allowed the charities to 'reduce the already moderate charges ... [and] deal with many more applications than was the case last year'.[97] Such terms were clearly very much appreciated; a female visitor, 'A Grateful Pilgrim', had approached several tourist agencies but found their prices 'far beyond my purse': one tourist agency had quoted her sixty guineas, whereas she and her mother made the trip with St Barnabas for £25 in total. The discrepancy outraged her sense of morality against those who appeared to be profiteering from a sacred need: 'The thought of such a thing being turned into a money-making business was unendurable to me.' In contrast, she pointed out that a three-day tour of Ypres could be undertaken with St Barnabas Hostels for £3 10s, which included full board and lodging on the continent for forty-eight hours.[98]

Appreciating fully the value and quality of the service provided by the charities, the IWGC recommended them to all enquirers, telling those needing financial assistance to contact the Church Army, which offered its tours for around £6 per person, 'as compared with the ordinary rate of £20'.[99]

A truly remarkable facet of the war graves charities was their ability to deliver a service of intense thoroughness tailored completely to the needs of the individual at a fraction of the costs demanded by private enterprises. Assistance in locating a grave, arranging passports and war graves passes, booking of accommodation and railways, and the provision of guides to take a person to the precise cemetery and grave were all a standard part of the charitable societies' operations.[100] 'Officers are especially set aside to render all possible service to these guests, making their pilgrimage easier, and proving of great help in cases where it might otherwise have been difficult to locate a grave,' noted the Salvation Army Year Book for 1921, adding, 'The personal touch for which The Army is well known in all that it undertakes has nowhere been of greater comfort than in this enterprise.'[101] The influential Canadian writer and journalist Mary Macleod Moore, who had provided much coverage of the lives of Canadian troops on the Western Front during the war, wrote a highly sympathetic account of the work of St Barnabas Hostels for the *Graphic* in August 1920. One of the first things that struck her was the comprehensiveness of the service provided. On arriving at the Calais hostel, she found that an entire itinerary had been devised for every visitor, complete with drivers booked to ensure transport to and from the relevant railway station and cemetery.[102] Macleod Moore's praise of St Barnabas Hostels was echoed by 'A Grateful Pilgrim', who told *Country Life* that Mullineux himself sketched out her itinerary.[103]

As Macleod Moore noted, the care and attention to detail was very reassuring for people who not only were undergoing an emotional challenge but, in many cases, had never travelled abroad before, which also implied something about the class of the average pilgrim.[104] The Church Army reported that the advice it gave on routes, trains, passports, visas, expenses, the booking of accommodation, 'and countless other details ... would puzzle and add to the troubles and anxieties of the inexperienced traveller'.[105] When the great cemetery at Lijssenthoek was completed and formally dedicated at Easter 1923, a large pilgrimage was organized by St Barnabas Hostels. The *Belfast News-Letter* wrote that it was 'for the exclusive benefit of those who were too poor, too timid, and too inexperienced to make arrangements for themselves'.[106] A report in the *Burnley Express* noted the trepidation of a group of visitors who had never travelled abroad before, were dreading 'all the discomforts and embarrassments usually associated with foreign travel', and were thus immensely relieved to find that everything had been arranged by the YMCA.[107] Arthur Hickman of the YMCA described a typical pilgrimage scene in the *Red Triangle* magazine. He asked readers to imagine the people coming down the gangplanks of the cross-Channel ferries at Boulogne. Mixed among them were those, in little clusters of two or three and carrying their photographs or some keepsake of their lost loved one, seeking out the YMCA triangle sign. As with many other commentators, Hickman thought it necessary to point out that many of them had never left Britain before this trip, were often elderly and sometimes painfully naive, and yet they felt impelled to make the journey in order to visit the grave of their beloved.[108] Innocents abroad they most definitely were.

Having personnel available at all points of the journey and overseeing each phase of the process reassured these timid visitors. 'A Grateful Pilgrim' stated: 'My mother and I can never forget the tender care bestowed upon us from the first moment to the last', just as those on the Burnley pilgrimage 'had guides placed at their disposal to conduct them' at every stage.[109] Everyone on the Ulster and Scottish pilgrimage of 1925 was impressed by the fact that 'each of the 900 pilgrims was personally escorted to an individual grave'.[110]

'Everything was made so easy for us,' wrote one pilgrim in a letter of thanks to the Church Army, adding 'it is a great consolation to visit the graves and makes one realise what a great amount of trouble is being taken so that the great sacrifice which our lads made is not forgotten'. The guide was especially thanked for 'his goodness, he never spared himself in the least'.[111]

The travel agent John Frame found the YMCA team in France extremely helpful and knowledgeable. He was especially struck by a Mr

Holmes, whom he deemed a 'man of extraordinary magnetism'.[112] L.M. Orton, who lived at Teynham in East Kent, told her local paper of her family's trip to her brother's grave. They had been wishing to undertake the journey for two years and had finally achieved it thanks to the YMCA. Despite a rough crossing on a crowded boat, everything went smoothly thanks to the efficiency of the YMCA's service. They were met at Calais, given a meal, and then taken to the station for their train to Amiens. She added that the guides 'kept us all "merry and bright" despite our sorrowful errand'.[113] Nellie Burrin's trip to see her brother's grave was greatly assisted by the YMCA. As her account makes clear, she was met and helped at every point of her journey. On disembarkation at Boulogne a representative smoothed the way through French customs; at Amiens, another took her and her friend to the YMCA huts for a much appreciated wash and cup of tea. YMCA huts also provided much-welcomed cups of tea at Albert after a long day tramping the battlefields. Prior to her return home the hostel staff at Amiens gave Burrin and her friend a 'good send-off', and another YMCA representative met them at Boulogne to ensure easy passage on to the steamer.[114]

WOMEN, WAR GRAVES VISITATION SERVICES, AND THE GUIDING OF PILGRIMS

For all these Christian organizations, the work was very much a sacred duty and as such related to the wartime rhetoric of *war service* rather than *war work*, a semantic, psychological, and spiritual distinction explicitly and implicitly prized by many of the middle and upper classes in particular.[115] As the war effort had largely been one of self-mobilization, so was the building of a commemorative infrastructure. Advertising its hostel at Ypres, which opened in August 1919 and was capable of accommodating forty guests, the YMCA reminded readers that 'during the war, eighty YMCA centres were maintained in the Ypres salient' alone. The implication was that the YMCA had simply shifted its welfare work to meet a new demand intrinsically linked to its wartime service.[116] Macleod Moore pointed out how some people 'foresaw the insistent longing of the bereaved to visit the graves of their beloved'. Among those who had come together to meet this need none was 'more imbued with the spirit of true sympathy and helpful forethought than the St Barnabas Hostels'. She carefully delineated the voluntary status of its core team, referring to the organization's reliance on 'private donations' led by former 'war-workers'.[117] Here, the term 'worker' was clearly meant to imply a dutiful response to a righteous call rather than a conscripted or financially motivated one. Such workers had made the transition to the new sacred cause of providing consolation to the bereaved.

Much of the work on the ground was carried out by women and, it was often implied, for women, commensurate with their status as chief mourner. The War Office and IWGC recognized the special place of women as organizers of services for the bereaved pilgrim from a very early stage. At its sixth meeting, on 19 November 1918, the IWGC noted an application from Lady Julia Drummond to build a hostel to cater for relatives visiting war graves 'who could not afford to pay their own expenses'. For the Canadian Lady Drummond, this was a very personal calling, as she had lost her son at the Second Battle of Ypres. Realizing the anguish that so many Canadian families must be suffering, she had established an information bureau on behalf of the Canadian Red Cross, as well as becoming president of the Maple Leaf Club, an organization that ran two London clubs for Canadian servicemen on leave or recuperating from wounds.[118] The IWGC was keen to explore the potential of such offers and Ware very firmly placed it within the realm of women, suggesting that 'a Committee of women should go into the whole question of visits to the graves in France and of accommodation for visitors there'. Sir Nevil Macready, the adjutant-general, concurred and added that this committee 'should be composed of women who had experience of war work in France and of other persons'. Harry Gosling, the representative of organized labour on the IWGC, agreed, being 'strongly of the opinion that this was a job for which women were eminently qualified', and that there 'should be no reluctance in using their services at the earliest possible date'. But, as a trade unionist, he was distinctly aware of the possible class implications and so offered to 'give the names of two or three working women who will be very useful indeed' to ensure the task did not become the preserve of bountiful women of good standing.[119] As it transpired, the organizations on the ground in France and Belgium were way ahead of the IWGC and were already getting on with the task of assisting relatives through their huge networks of highly motivated women.

Mary Macleod Moore strongly perceived the special role of women as both mourner-visitor and guide. The tone of her published account was set in the opening lines of her article on war graves visiting:

'The hearts of many British women are buried in the soil of France, where are the graves of their sons.' Her pilgrim searching for a grave was there not to find a husband or brother, but a son: the essential pilgrim was not just a woman but a mother. Bereaved mothers were presented as sharing a universal desire to see the graves of their sons. For Macleod Moore, the highland Scottish woman wailing for 'My bairn, my bairn!' had an intrinsic affinity with the woman in 'distant Devon' mourning 'for her first-born son'. Her article then followed the journey of a woman whose sole intention of seeing her son's grave was achieved thanks to the skill and dedication of St Barnabas Hostels staff, and in particular, its female representatives.[120] Macleod Moore presented the entire team as not only strictly practical and administratively efficient but deeply human, with a particularly feminine sympathy: 'From the time a traveller starts she finds friends', treading a 'path made easy for the feet of sad travellers' by the women of St Barnabas. Reassurance was instantly imparted to each anxious pilgrim cautiously walking down the gangplank at Calais by these 'gracious women' wearing their badges bearing the organization's motto, 'Consolation'. Bathed in this feminine aura of care and consideration, the St Barnabas team offered the pilgrim an atmosphere in which they could release their emotions: 'It is an unobtrusive sympathy of which one is sensible as of a perfume. Into the ears of these hostesses have been poured many stories of loss and sorrow, while fervent prayers have been breathed in the hush of the small chapels to be found in each hostel.'[121]

Many of St Barnabas's daily services were overseen by Eleanor Barker. The daughter of a doctor who served in the Crimean War, and wife of a vicar, Barker had served in the Voluntary Aid Detachment during the war, helping to run a club for non-commissioned officers and petty officers in London before serving with the Church Army in Cherbourg. She became co-ordinator of Church Army activities after moving to Calais, and there met Mullineux, joining him on his establishment of the first St Barnabas Hostel in the town.[122] Macleod Moore described her as the 'lady superintendent', supported in her work by her team of 'hostesses'.[123] Barker's immense contribution was also recognized by 'A Grateful Pilgrim', who told *Country Life* that St Barnabas's success lay in her 'great organizing powers'.[124]

The War Graves Visitation Department of the Salvation Army relied upon the tireless Commissioner Catherine Higgins. During the war she had managed the organization's supplies of comforts to prisoners of war. After the war, on completing a global tour of Salvation Army branches as International Headquarters Representative, she was made head of the War Graves Visitation Department, overseeing its work for eight years, most of them spent in constant rounds of inspection in France and Belgium.[125]

Another woman with much administrative experience of running wartime services for soldiers was Helen Bax-Ironside, a former superintendent of the Officers' Club at Calais, who was appointed director of the hostels of the Fields of Honour Society, an organization designed to meet the needs of the bereaved.[126] Despite her skills base and track record of wartime charitable work, Bax-Ironside's operation created a grey area. An important part of the internal culture and external messages of the Christian charities had been their distinctness from the commercial organizations. But the formation of the Fields of Honour Society in January 1920 offered the potential for confusion through its hybrid operating model and status. Announcing the formation of the society, the *Daily Mail* told its readers that it was being formed with funding from the British and French governments and the allied Red Cross societies. The Society was said to be building camps at Arras ready to receive and house visitors for two days at an all-inclusive cost of £8 per person.[127] In fact, the Fields of Honour Society was a registered company, in which Bax-Ironside was listed as a director along with two other partners. This new venture was stillborn; it never made a formal business return to the Companies Registration Office and was formally wound up in August 1923, having been declared 'defunct never commenced business'.[128] The problem the Society may well have faced is that of trying to straddle the market between the pilgrim of modest means and the wealthier visitor prepared to travel independently or to commission the battlefield visiting services of an established travel company. Lacking infrastructure in France and Belgium, and very probably lacking the capital to make any kind of significant investment to secure it, the Fields of Honour Society had no ability to enter an environment already divided between specific types of experienced operators.

Further, the Fields of Honour Society was highly unlikely to gain any kind of comradely sympathy from the charitable organizations, which were probably highly suspicious of its nature. Indeed, the Reverend Mullineux maintained a judgemental eye on his fellow charities and was not above passing direct critical comment. In 1923, when mechanisms for distributing the profits from the sale of *The King's Pilgrimage*, a book detailing the king's tour of the battlefields, were circulated, he announced his displeasure at the proportions suggested. He was particularly aggrieved that the YMCA was to be given a share, declaring it a profit-making organization. Writing to an IWGC official he complained about the overly expensive YMCA war graves visitation scheme. 'I will say nothing further about this commercial Society beyond this. I have just seen a report "For Official Circulation Only." In 1921 this Society took 999 assisted Pilgrims at a cost of £15,000. 0. 0. With our £5000 we took nearly 1200 Pilgrims. I am all against putting anything into the coffers of the YMCA. It means an abuse of what is practically Trust Money.'[129] It was probably operations such as the YMCA's Red Triangle travel service and its offering of a three-day battlefield tour for £10 10s by the summer of 1921 that provoked Mullineux's ire.[130] Mullineux's feelings were shared by his IWGC correspondent, who agreed with his comments 'regarding a certain organisation'.[131] Given the massive role of the YMCA in war graves visitation, and the high levels of appreciation for its work, it seems likely both men had confused the YMCA's commercial arm for its charitable one. In turn, the commercially run battlefield tours must have created profits capable of cross-subsidizing the programme for poorer visitors.

On the ground in France and Belgium the sharpness of the distinction between the worlds of the travel industry and the charitable services was revealed most strongly in the choice of accommodation; nothing delineated the class, and perhaps also motivations, of the traveller more than where they laid their head at night. As the French authorities were well aware, trying to accommodate large numbers of visitors in the devastated region was a major problem, placing a great strain on what few signs of civilization remained. Nonetheless, plenty of French and Belgians recognized the profits to be made by catering for visitors, which led to rapid restoration of pre-war businesses, as well as new ones improvised from scratch.

Those cities sufficiently behind the lines to escape serious damage but still within easy reach of the battlefields held an advantage for those seeking the usual levels of comfort. It was for this reason that Sommerville Story recommended the 'excellent hotels' of Lille.[132] However, being beyond the range of the gunfire was no guarantee of fine surroundings. Marjory West, who had worked for the YMCA in Le Havre during the war, took the opportunity to visit the battlefields in February 1919, accompanied by her sister. In supposedly luxurious Lille they found a nice hotel, but it was bereft of metal fittings, the Germans having removed them all for conversion to munitions. As a result, their bedroom had no bolt, and the water supply to their bathroom had been cut, as pipes had been confiscated. Unlike many visitors, however, the sisters seem to have taken this in their stride, with Marjory noting, 'This was one of the few pieces of comedy we found resulting from the War.'[133] For those wishing to see the Somme, Amiens was recommended. As a fine city with a long history, good facilities, and minimal war damage, it was the ideal base. Writing in the *Pall Mall Gazette*, 'Old Brigade' praised Amiens for its banks, good hotels, and restaurants, the latter allowing the discerning visitor to obtain a good lunch basket for the long day's exploration with no difficulty.[134]

Closer to the battle zones, standards could be less predictable or stable. The guidebooks provided information and also tried to shape expectations. According to the Reverend J.O. Coop, former chaplain to 55 Division and author of an early guide, the visitor had to expect 'to put up with some little discomfort and inconvenience in the matter of hotel accommodation', and he urged all arrangements be made well in advance due to demand.[135] The visitor was told not to expect 'pre-war comforts and pre-war conveniences' but would be amazed at 'how even the present accommodation offered has been improvised'.[136] Not everyone was impressed. Staying in Amiens in early 1919, the travel agent John Frame did not think much of hotel standards, but added, 'But what we thought was a poor hotel then was nothing to what we encountered later in our tour of the battlefield area.'[137]

Stephen Graham, a writer and journalist before his wartime service in the Scots Guards, came across the evidence of a rapidly improvised hotel business in the devastated area during his 1921 visit to the battlefields. On the road into Ypres, he saw the long line of temporary wooden buildings serving as hotels, each subdivided into tiny cubicle bedrooms.[138] Trying to appeal to all types of visitors, but with a firm moral guide in its title, *The Pilgrim's Guide to the Ypres Salient* provided clear information on what kinds of hotel could be found in the devastated city. 'Situated outside the Menin Gate there are several restaurants newly built of wood, which have rooms to let at from 5 to 10 francs per night,' it stated before running through

Figure 2.1 The Excelsior Hotel, Ypres, c. 1920.

an annotated list. The Splendide was deemed 'rather expensive', the Ypriana, 'cheerful, clean, cooking good', while the Metropole served 'excellent cooking of a French type' with 'bedrooms clean and comfortable rooms nice, but small'.[139] This was possibly Stephen Graham's Ypres base, for he wrote of trying to 'sleep in a little bed in a cubicle with [a] tiny doll's house window'.[140] For those looking for more comfort in the old Ypres salient, *The Pilgrim's Guide* recommended the former rest, recreation, and staging post centre of Poperinghe, some eight miles east of Ypres. Skindles Hotel was particularly recommended for its 'very comfortable rooms, excellent cooking, hot and cold water in bedrooms' (see also chapter 11).[141] So lucrative was the business that a second branch was opened in Ypres, on the station square, in July 1920.

The wealthy Baroness Campbell, who was also a pioneer female motorist, certainly intended to counterbalance the difficulty of navigating the appalling roads of the battle zone with a bit of luxury. Having gained a list of recommended hotels from the Royal Automobile Club before setting out, she followed its advice wherever possible. Requiring lunch in Ypres, she stumbled upon one of the hotels listed in *The Pilgrim's Guide,* a wooden building with 'the inappropriate name of "Le Splendide"'. Having been unable to follow the Royal Automobile Club's recommendations in Ypres, she was delighted to find it proved absolutely correct in its assessment of a particular hotel on the Albert-Doullens road having comfortable rooms, hot water and 'first rate food'. In complete contrast, at Amiens she experienced immense disappointment in the very expensive Hotel du Rhin. No hot water was available, dinner was 'very poor', and matters were compounded by the staff's failure to garage her car properly, resulting in a dented back panel and damp interior caused by heavy rain.[142] (The *Gloucestershire Echo*'s correspondent was either luckier or less exacting, for neither he nor the author Elizabeth Braithwaite Buckle, visiting en route to her son's grave in the summer of 1919, passed any negative comments on the same hotel.[143]) Continuing Baroness Campbell's catalogue of complaints, the Hotel du Grand Cerf in Tournai was deemed a particularly rough and ready affair. It had no carpets, curtains, or lights, and the party's rooms contained no windows. However, she did explain that this was largely because the Germans had entirely stripped the place before quitting the town. In Arras her hotel room lacked a proper ceiling. As a result, she found it very difficult to get to sleep while plasterers remained at work in the room above late into the night. At two o'clock in the morning she implored them to stop, which brought the owner to the scene. He promptly told the men to knock-off before explaining that there was so much work to do that night shifts were common. At St Quentin she experienced 'the worst accommodation we had found, and the most knocked-about hotel'. In addition, 'sanitary arrangements were as in Japan in the old days', but at least the food was good.[144]

Such judgements were by no means confined to Baroness Campbell and her rather exacting standards. At the Hotel Folkestone in Boulogne, Olive Edis was 'staggered' by the bill for her very simple breakfast. Things were little better in Vitry-le-François, where she was most disappointed to find a room 'anything but to my liking'. The bed contained only a thin sheet and a grubby coverlet. To stay warm, she had to supplement the bedclothes with a towel, her dressing gown, and travel rugs. The hotel at Commercy was described as 'little-loved' when she moved on.

By contrast, Bruges offered comfort, and after a freezing day on treacherous roads, Edis and her chums had a stroke of luck on discovering the Hotel Paris in Compiègne. It turned out to be highly luxurious and 'the memory of the glorious supply of hot water which drowns all other memories of that evening, when I retired from public view for a few hours of real leisure, the like of which we had not known since we started our tour, remains like an oasis in a desert'. On arriving at the Hotel Metz in St Menehould, 'Daddy Blow', the Quaker relief worker accompanying Edis and her friends, was delighted to get their car into its sole garaging space. 'Indeed, ours was the only car in shelter,' Edis noted. This was no minor sybaritic detail, but a crucial point: with most cars being soft-topped tourers, getting them out of the weather, particularly in the snowy conditions of spring 1919, stopped road travel from becoming an even more miserable affair, as Baroness Campbell also realized.[145]

Once rebuilding had commenced in the devastated zones, the infrastructure improved, as did the standard of accommodation, and some hotels gained a suitably enhanced reputation among pilgrims and battlefield tourists. In Ypres, the Hotel Splendid et Britannique was used regularly by the Ypres League. The owners, Monsieur and Madame Kock, were enthusiastic collaborators with the League and worked hard to ensure all visitors were entertained with great hospitality.[146] A party from the Whaley Bridge British Legion left very much impressed by the care and attention shown by the owners, thanking them for their 'unfailing courtesy' throughout their stay.[147] The brisk efficiency of the Kocks was the subject of gentle comedy in an account of the 1928 Easter pilgrimage: Madame Kock was described as making 'billeting arrangements in her own masterful style, allotting bedrooms in a reckless manner and without any regard for the private wishes of the victims'.[148]

For those visiting the Arras and Somme battlefields, the popular Hotel de l'Univers in Arras, with its reputation for good food, quality, and style at affordable prices, was 'thoroughly recommended' by W.A. Francis, honorary secretary of the 2 London Regiment association.[149] The men of the 1/7 Northumberland Fusiliers were equally impressed by the 'good food and comfortable bedrooms [which] materially conduced to the success of the tour', as were their fellow countrymen from the Northumberland Hussars old comrades association.[150] The 5 London Field Ambulance association had a 'first-rate lunch' in the hotel, and the Whaley Bridge British Legion party 'enjoyed an excellent dinner'.[151]

Exploring the battlefields in the summer of 1938, the veteran A.W. Keith was delighted by the wonderfully warm welcome shown to him by the Australian ex-servicemen who owned the Hotel Lamartine in Amiens.[152] For those facing an emotional and physical battering while visiting the battlefields and cemeteries, the quality of the accommodation was crucial. Often exhausted by evening, the ability to rest and relax in convivial surroundings was paramount. Whether a spartan but cheery YMCA hostel or a fine hotel, where a person bedded down significantly affected their ability to cope with the battlefields, particularly in the years immediately following the war, when the former Western Front was a wasteland.

*

Travelling and finding a place to stay in France and Belgium, so easy before the war, was a difficult, confusing challenge by 1919. For those suffering the agony of loss whose only desire was to see the grave of their loved one, the challenges in getting to and from the former fighting zones and locating accommodation as quickly and efficiently as possible must have seemed dreadful. For those of modest means, it must have seemed not so much a challenge as an impossibility. Recognizing the poignancy of their plight, the charitable organizations devised a workable system with amazing rapidity, while the commercial travel industry ran in parallel, exploiting the new opportunity with equal skill and ingenuity. By 1921 the world of the old Western Front was swarming with a range of British visitors. Some were pilgrims on progress, while others were traipsing tourists, but both kinds of visitor felt a hypnotic draw to a world of devastation, and both sought to walk with ghosts in the land of death.

3

Postcards from the Road

Relatives wishing to visit graves in the Ypres salient in the least possible time can go direct to Ypres viâ Dover and Ostend, the journey occupying about nine hours.

    Ward, Lock and Co., *Handbook to Belgium and the Battlefields* (c. 1921)

From Beugny I walked along the Cambrai road towards the line. A lorry laden with bricks and tiles overtook me and splashed me with mud. There was nothing very new about that, I reflected.

    Arthur Behrend (1921)

I turned back from Combles and made a diagonal at right angles to the direction in which I had been going, in order to run along the back (roughly) of the second German line, and hit the Albert-Bapaume road, almost in the centre.

    R.H. Mottram, *Journey to the Western Front* (1936)

Accessing the cemeteries and battlefields was usually achieved using a combination of transport methods – rail, ship, motor transport, and walking. Once on the battlefields, the primary mode of movement to be adopted was a topic of much discussion and consideration. This discussion went beyond utility and convenience to encompass moral and spiritual implications. How a pilgrim progressed was as important as where they progressed to.

EXPLORING BY CAR

Wealth and status were, of course, great determinants of precisely how someone got about. For those of modest means, the usual way was by train to a channel port, followed by another train journey on the other side, before the final link with the ultimate destination was made by motor transport of some kind.

The main railway stations along the former battle zone soon spawned local taxi services specializing in battlefield and cemetery visiting. As with the battlefield tourism packages offered by UK-based operators, many of the local drivers emphasized their credentials to ensure trust and confidence. In Ypres Captain P.D. Parminter, founder of Wipers Auto Services, could arrange to collect people from ports or railway stations, organize guided tours of the battlefields, and photograph graves and lay wreaths on behalf of relatives.[1] Malcolm Cockerell's guiding service was based at Station Yard, Albert, on the Somme. His letterheads and advertisements sought to gain the trust of visitors by announcing his status as a former captain in the Royal Army Service Corps. The list of services Cockerell advertised was typical of those provided by the battlefield guides based in Belgium in France. He proudly stated that he could arrange for visitors to be collected from railway stations, arrange battlefield tours led by British drivers fluent in French, photograph graves and cemeteries, and supply and lay wreaths on specific graves.[2]

There were plenty of others in the same mould: Captain Stuart Oswald operated from the Hotel Carlton, Amiens, conducting 'day and half-day tours to the battlefields in private, comfortable motor cars, and employ[ing] British ex-service driver-guides',[3] while Philip Vyner, formerly of the Army Cyclist Corps, was proprietor of the British Touring Service, based on the station square at Arras. Vyner was also the representative of the Ypres League in the area, and he, too, offered tours led by British driver guides.[4] Visiting Albert in the late twenties, Captain H.A. Taylor identified a number of British expatriates making 'a living by driving British visitors to war cemeteries'. All were elevated above the ordinary by their veteran status: 'Generally, they are very good fellows, these former soldiers whose British countenances make an odd contrast with their French clothes.' An IWGC official told Taylor these ex-servicemen were a fine crew. He knew of drivers prepared to take poor women to their destination gratis with a '"Jump in, Mother, and send the fare some day when you can spare it."'[5]

By contrast, the most affluent and adventurous could act independently, taking their own cars across the Channel. This was a clientele many of the earliest battlefield guidebooks recognized. To promote smooth passage through customs, the Reverend J.O. Coop warned motorists to get their paperwork in order before crossing the Channel,[6] while *The Pilgrim's Guide to the Ypres Salient* provided a full page of 'notes on the conveyance of cars and motor-cycles to the continent' including the current customs rates due at the French and Belgian borders; similarly detailed information was given by *Muirhead's Belgium and the Western Front*.[7] As a pioneer of motoring in Britain, and author of a 1913 work, *My Motor Milestones*, extolling its virtues as a practical and flexible form of transport, Baroness Campbell was determined to explore the battlefields by car.[8] Her full title was actually Baroness Campbell von Laurentz, which she gained through her husband, Captain Edmund Kempt Laurentz Campbell, who served with the Prussian Lancers during the Franco-Prussian War and was for many years equerry to Duke Ernst at the court of Saxe-Coburg-Gotha. By 1919 she was a widow, which may have allowed her to downplay her association with a German princely family. She put further distance between herself and Germany by dropping the German half of her name early in the war.[9]

It was therefore as 'plain' Baroness Campbell that this confident and experienced driver set off for the battlefields in the autumn of 1919. The inspiration for her trip came after meeting a friend who had just returned from the battlefields, and the prospect was made all the more appealing thanks to the recent demobilization of Campbell's chauffeur. Her first act was to contact the Royal Automobile Club to obtain information on hotels and the condition of roads. Next, she made the necessary customs arrangements, gained a passport, and obtained a bankers' letter of indemnity for insurance purposes.

A further indication of her status was revealed by her decision to take at least three cars to France, but, due to poor weather, only one could be safely loaded on to the ship. Once in France she met up with a friend, who happened to be a general. This senior officer proved his worth instantly by taking her to lunch at General Headquarters, where 'much valuable information was gained, and all trouble with passports and landing of the car taken off our hands'.[10] Very nicely set up by these encounters, she set off along the 'splendid road' from St Omer to Cassel. The equally well-connected Lady Norman commenced a battlefield tour in early 1919 along with Olive Edis, an official photographer. The mission of the two women was to obtain material for the Imperial War Museum's women's war work collection. Although Lady Norman's position ensured the services of a touring car and chauffeur, the latter proved far from content. Dismayed by the bulk of Edis's equipment, he insisted that they break open the boxes and empty them to reduce the weight. The two women were content to put up with this grumbling; the 'comfortable Vauxhall car' to which they were transferred was a much better option than the rickety old ambulance they had been forced to use at the start of their tour.[11] Women, particularly women of a certain status, were therefore among the pioneers of battlefield touring.

For visitors attempting to traverse the battlefields independently like Campbell and Lady Norman, ensuring petrol supplies was an important issue, and many accounts designed for public consumption stressed this piece of advice. For Campbell it was a constant source of anxiety. At the start of her journey, she secured 'eight *bidons* of *essence*' at a garage in Cassel. A few days later in Lille she was told that only very limited supplies were available, but thanks to her influential connections she managed to secure a further twenty-five litres. However, by the time she reached St Quentin she was again running desperately low. This led to an anxious moment when the only person able to resupply first had to fulfil the orders of his regular customers. Fortunately, she was then able to get all she needed.

In Amiens she had to commence the search all over again, while at Ostend, after much searching, she discovered someone selling petrol from his house. 'We took the car round, and he had many two-gallon "Shell" cans filled for us from evidently a large store of petrol in his back premises.'[12]

Also writing in 1919, Captain Atherton Fleming informed readers of his battlefield guide that the motorist should always take plenty – 'and by that I mean plenty' – of lubricating oil and petrol, as obtaining it locally might be very difficult.[13] Similarly, Coop's 1920 guide warned that once the driver had left Amiens and reached the Somme battlefields, obtaining fuel became a very difficult task.[14] One person exempt from this problem was Lady Norman. She was fortunate enough to be touring the battlefields so soon after the end of the war that they were still very much militarized and populated with troops, and this happenstance, combined with use of her official pass, helped ensure a ready supply of petrol from official stocks. She noted that the crucial liquid was 'freely offered to us wherever we went' and there was no need for any wangling.[15]

Road conditions too had to be taken into account. 'Owing to the state of the shell-pitted roads in the war area travelling by motor car is a slow and precarious business,' wrote a Press Association journalist in June 1919.[16] L.M. Orton, visiting her brother's grave in October 1921, was surprised by the charge of FF100 per person for the car journey from Bapaume to Ypres, but having seen the difficulty of the driving and the condition of many of the roads, she quickly understood.[17] Wilfrid Ewart, officer-veteran of the Scots Guards, tried following the Flers-Longueval road on the Somme. Although 'so clearly marked on the map', it disappeared into undergrowth and debris and then reappeared every so often.[18] As Coop warned his readers, if bad weather descended, the going could become treacherous.[19]

In the bitter winter of 1918–19 the roads were indeed treacherous. On the snowy drive back to Boulogne with Lady Norman in early 1919, Edis was pleased to bump into an officer requiring a lift, as he knew every route and could help navigate them expertly around every blocked street.[20] Setting out from Doullens for Mons on a freezing February Sunday morning, YMCA worker John Hastings Eastwood had to traverse the full width of the Somme battlefields. The intense cold combined with the appalling roads made the journey a trial of patience and endurance. A cracked radiator, doubtless the result of the endless jolting on the unforgiving surfaces, required constant refilling. The mains supply in Doullens being frozen solid, Eastwood had to make five trips to the river before he could even begin to start the laborious process of firing up an engine reluctant to kick into life on such a cold day. Nursing the leaking radiator, with regular stops for top-ups, meant that progress was magisterial at best.

With every inhabitant of the devastated zone clinging on to a fragile existence, obtaining help when things went wrong was no easy task. One farmer allowed Eastwood to use his well, but would not lend a bucket (a precious piece of equipment in that devastated region). Thankfully for Eastwood, a farmer at a subsequent stop was prepared to supply one and even helped to lower it down and wind it up. The one saving grace was that it was a glorious winter day of bright sun, blue sky, snow everywhere, and 'roads like glass'.[21]

Travelling at much the same time, Olive Edis and Lady Norman were driven along roads 'which in some places were a sea of mud', leading to 'several very lively skids'. All the time they drove down the Belgian front 'the mud became deeper and deeper, and our axles in seas of water'. Moving out across no man's land, they used 'a straight and shiny tract of watery road … and again we plunged and jolted through lakes and shell holes, till it was a marvel our axles did not break'. At Hooge, on the Ypres battlefield, their car got its third puncture, 'and Lady Norman and I scrambled about it in the mud, about which one had to become absolutely reckless, plunging in it up to one's ankles'. At Lassigny, the road was completely blocked by debris and yet more mud. Taking a different route, the way was blocked by a wrecked lorry. Carefully negotiating their way round, and as a result slowing progress down considerably, they finally reached Péronne, 'almost frozen in our seats'. Travelling in an open touring car in 'the appallingly keen' wind made them very 'thankful for all our warm rugs'. Heading for Toul in driving snow and an equally cutting wind made Edis equally thankful 'not to be responsible for the driving' in such conditions.

The roads near Rheims then proved an extreme trial for the car. With the guide declaring the tyres to be in such a poor state that 'it would be absolutely impossible to get further that night', he advised finding shelter quickly. Although they were offered hospitality by a local woman, the terrible conditions in which the poor woman lived dissuaded them from staying and they determined to push on. Fortunately, they managed to navigate the treacherous pavé and reached Compiègne.

During their travels they were joined by an experienced Quaker relief worker, nicknamed Daddy Blow, who accompanied them for part of the tour. When he found that they intended to visit the Verdun forts, he did everything in his power to dissuade them due to the dire state of the roads. Nonetheless, they remained determined to proceed, and as the going became increasingly difficult, 'Daddy Blow grew more and more rebellious'. However, he remained loyal and proved invaluable to the party. Despite his fears, warnings, and protests at their itinerary, he successfully navigated them through much of the trip, and brought them back safely. They were left convinced that their original 'professional chauffeur assuredly [would] have arranged accidents or punctures at many critical points' in order to abort the excursions.[22]

With the going so difficult, expert advice such as that provided by Daddy Blow was extremely important. Sommerville Story's 1920 battlefield guide informed the reader intent on detailed exploration to ensure 'he has someone with him who can give him information'.[23] Like Edis and Lady Norman, Baroness Campbell had the good fortunate to be travelling with a companion as dedicated as Daddy

Blow: the general utilized his connections network to determine the best routes around Ypres. But even this did not ensure a completely smooth passage. Although an expert driver herself, Campbell was pleased to hand over to her chauffeur having been 'shaken to pieces crawling at five to seven miles an hour' navigating the 'detestable' roads around Ypres. She told her driver, 'Now you shall take her, for nothing will induce me to drive over that vile road for a second time.' A day later her party travelled fifty-four miles, 'but such miles! ... over a truly awful road – bump, bump, all the way, dodging in and out of holes'.[24] On another day the wheel was 'nearly wrenched out of my hand several times'. As a result of this battering the radiator developed a leak, but, to Campbell's joy and amazement the suspension springs held up, which she saw as a credit to the quality of a Rolls-Royce. (Indeed, Rolls-Royce must have had similar confidence in its product, for it took out a four-page advertisement spread in *The Pilgrim's Guide* extolling its limousine as 'the zenith of motoring luxury'.[25]) The tribulations went on and on, and there was never a day of easy driving.

The devastation of the Somme proved exceptionally trying, especially when two of Campbell's companions decided to explore a wood in more detail and asked her to meet them on its farther side. After having experienced a 'truly awful road' near Chuignes that ended in a pile of planks, she was not confident, and told her friends that if she attempted it, 'I shall never see you again, for I have been warned never to try the side roads'.[26] She was quite rightly sticking by the advice of the reputable guidebooks. The *Pilgrim's Guide to the Ypres Salient* solemnly warned, '"Short Cuts" should be avoided, as in this devastated country they are difficult to identify.'[27] In a similar vein, Captain Fleming advised both walkers and motorists to avoid the tertiary roads, especially at night.[28] Nonetheless, Campbell attempted to navigate the route after a local woman said it could be done. Here again, she was rigidly in line with the guidance. 'Travellers are warned that all roads and tracks, other than the main roads, present difficulties and, indeed, occasionally become indistinguishable,' Coop's guide declared, adding, 'In many cases the traveller will probably pass his landmarks without recognizing them, and may very easily get lost.' He therefore advised relying upon local expertise, urging visitors to make enquiries before travelling on tracks or side roads, and ideally to be 'accompanied by some person thoroughly acquainted with the locality'.[29]

Lady Campbell had followed this advice closely by carefully enquiring about the conditions from a local inhabitant, but then she met another who warned that a car would never make it through. At this point she despatched her driver to investigate. 'Quite impossible, the car will sink in', was his response. Close study of the map revealed a safer route, but this took a long time to complete. After an eight-mile detour Campbell and her driver caught up with her friends, who expressed concern at their long absence. Left alone, waiting for her return had caused them to wonder whether they would be forced to shelter for the night in a nearby abandoned hut.[30]

Alice Knight, a young woman from Kent who crossed to France in the early summer of 1919 to work for the American YMCA, was also forced into strange adventures due to the war-damaged roads. Along with a party of others she was taken on a battlefield tour of Belleau Wood. On alighting from the train at Château Thierry, the group clambered into lorries with French drivers. Progress was miserable as the lorries broke down continually with the usual problem of radiator trouble caused by the constant buffeting on the poor surfaces. It was probably a radiator leak which gave Knight the sudden shock of a spout of warm water soaking her clothes. With time running short due to the constant stoppages, and fearful they would not get to see the wood, some members of the party dismounted and started to walk. Knight had not intended to join them, but the lorry then set off with a start, catapulting her over the tailboard; fortunately, she was caught by two sailors. Totally oblivious to this loss of his passenger, the driver went ploughing on.

Knight was now left with no choice but walk into the wood alone. This was a disconcerting experience, as she told her parents: 'I cannot explain to you my feelings stood there absolutely alone in Belleau Woods by the side of the little cemetary [sic]. All of a sudden I spotted a soldier who I had met at Reims and it appears he was on another lorry and the same thing had happened to him. Really although I felt so nervous I was jolly glad to see him. I asked him what could I do, and he said "Never mind little one. I'll look after you or take you back to Chateau Thierry if I possibly can."' She seemed to have realized that she was in a potentially awkward or even dangerous situation, adding her relief at not having 'Wrightie', her slightly primmer friend, with her, who would have been made very wary by the soldier's manner.

Figure 3.1 Plank road near Ypres, summer 1919. Left to right, Miss Butcher, 'Curly' Allen (note the shell-case souvenir), and Miss Ellis.

The soldier went on to say, 'I know some of the Americans are awful rotters but trust me and you'll be alright.' Heading in the direction of Château Thierry, the two bumped into another woman with an American officer and they joined up. To keep their spirits up they sang songs on the way while joking about which of the ruined houses they might be forced to bed down in. 'I might say some of them only possessed a piece of a wall, you could see there had been an upstairs by the shape of the window standings. The majority of them looked as though one could blow them down. Oh my! The Germans absolutely wrecked the whole place.' Fortunately, after some time, a car, driven by some YMCA women workers who had been sent out to find her, stopped to pick them up. Safely back at the YMCA offices in Château Thierry, all three weary travellers were given dinner and told to make themselves comfortable for the night, but Knight did not sleep well because she was very nervous about missing the 6.20 a.m. train back to Paris. Despite the saga, she told her parents, 'I shall never forget that experience so long as I live. The whole thing is a great experience from beginning to end.'[31]

A slightly less dramatic, but equally challenging situation was experienced by a party of veterans of 49 Division in the summer of 1921. They found the roads across the Somme mostly impassable except on foot, and so punishing was the experience of trying to navigate them by car that during the course of the day their transport sustained three punctures, a smashed windscreen, a broken spring, and a burst radiator. Perhaps unsurprisingly, the car then broke down completely as they began their homeward journey to Amiens. Unable to make a repair, their ex-service French chauffeur shoved the car off the road, and the veterans began the task of seeking a lift in the gathering gloom. Having no luck, they walked into Villers-Bretonneux, caught the train to Amiens, and just managed to get back in time for their connection.[32] A correspondent for the *Scotsman* exploring the battlefields by motorcycle in the spring of 1923 was another who almost faced defeat due to the roads: 'There was no need of traffic noticeboards calling for slowing down of speed around Messines. Every yard of the way had a jolt in it,' he noted. When considering the best route across the border, he was told to make for Neuve Eglise rather than proceed via Ploegsteert for the simple reason 'there was no more road'.[33] Aware of the difficult, and occasionally treacherous, conditions to be found in the old Ypres salient, the Ypres League's 1925 map delineated between 'roads passable for motor traffic', those requiring 'very careful driving', and those 'generally unsuitable for motor traffic'. Three years later, a full ten years after the Armistice, some visitors found the condition of the roads remained poor. W.J. Baumgartner believed many of the roads to be the only things left in their wartime condition.[34]

According to its enthusiasts, a form of transport far less likely to be held up by such conditions was the bicycle. In the summer of 1922, Captain Mee took his bike over the battlefields. Writing up his experiences for his local newspaper, he thoroughly recommended the bicycle as a means of exploring, owing to its flexibility, which gave him the ability to 'get off the highways and to the out-of-the-way yet famous corners'. He proved this contention by completing a 330-mile round trip from Zeebrugge down to the Somme and back entirely by cycle.[35] Another veteran, F. Hermen, also recommended the bicycle. He told readers of his local newspaper 'to hire a cycle at Ypres. One can then leave the main roadway and search out that old dug-out which, if of concrete, will still be there; or that ditch that caused so much bad language during a raid.'[36]

Figure 3.2   Menin road at Hooghe, c. 1921.

MOVING BETWEEN FRANCE AND BELGIUM

A major annoyance for all visitors, whether they defined themselves as a tourist, pilgrim, or returning veteran, was the customs procedures of the Belgians and French. These were especially likely to affect those travelling by motor vehicle and those in large parties. Few comments were made in the guidebooks about the initial point of entry at the Channel ports, which implies that all formalities were relatively efficient and, crucially, expected. On arrival in France and Belgium, veteran groups could also play on their status to help speed them through customs. A party of veterans from the Manchester Regiment arriving in the summer of 1936 certainly believed they had a magic touch. On disembarking at an unnamed French port, customs were smoothly negotiated, as "'42nd Division" again proved to be an "open sesame"'.[37] In complete contrast to the customs procedures at the ports was the exasperation experienced by travellers attempting to cross, or re-cross, the border between Belgium and France. Many were annoyed by a process that seemed overly zealous, officious, and lacking any sense of general procedure or order.

On leaving Belgium at Ploegsteert, Baroness Campbell found the Belgian border post unmanned, but on arriving at the French equivalent, was turned back for not having the correct Belgian exit documentation. Her driver was instructed to take a different route through the nearest Belgian town where the post was likely to be in operation. At the customs point Campbell found the 'Belgian people most tiresome. [They] would not believe that a Rolls-Royce has no engine number, and examined the car all over.'[38] As the early guidebooks make clear, Belgian and French regulations for driving foreign-registered cars were extremely complex and must have stumped many.[39] The veterans of the 1/5 Gloucester Regiment association making a trip from their base in Albert found that only one of the three vehicles and drivers had the correct paperwork to cross into Belgium, and so the party dismounted, left the bus behind, and walked to Ploegsteert Wood, where they visited the cemeteries that included some of the earliest casualties from their battalion.[40]

A tour for the Christian organization Toc H (see chapter 11) in 1929 was also left frustrated by the problem of paperwork. Heading for France in a fleet of buses, they drove via Reninghelst, over the foot of Kemmel Hill, to the border at Neuve Eglise. Here the French frontier guards argued that there were discrepancies in the paperwork of the Belgian drivers. 'Much picturesque rhetoric flowed – as well as time and watery beer', as the pilgrims watched a lengthy wrangle unfold. Finally abandoning hope, they drove to Menin, where the French officials made no demur at all, and they went on. Nonetheless, they took an hour to get through the customs barrier as everyone and everything in the long queue of people waiting to pass was checked. To while away the time the Reverend 'Tubby' Clayton, co-founder of the Toc H movement during his time as an army chaplain during the war, engaged everyone in a sing-song. The chronicler of the tale at least managed to find wry humour in the whole thing: 'To the citizen of Menin, who suffers these inconveniences every time he wants to walk down the street, the talk about brotherhood of nations at The Hague must seem a little academic!' The disruption played havoc with their timetable, as they got to Lens at five o'clock in the afternoon, rather than one o'clock as originally scheduled.[41]

The Ypres League August Bank Holiday pilgrimage of 1935 had much fun on the Franco-Belgian border with the Belgian customs officer, who had spent time in Yorkshire during the war and was keen to have a chat with them. However, on the French side things were very different, as everything was unpacked and inspected by 'a pompous officer [who] made enquiries if we possessed tobacco, cigarettes, chocolates, etc'. On the way back, the same Belgian customs officer was on duty and greeted them with good humour: 'Of course, you have nothing to declare! Oh no!' and he promptly lifted the barrier.[42] The Reverend T.B. Stewart also found humour in the customs procedures. Despite being bundled off the train while officers searched the carriages, he still managed to grin when they returned with two boxes of undeclared cigars and confiscated them leaving 'two Edinburgh husbands ... still mourning their loss – if their wives have had sufficient courage to confess it'.[43]

Henry Williamson was far more cynical about the process. Wracked with disillusion, which expressed itself in a highly jaundiced view of Britain's former allies, he deemed most Belgians and French swindlers and charlatans. The customs officials were thus portrayed as prime examples of the mean-spirited nature of both nations. Williamson and his travelling companion, a fellow ex-serviceman, were accosted by a surly French border guard near Ploegsteert. Opening their packs, the guard promptly pocketed their cigars. Both men protested and were told they would have to pay 160 francs duty. Stunned by this outrageous demand, they said they would go back into Belgium and smoke them before returning. At this point they found themselves in an ironic bind worthy of many tales of the war, as the guard told them that they were now on French territory and so could not go back with their cigars. More discussion followed, in which it transpired that three cigars could be taken across the border duty-free and, with each man having lit two cigars and stuffed them into their mouths, they crossed over the border.[44]

The whole scene is presented as a farce driven by petty-minded, unthinking, and unsympathetic bureaucracy. By so pointedly ensuring that the key identifier was the man's nationality, Williamson seemed bent on portraying not the world as enslaved to such people so much as the national sentiments behind them. Although he did not say it clearly and unequivocally, his vituperative tone almost implies that Britain fought the wrong enemy. Totally different in his outlook was Stephen Graham, who remained completely blasé about his experiences. Making the crossing between Belgium and France near Locre in 1920, he found a rope strung across the road. 'The customs gendarmes will examine you if you are coming into Belgium, though they will pay little attention if you are going out.'[45] Coop also accepted it as a fact of travel between the two countries, worthy of little additional comment. Referring to the customs posts at Abeele (the route many would have taken on the train from Calais or Dunkirk via Hazebrouck) and Locre, he deemed the customs officials 'polite and suave, but they are firm – especially the French'.[46]

WALKING AS THE SPIRITUALLY AND MORALLY APPROPRIATE METHOD OF EXPLORATION

For walkers like Williamson, less likely to be burdened-down with taxable objects, despite his complaints, the process was usually easier than it was for those in motor transport. It was just one of the advantages that walking offered. Of far greater importance to most chroniclers of battlefield visiting was the ability to engage fully with the landscape when the visit was made on foot. Writing in a very early battlefield guide published in the autumn of 1919, Fleming ranked the 'walking tour' as the primary way of seeing the battlefields, but also acknowledged that this was for the person 'with plenty of time on his hands', and therefore perhaps also with plenty of money. The second means was to take trains to the hub of each region and then undertake walking tours from those points. Third was use of the motorcar. Finally, there was the 'possible fourth, and that, there is no doubt, will be the inevitable "conducted tour"'. Here Fleming made some subtle value judgements. He clearly believed that certain ways of visiting the battlefields were spiritually more apt and thus more enriching. Although he believed the dawn of the conducted tour was inevitable, his resignation to it did not mean he thought it a wholly sound or decent development. Rather, he saw in it the likelihood of a tick-it-off-the-list approach, which was wholly unsuitable for a deeper, spiritual, and intellectual engagement with the battlefields. Only walking could achieve the right understanding, and 'this in my opinion, is the ideal method', Fleming added.[47]

Writing for wealthier, particularly wealthier American, visitors, Sommerville Story provided the option of travelling on foot, but noted that it required 'a good deal of time and patience and will be found highly fatiguing under the circumstances'. For such visitors walking was to be regarded as a small optional extra to provide a little bit of extra depth, rather than an integral part of the experience. 'Plenty of opportunity is given on this trip (as on others I have described in the previous volumes of this series) for a close examination of various points on the route for a little walking.'[48] Story clearly felt a little walking was the most his particular readers would be prepared to countenance. Others agreed with Fleming and encouraged exploration on foot as the only way in which the battlefields could be truly understood. Lieutenant-Colonel T.A. Lowe, veteran and author of one of the earliest guidebooks to the battlefields, was a definite proponent of this means of travel, and declared reliance on foot the best way of seeing the battlefields. Making an explicit allusion to the wealthier class of visitor who could afford a car or hire a chauffeur-driven service, he added that 'the poor man will have the consolation of knowing that by walking he is seeing a great deal more of the country than the motorist could possibly see'.[49]

Given the difficulty of finding and accessing many places, walking was also a practical necessity and the traveller had to be prepared to undertake a good deal of exploring on foot. Generally, it was possible to get between the major towns and cities by car, but any kind of excursion from the main routes could prove extremely difficult, while finding remote cemeteries and memorials was almost impossible unless undertaken on foot. As one newspaper reviewer remarked of a battlefield guide, whatever moral benefits there might be in walking the battlefields, the visitor would find a bit of 'foot-slogging' impossible to escape, given the nature of the roads in the immediate aftermath of the war.[50] In his 1920 battlefield guide, Coop admitted that roads had improved a good deal since the Armistice. However, many remained very difficult for cars and horse-drawn traffic and in wet weather were 'likely to be impassable'. As a result, the visitor 'must be prepared to cover a great part of the ground on foot'. By walking, 'the traveller is less confined to any beaten track and can explore [the] important portion of the battle-field at his own will and in his own leisure', Coop declared.[51]

Whether by necessity or choice, many visitors walked many, many miles. In the earliest days of visiting, the amount of ground covered on foot was at its highest, in large part due to the lack of alternatives. Presumably because other transport simply was not available, Nellie Burrin spent a good deal of time walking during her battlefield pilgrimage in the summer of 1920, as her diary attests. Along with a companion, she walked from Villers-Bretonneux to the Heath British Military Cemetery to see her brother's grave. Two days later, wishing to see the grave again, they completed a similar trudge 'for some distance along a straight, dusty, never-ending road to Heath Military Cemetery, arriving at 11.15 a.m. – about an eight-mile walk'.

A day earlier they had walked from Albert to Pozières and back, covering nine miles in total. Facing such distances, it was little wonder they were very pleased to accept lifts from passing cars when the opportunity arose.[52]

Alice Knight was a similarly grateful recipient of lifts from passing vehicles. On a freezing day in 1919, she set out to find the grave of a family friend on the Somme battlefields. Accompanied by her chum, 'Wrightie', Knight alighted from the train at Albert, where the two young women commenced the long walk to Delville Wood. Presumably, Knight must have been given a map and directions, for the two women seemed to know which roads to take. 'It was awfully cold and the sky looked full of snow,' Knight wrote in a letter to her mother. The snow started an hour in to their walk and continued the whole day. Knight and Wrightie were determined to press on: 'We didn't trouble any about it. We just buttoned our coats up round our necks, and we had got some good shoes on, and we walked merrily on. We sang, and we laughed, and we talked.'

Fortunately for them, when they had covered just over half the distance, some British soldiers came past in a car and gave them a lift to Montauban. After a stop for a meal in the YMCA hut, the two women marched on to Guillemont before finally arriving at Delville Wood. On the return leg they had a similar stroke of luck, as at Mametz they were picked up by passing soldiers and taken back to Albert. They estimated they had walked about twenty miles in total but were told by the soldiers it was more like twelve, which the women doubted. However, the soldiers' estimates were probably about right, as it is just over eight miles from Albert to Delville Wood, and the lifts the two were offered meant they did not cover the sixteen-mile round trip entirely on foot. However, it was still a considerable achievement given the dreadful conditions and the nature of the roads.[53]

As transport links improved, walking became a deliberate option. When veterans returned, the allure of foot travel was immense, and it proved a standard approach for those commissioned to write up their experiences for newspapers. Ferdinand Tuohy walked the old line from Albert to Ypres in the summer of 1927 and published his reflections in the *Sphere*.[54] Eleven years later, A.W. Keith undertook a similar epic meander across the battlefields, which inspired ten articles for his local newspaper.[55] Henry Williamson, although he used the train to connect between the major battlefields, undertook the detailed explorations on foot and eventually compiled his newspaper articles into a book, *The Wet Flanders Plain*, published in 1929. Many other veterans were equally convinced of the value of walking. When W.A. Allinson returned in the summer of 1926, he walked through the Menin Gate, up the Menin Road, turned off at Birr Cross Roads Cemetery, and followed the road up to Zonnebeke. By the time he and his companion returned to Ypres, they had covered the best part of eleven miles.[56]

The appeal of walking was not confined to veterans. Toc H pilgrims always took the opportunity to walk: their particular identity as a Christian organization that had emerged from the blood-soaked fields of the Ypres salient, as well as their sense of community, demanded it. In 1929, they walked back to Ypres from Kemmel, a journey of six and a half miles, arriving in the city as dusk fell.[57] A year earlier, a party of Ypres League's pilgrims elected to walk across the battlefields rather than take the bus or steam-tram.[58]

Those who could not find the time to explore by walking seemed to feel the need to explain their actions. A party visiting Ypres in the summer of 1923 felt that 'such a pilgrimage should, most certainly, be made on foot', but pressures of time forced them to take a motor tour with a guide.[59]

For the non-veteran, walking was thought to be a way of gaining insight into the experiences of soldiers. Lowe was a firm believer in this claim: the visitor 'will feel that he is tramping as they were obliged to, minus the burden of a rifle, equipment and heavy pack – details which are well calculated to minimize the delights of a walking tour'. However, he also said the potential traveller ought to assess themselves physically, as this method would be hard going.[60] According to Ferdinand Tuohy, the battlefields were a landscape best appreciated on foot with the 'occasional lorry hop', as any Tommy would have done behind the lines during the war.[61] For Fleming, the only way to get a real understanding of the Arras front was to be prepared to venture off-road:

> I am afraid that the earnest student of the battlefield area
> will miss very much if he sticks only to those routes which
> are in good condition. 'Foot-slogging' is the only way to
> see the real points of interest. Make an early morning
> start, with some good sandwiches and a bottle of wine
> in a haversack, a good stout pair of boots, and the fixed

determination to see as much as you can in the time, even if you do finish the day weary and dirty, and you will see more in two hours than I could describe in two volumes. You will see a lot more of the country than Tommy did, because you will be able to put your head up without being sniped at.[62]

Fleming deliberately provided no particular advice to the visitor on what to see and do around Ypres other than to soak-in the landscape: 'The battlefield of the "salient" has been the scene of so many Homeric fights that it is extremely difficult to advise the visitor what to see and what to miss. The only satisfactory way to "do" this district is to walk it – or ride if a horse can be obtained. There is not a square yard between Langemarcke [sic] and Hollebeke that is not noted for some episode or other.'[63]

Setting out to walk and adapting to a necessarily more relaxed timetable freed the visitor from the tyranny of the tourist charabanc itinerary and opened up the possibility of a genuine relationship with the country. A British Legion member deemed this especially true of the tourist hotspot that was Vimy: 'All visitors to Arras make the pilgrimage to Vimy, but if one is on foot and has plenty of time to spare one can wander by secondary roads and footpaths down the western side of the ridge to the villages of Souchez, Ablain-Saint-Nazaire, and from the latter make the ascent of the ridge of Notre Dame de Lorette which overlooks the valley and dominates the country around.'[64] And wallowing in the Vimy landscape is precisely what some members of the British Legion's great pilgrimage of 1928 did. Having been dropped at Vimy station, Mrs E.A. Smith and others set off to walk up the ridge. She wrote in her diary: 'There seemed an air of quiet happiness everywhere and mothers and wives seemed to me to forget their sorrow in their pride at being there, just walking along where many hundreds of lads had walked from 1914–18.'[65]

'Tubby' Clayton led Toc H pilgrims up the slopes of Mount Kemmel as part of their summer 1929 pilgrimage to Ypres. His short cuts took them across fields still littered with clusters of barbed wire, but as was noted this helped them all appreciate the problems soldiers faced. The sense of almost re-enacting the experiences of the troops was reinforced by the arrival of a specially booked lorry laden with tea and other refreshments 'worthy of the best ration party known' during the war.[66]

Walking across the Somme battlefields and seeing the remains of trenches, noting that German lines were often in better positions in the process, inspired the *Aberdeen Daily Journal*'s correspondent to 'give tribute – more hallowed and more sincere than words can express – to the four million Allied soldiers who lie sleeping under that greyness'.[67]

A remarkable attempt to come to terms with the landscape by walking it was made by Charles Jones and his wife in the summer of 1920. Every place their son had served was visited before they arrived at High Wood, where he had been killed. Jones seems to have taken a copy of Atherton Fleming's guide with him, and prefaced his diary account of the pilgrimage with a direct quote from the book: '[a] very good view over the battlefield can be obtained from High Wood, which can be reached by walking across country either from Flers or Martinpuich, if the visitor is energetic enough … [a walk across the battlefield from this point exposed] the usual battlefield flotsam and jetsam, if one maybe allowed to use the term', and 'if the day be fine and clear, a very good view may be obtained'.[68] The account then unfolded with a profusion of topographical details in which every location was very carefully described and annotated by reference to other sites. Typical of the many orientation remarks was his comment on spotting Le Sars with 'Pys a little to the right and came opposite Eaucourt l'Abbaye, a little further on was Courcelette and the debris of its sugar refinery'.[69] This level of detail perhaps reveals Jones as a man desperate to authenticate and verify in order to achieve absolute knowledge about the conditions in which his son was killed. In adopting a spirit of Olympian detachment he was perhaps also attempting to make the pain bearable. Somewhat ironically, Jones's pedantic precision and determination to understand, interpret, and contextualize the landscape in which his son was killed by walking across it gave him far more insight into the battle and its geographical environment than his son ever achieved locked into the labyrinthine world of the trenches.

COMMUNING WITH THE LANDSCAPE AND UNDERSTANDING THE GEOGRAPHY AND GEOLOGY

Charles Jones's ability to understand in forensic detail was just one of the advantages walking provided. For many others, walking pathways across the battlefield opened up a spiritual relationship of true communion with the landscape and its significance. H.A. Taylor, formerly a captain in the Royal Fusiliers who had also served on the general staff during the war, produced his book *Good-Bye to the Battlefields: To-day and Yesterday on the Western Front* in 1930. Much of the book was drawn from his series of articles printed in a number of newspapers between 1926 and 1928. Taylor adopted the pseudonym 'Raymond Bridgeway' for his journalistic work, and this nom de plume reveals much about his experience of the Western Front and his understanding of pilgrimage. 'Raymond' derives from a name of Germanic origin meaning protecting or guiding hand, and the surname has strong resonances of an ancient trail in Southern England, the Ridgeway. The Ridgeway is at least 5,000 years old, and originally stretched from the Dorset coast to the Wash on the Norfolk coast, but its core ran across the North Wessex Downs of Wiltshire, connecting the sacred sites of the Neolithic peoples. By adopting his pseudonymous first name, Taylor perhaps implied that he had played the role of wartime protector and his experiences as a veteran gave him the rightful status as guide and interpreter of the battlefields, while 'Bridgeway' also conjures up a mystical association with landscape and place, implying that the weight of experiences pressed down into the land of the former Western Front had telescoped time, making it seem like a trail already thousands of years old. At the same time there is an element of prophesy: this way will be trod for thousands of years to come; it will become as embedded in the landscape as the Ridgeway itself. The name also makes a statement about the way of moving across the Western Front: only the walker would truly appreciate its significance and meaning. Finally, Taylor made an intrinsic link between the Ridgeway and his personal Bridgeway, with he and his experiences acting as the bridge linking the histories of Britain and France.

Unlike most guidebooks and meditations on the battlefields, Taylor eschewed Ypres as a starting point, and launched his book with an exploration of the Somme. Like Wiltshire and the southern coast of England, this area of Picardy is part of a great streak of chalk, creating a rolling landscape of ridges and downs. It was a connection made during the war itself. Raymond (that name again) Asquith described the Somme as 'a rolling down country, rather like the uplands of Hants or Wilts'. John Buchan agreed, comparing the Ancre to the streams of Wiltshire.[70] For Charles Carrington, it was an 'open chalky downland ... which resembles Wiltshire, with its long bare slopes and occasional deep fertile river valleys'.[71] The land between Arras and the Somme resembled Salisbury Plain, said C.E. Montague, adding, 'the ground has the same large and gentle undulation; and these great rollers are made, as in Wiltshire, of pure chalk coated with only a little brown clay'.[72] B.S. Townroe, a great admirer of Montague, was clearly much struck by this definition of the landscape, for his work, *A Pilgrim in Picardy*, also compared the countryside to Salisbury Plain, and spoke of it having 'the same stretches, that seem to extend to infinity, of rolling chalky ground, covered with a thin coating of brown clay'.[73] Taylor's entry point to the battlefields was the geological (b)ridgeway of chalk linking the sacred sites of ancient England to its new sacred way in France.

Graham Seton Hutchison, a veteran of the Western Front who drifted into extreme right-wing politics by the end of the 1920s, shared the fascist belief in the mystical relationship between humans and the land: the racially pure were rooted in the soil and rural landscape, and in turn realized their roles as guardians and promoters of it.[74] Unsurprisingly, Seton Hutchison was a firm believer in stepping out across the former battlefields, seeing this as a method through which the veteran would reverse time and find affirmation in his former self:

> Go back to Ypres if you wish to recapture part of your former self. Tread the pavé roads. Plant the heel firmly in the muddy soil. As you hear the squelch, or your nailed boot echoing upon cobble stones, where you stand will become peopled, and your horizon broken by a forest of rifles and tin hats all askew. Walk swiftly across the fields as if expectant of a barrage of gas or a five point nine morning hate. Then slip the fingers beneath the shirt buttons on the breast. You will feel animal sweat; and you will wonder as you withdraw your fingers, as you did before, if this warm wetness may not be that of blood. The Salient is very silent now. But if you stick your fingers in your ears you hear again all those sounds, so similar, yet so distinct, which for the initiated spelt death close at hand, or some mighty metallic atom hurrying safely overhead. And in the darkness, or with closed eyes, you will see visions – British soldiers huddled close for physical warmth and spiritual reinforcement; men from the blue haze of an

English countryside, wrestling with death in Battle Wood; bare-legged boys from shingle coves playing in the waters of Dickebusch; Canadians breasting the Abraham Heights; 'Aussies' in Cameron Copse, 'Tac Heels'[75] at Zillebeke; old soldiers who remembered Darghai and Magersfontein, and men who, as youths, knew only Plumer's final drive; those who gripped hands at zero hour, and those who fell in mud and dust and rose no more.

E.F. Williams similarly urged his fellow veterans to take to the road:

Go out along the Boesinghe road past Salvation Corner, with its adjacent shattered ruin, which still remains in the state Fritz's gunners left it. Explore again the Canal Bank; try to find your old dug-out among the few that can still be noticed from the road ... Gaze along the Canal Bank up to Ypres, or along to where Blighty Bridge once stood ... Walk out through the magnificent memorial arch now nearing completion at the Menin Gate, up past Hell Fire Corner to Hooge with its famous crater ... pay a visit to all the famous spots you knew so well.[76]

Many veterans felt walking did indeed inspire memory. 'Mr X' found that as he walked across the Messines battlefield his 'thoughts turned to that delightful commander of the 16th Division, Major-General Sir W.B. Hickie, whom I recall standing on a lot of ammunition boxes congratulating us after victory'.[77] Similarly, it was only by walking that 'J.B.M.' found himself able to reconstruct his old world, especially once he started exploring Thiepval Wood. Here the memories came thickly and rapidly allowing him to contact again the 'Deathless Legion' of his comrades.[78] Walking up the Menin Road inspired a flood of memories for F.J. Lineton and J. West, and they recounted a host of anecdotes and vignettes inspired by their route march.[79] An ex-serviceman who returned to Ypres on 31 July 1937 did so with a very clear intention in mind: he wanted to walk the ground over which he had advanced on that day twenty years earlier, at the opening of the Third Battle of Ypres.[80] The veterans on the Ypres League 1934 August Bank Holiday pilgrimage 'walked out [through the Menin Gate] to their various old battle haunts' reported the *Ypres Times*.⁸¹ Despite complaining about the scale of reconstruction, Ferdinand Tuohy found it fascinating to walk the old front line from Albert to Ypres in the summer of 1927. Walking allowed him to blot out the present as 'great battles, great crises, came back from afar as one trod this and that vantage point – history lived again'.⁸²

THE IMPORTANCE OF APPROPRIATE CLOTHING

Whether the visitor travelled by car or foot, the conditions in the devastated areas required clothing to be considered most carefully. Robust and practical rainwear and shoes were vital. 'Above all things the traveller must be weather-proof,' warned Coop, advising the visitor to be 'well prepared against mud and rain'. He then alluded to the dire condition of the countryside ('it is perfectly astonishing what a change a shower of rain will bring about – especially in Belgium') before making the lesson absolutely explicit: 'A good portion of the battle area, even in good weather, is difficult to negotiate; in or after wet weather it is indescribable'.⁸³ Given the likelihood of rain and absence of sheltering places on the battlefields, a waterproof coat was an absolute necessity according to Lowe, 'except when there is no doubt about the weather'. Once again, his service perspective came through. He advised carefully rolling a mackintosh and carrying it on the back, as a soldier would; stowed thus it would give 'little trouble'.⁸⁴

Getting across the rough ground made a walking stick advisable, as were thick-soled, stout shoes or boots.⁸⁵ 'The only real way to see the Salient is to hike it,' wrote veteran R.J. Mason, but he added, 'It is necessary to wear thick soled shoes.'⁸⁶ Having experienced the cut-up and swampy ground of the battlefields, Olive Edis bought herself 'a good stout pair of shoes ... as one pair of mine were not likely to recover from the mud of the Menin Road'. She was then most grateful to Lady Norman's son for giving her an army mailbag, which was the perfect place to store her muddy shoes and gaiters. (The bag then proved very useful for collecting souvenirs.⁸⁷) As for Lady Norman, her shoes became so dishevelled, deformed, and mud-caked after walking the battlefields that she struggled to get her feet back into them.

Lieutenant-Colonel T.A. Lowe was equally forthright in his advice on clothing. The class perspectives as an officer underpinned his recommendation to wear an old golfing suit combined with stout boots and leggings, so as to avoid snagging on the copious amounts of barbed wire still scattered across the battlefields. He additionally made an explicit comment about female clothing, recommending strong boots, thick woollen stockings, and short skirts, combined with a good woollen jumper 'if they really wished to enjoy the tour'.[88]

Having identified the dangers posed by rusty wire, Lowe advised a first aid kit with antiseptic as a useful addition to the daypack. This was indeed wise advice, for it was almost impossible to avoid wire. The horrors of it significantly affected the progress of Scots Guards veteran and writer Wilfrid Ewart and his companion as they trudged across the Somme battlefields from the Butte de Warlencourt to Flers in the autumn of 1919. Here the ground was a shattered mess, and as a consequence they found the going very difficult, especially as the shell holes were often camouflaged by thick growths of weeds, resulting in regular falls and tripping on remnants of wire.[89]

NAVIGATING THE BATTLEFIELDS

Another issue facing all visitors, regardless of the form of transport adopted, was the need to know precisely where they were going, which made good maps crucial: Lowe declared maps an essential item for any battlefield tour in the first paragraph of the first page of his guidebook.[90] As 'a great deal of the success of one's visit depends upon the acquisition of just that information that the expert guide or well-informed guidebook can afford', the *Dundee Courier* advised its readers to make the investment in good maps and books. The Michelin guidebook was recommended, along with the histories of Conan Doyle or Philip Gibbs's collected war journalism to help flesh out the details.[91] Muirhead's 1920 guide to *Belgium and the Western Front* recommended the British Army's General Staff maps of the region available from Edward Stanford and other specialist map suppliers. As an experienced driver, Baroness Campbell knew all about the difficulties of navigation. She therefore took with her war maps produced by the *Times*, marking them with coloured pencils to help her understand the positions of the front lines at various times.

But the challenge of finding remote graves often made even excellent printed maps of little use. A correspondent for *Answers* reported a fruitless four-hour search for a grave on the Somme despite detailed instructions supplied by the army's Directorate of

Graves Registration and Enquiries (DGRE) office. The key to the situation was a sketch map prepared by a friend who had already thoroughly explored the area. 'Treasure all such records,' was the advice given.[92] Writing very soon after the lifting of travel restrictions in August 1919, a correspondent of the *Gloucestershire Echo* who was recently returned from the battlefields provided advice on interpreting British army maps (these were vital for navigation, but tricky for the uninitiated).[93] By the same token, the fluidity of the situation in the devastated zones meant that maps rapidly became outdated, as Coop warned his readers: 'Maps and other references, which may have been quite correct even a few months ago, may now be altogether misleading.'[94]

Knowing the desire of relatives to find the precise spot where their loved one was buried or commemorated, ensuring those sites could be accessed with as little difficulty as possible was of great importance to the IWGC. Unable to play a direct role in the transit and guiding of pilgrims, the organization worked hard on the elements it could control. With the wrapping up of the DGRE in the autumn of 1921, the responsibility for providing directions fell solely on the IWGC and its enquiries department.[95] In turn, this made clear and effective signage an important issue for the IWGC, as it was realized that navigation in a devastated landscape could be tricky and anything assisting the outsider was highly valuable. However, as its attention was devoted initially to cemetery construction, it took the IWGC until the autumn of 1923 to consider signage more closely. Even then, a quick solution was not obvious. Major W.B. Binnie, deputy director of works in France and Belgium, believed the devastated regions were changing so quickly as to advise caution on implementing a general policy. He wanted to wait until the local authorities had fixed their plans for road construction. Referring to the eleven cemeteries in and around Beaumont-Hamel on the Somme, he claimed to 'only know of two where the road has definitely been fixed'.

At Polygon Wood on the Ypres battlefield, the Australian government wished to put in a new memorial road 'which will completely alter the design of Polygon Wood Cemetery, as we will naturally site the main entrance of the Cemetery in relation to this new memorial road'.

Ploegsteert Wood was another site extremely difficult to navigate. Despite the six cemeteries in and around it, the wood was still riddled with numerous, intersecting tracks and trenches and dotted with clusters of its original trees that had survived the bombardments. Binnie and Captain A.L. Ingpen, a senior Commission official based in France and Belgium, conceded that 'the very thickly grown' wood meant 'even Commission officials who know the site of the cemeteries fairly well have difficulty finding the roads'. To solve the issue, the two men suggested a rubble cairn about seven feet high with a granite direction plaque. In Binnie's opinion, any other form of signage would 'be quickly overgrown with trees and will be difficult to find'. He drew up some sketches and submitted them to Colonel H.T. Goodland, director of works in France and Belgium, for his thoughts. As an alternative, Ware suggested markers fixed or blazed onto trees. Goodland was less convinced by this idea, as the site lacked mature trees, and was instead much more of a thicket of undergrowth. Further, as the wood was in private hands, he was loathe to do anything that might cause offence or concern to the owner. A conversation between Major W.S. Browne, the area superintendent, and the gamekeeper had added to the concerns, for Browne had reported that the owner was considering closing all paths in the wood as visitors had disturbed his shooting. As a temporary solution Goodland ordered more wooden signposts to be placed in the vicinity. Regarding the broader issue of direction signs, Goodland backed Binnie's recommendation to refrain from erecting anything more durable or permanent until the road plans had been finalized. In the meantime, the current wooden boards were to be maintained and kept in good repair.[96]

Things were not necessarily easier in the old rear areas far from heavy enemy activity. Mr K.T. Gemmell wrote to the IWGC informing them of his adventure searching for the British plot in the Boulogne city cemetery. Despite speaking 'French like a native' and being in a town of significance, he still ended up at the wrong spot, having marched on to Terlincthun British Cemetery some mile or so outside the city limit. As the locals did not know the difference between the two cemeteries, they had immediately assumed he meant the largest one and sent him that direction. If there had been clear signs with the cemetery names, he would have been able to find his way much more effectively, he wrote.

A visitor searching for Landrecies British Military Cemetery managed to find it by asking in the town, as there were no obvious direction markers or boards. According to F.J. Kirby, who was seeking a grave in the Villers-Bretonneux cemetery in the summer of 1926, the signage needed to start at the railway station. Relying only on local help, he found himself visiting a host of other IWGC cemeteries in the region rather than the one he wanted. In turn, this forced him to incur a further day's expenses, as he had walked nine miles and missed his return train. He then brought in a gender element, for he asked them to imagine the effect of such an extra tribulation on poor mothers. Alfred Whittle, visiting his brother's grave at Roeux cemetery, did not have to imagine the plight of distracted women. While at the cemetery, which he had struggled to find due to the lack of directions, he met 'two ladies from Edinburgh visiting the grave of a Royal Scot buried there, and they complained bitterly that they had been searching for many hours and had visited many cemeteries before finding Roeux'. But, as with Mr Kirby, much of the exasperation at the search had been lessened by the 'splendid way you attend to the graves'. Female visitors were also highlighted by Mr C.H.W. Cook, who, on his annual pilgrimage to his brother's grave, felt Dozinghem cemetery was not well signposted despite previous correspondence on the subject. On his 1930 trip he gave a lift to a Canadian widow unable to find the cemetery.[97]

In considering options for signage, the IWGC had to take into account the plans of other stakeholders. Since the expansion of motoring at the turn of the century, car and component manufacturers and driver interest groups had encouraged touring by investing in every element of the supporting infrastructure.[98] Among the important services provided was signage. Binnie and Goodland recognized this and wondered whether these organizations might solve the issue for the IWGC. For example, Citroën had already added the British military cemetery at Bois-Guillaume near Rouen to its enamel signboard. Alternatively, Goodland wondered whether it might be better to approach the French government and request permission to place IWGC plaques next to its existing route signs. However, as was also recognized, integrating its systems with that of the French state would not provide a comprehensive solution, as so many of the cemeteries were in isolated locations a long way from the main network of roads and accompanying signs. Whatever was decided, Goodland noted, the IWGC's design and lettering should be sympathetic to or in keeping with the official signage: the corners of the foreign fields forever England would at least have Franco-Belgian direction aesthetics.[99]

As the Citroën example shows, external organizations had noticed the lack of provision and were responding to it. An important player was the Ypres League. For the League's founder, Henry Beckles Willson, clear signboards were not just practical instruments, they were also a way of ensuring the sacred memory of the salient by maintaining its Britannic stamp regardless of the work of nature and the locals. Writing in the *Ypres Times* in 1924, he said visitors could not find 'the sites of great exploits once so familiar to us all, largely owing to the natural revision to local names of those places which must ever be immortal in British military history', before listing a range of iconic sites including Hellfire Corner and Sanctuary Wood.[100] For Beckles Willson, simply indicating the site was not the optimum, but a minimum. He also suggested accompanying information plaques and produced some examples he had devised to illustrate his point. Breathless, heroic, and awe-inspiring they were, too:

> On the Menin Road I should like to see reared the following legend: –
>
> ON THIS SPOT ON THE 31ST DAY OF OCTOBER 1914, AT A CRITICAL MOMENT IN THE FIRST BATTLE OF YPRES, BRIGADIER-GENERAL CHARLES FITZCLARENCE, V.C., ORDERED ON HIS OWN RESPONSIBILITY A BATTALION FROM ANOTHER DIVISION, THE 2ND WORCESTERS, TO ADVANCE AND HIMSELF RALLYING A RETREATING LINE, SAVED THE DAY FOR THE BRITISH ARMS.

For Hill 60, he ventured:

> THE ELEVATED GROUND SURROUNDING THIS SPOT KNOWN AS HILL 60 WAS THE SCENE OF SOME OF THE FIERCEST FIGHTING IN THE GREAT WAR. IT WAS THRICE GALLANTLY ATTACKED AND AS GALLANTLY DEFENDED, AND AFTER EXACTING A TERRIBLE TOLL OF BLOODSHED, WAS FINALLY MINED AND BLOWN TO BITS.[101]

A utilitarian orientation marker became a didactic instrument in Beckles Willson's vision. It was one in which individual agency mattered: senior officers appraised situations and gave firm orders; men acted on them and changed the course of history. Men were not merely brave, they were gallant, in the terminology of Tennysonian chivalry, and this even made the annihilation of the landscape – 'blown to bits' (not the men, it will be noted) – romantic. Beckles Willson did not depart broadly from representations of the war in British popular culture of the time here, as the hugely popular battle reconstruction films made by the British Instructional Films company show.[102]

As Beckles Willson told his fellow members of the League, temporary signs could be put up very quickly and inexpensively and replaced later with something more permanent. Each should carry the statement '"Erected by the Ypres League," and it might well be for the units immediately concerned to undertake their perpetuation', he added.[103] Seemingly inspired by this appeal, Brigadier-General Fitzclarence's widow announced her desire to erect a noticeboard at Gheluvelt to commemorate her husband.[104] Fired on by Beckles Willson's enthusiasm, and never an organization lacking ambition, the League quickly identified a series of sites around the salient and commenced the erection of signposts during the summer of 1924.[105] By the following spring forty had been erected across the salient.[106] And they embodied his original vision, for the iron posts erected bore the 'place-names given to them by the troops which became so familiar, and will go down to history'.[107] The British presence was made indelible. The War Office had also considered making the British military footprint permanent. Reacting to enquiries from the Belgian government regarding the preservation of battlefields, it instead proposed a series of markers in 'simple concrete work with a commemorative tablet' covering the actions 'of all British and Dominion troops'. The sites it believed important were 'the deeds of the original Expeditionary Force'; those 'identified with the second battle of Ypres and Dominion troops'; 'the inside ring' of the salient, which was defined as Pilckem, St Julien, Hooge, Hill 60, St Eloi, and Neuve Eglise; 'the successes of 1917'; and 'the defence of 1918'. These would 'show to anyone the inner and outer limits of the salient and identify the actual salient for future generations'.[108]

Reviewing its own progress on signage and direction markers in the autumn of 1924, the IWGC noted how the maturing of the cemetery construction programme was creating anomalies. Dury Mill cemetery had the headstones on site, the Cross of Sacrifice had been erected, but the access road was 'very bad' and in poor weather 'almost dangerous for mechanical transport' with an unmarked footpath over a hundred yards long and 'overgrown with beetroots'. Nearby Dury Crucifix cemetery was 'in beautiful condition horticulturally' but had 'no direction board indicating the way to it at the fork of the road ... and may easily be missed'.[109] Following a tour of the French and Belgian battle zones in the spring of 1925, Sir Fabian Ware, the vice-chairman of the Commission, produced a memorandum for internal consideration. During the tour he had tried to place himself in the position of a member of the public who wished to see the memorials and cemeteries. At the same time, he also tried to imagine himself as someone seeking the grave of a lost loved one. In adopting these dual roles of general visitor and pilgrim, he made a number of observations, chief of which was the absence of direction boards. He provided some definite examples. At Dunkirk town cemetery he found no markers of the IWGC plot. Adinkerke churchyard extension lacked any indication as to the route to the cemetery gate; he failed utterly to find Adinkerke Military Cemetery.

In St Quentin, 'I again adopted the attitude of the ordinary tourist' and while having lunch in a local hotel restaurant 'still playing the innocent tourist', he enquired of the staff where the British cemetery could be found. He was given much detail on a huge cemetery containing some 30,000 graves. As he well knew, they were describing the French national cemetery in Vermont. For Ware, the matter was now urgent: 'We cannot contemplate the Summer's passing, and nothing being done while we are waiting for the erection of permanent indication boards.' He urged the production of maps, to be distributed to hotels across the former battle front, allowing staff to assist visitors. Temporary signage had to be made a priority, as did information boards at the cemetery entrance. At Bray Road Cemetery, there was no way of understanding the cemetery's significance as a mass grave, as the explanatory tablet embedded in the lawn was not obvious. 'We shall be very seriously criticized for what can only be called stupid work of this kind.' Believing the issue of signage had been a peripheral one for many departments, he called for a new interdepartmental committee to come up with a plan.[110]

A network of temporary direction boards was then rapidly produced, allowing the IWGC's fourth annual report to record the erection of 598 road signs during 1922 and 1923.[111] Although these must have been helpful, the lack of coverage was often a frustration, and was something that could be unfavourably contrasted with the cemeteries themselves: Mr Paul Taylor of Kettering wrote to the

IWGC to express his wonder at their beauty and the quality of care. He had found his particular destination easily thanks to the clear instructions he was given and the direction signs. However, many of the signs were found to be placed off the main roads, and in a number of cases were 'only temporary affairs and are occasionally in rather inconspicuous places. Occasionally we came across one that seemed to have been knocked down and fitted up temporarily.' While he understood that the overall project was 'incomplete and that probably something permanent has been planned', he still thought it worth bringing to their attention. IWGC officials tried to keep a close eye on details. When it was noticed that the 51st Divisional Cemetery had been inadvertently marked in an incorrect military citation style as the 51/Divisional Cemetery, it was immediately noted, and the slip condemned as 'deplorable'.[112] Here, the IWGC's nature, as an organization filled with war veterans anxious to maintain military nomenclature accuracy and to advertise that principle to ex-service visitors, made itself felt strongly.

Of course, when it came to the creation of permanent markers, the IWGC stuck with another of its principles, insisting on fine design; utility and aesthetics were combined. Some initial designs were requested from Charles Holmes, formerly director of the National Gallery and an adviser to the IWGC since its earliest days. Internal discussion over their merits was extensive, and the IWGC's usual oracle, Sir Frederic Kenyon, was asked to adjudicate.[113] The final version was a cast-iron sign painted in the IWGC's apple green livery, with white lettering, mounted on a hooped green and white cast-iron pole, and deliberately echoed the style of French and Belgian road signs.[114]

As with cemetery fixtures and fittings, an impression of sturdiness and solidity combined with delicacy and beauty was the overall effect.[115] So proud was Ware of the innovation that he not only flagged it up in his preface to Sidney c. Hurst's 1929 guide to the war cemeteries, *The Silent Cities*, but also included a photograph of one in situ. 'They are effective and dignified,' he wrote, and 'besides fulfilling their purpose will in future remind all travellers on the French and Belgium roads of the sacrifice which the British Empire made in the Great War.'[116] Production and erection commenced in 1927, and by early summer 240 were in place with sixty in hand.[117] Equally carefully planned was the precise siting so as to create a sense of itineraries and routes through the landscape, and Holmes himself accompanied Ware to oversee the initial installation phase.[118] As with so much else about the British and imperial relationship with the battlefields, the Ypres district appears to have been given the highest priority. Goodland reported the establishment of sign route clusters starting at Abeele, on the Franco-Belgian border, a further set on the main road into Ypres and then out into the salient on the St Jean and Potijze roads as well in as the neighbourhood of Boesinghe and Pilckem, thus providing comprehensive coverage of the central and northern sections of the Ypres battlefields. Matters progressed more slowly in France, where additional work with local and national government officials was required, as well as greater compliance with regulations.[119]

All direction markers appear to have been greatly appreciated by visitors. For the Ypres League, its additions were a badge of pride. The Whitsun 1925 pilgrimage passed many of these markers, 'and the idea of permanently identifying all the old spots by their war names was unanimously applauded'.[120] A scout troop marching back to Ypres in pouring rain 'were cheered to see signs of the Tommies' undaunted spirits still standing, in the shape of Ypres League signboards marking "Hell Fire Corner," "Clapham Junction," etc'.[121]

For veterans such signs proved crucial in the rapidly changing landscape and, in their very physicality, reminded them of the original trench signboards so crucial to their wartime sense of place and orientation. E.F. Williams told his fellows the signs were extremely useful: all the famous spots could be visited as they had 'been well marked out by the League and won't take much finding'.[122] Writing for the *Ypres Times*, W.G. Mowle, who served with 50 Division during the war, was equally enthusiastic and appreciative. Although very familiar with the salient, he was a bit disoriented by the restored landscape, which 'somewhat taxed the memory and imagination, and it was good to have confirmation (by means of the League's signs) that places one suspected to be in the vicinity were really so!'[123]

It was a problem W.H. Arthur Duncan similarly faced on his return to the battlefield. Declaring the old Ypres 'we knew ... gone, the change is a revelation', the only way he could identify 'the old historic places such as Shrapnel Corner, Salvation Corner, Hell Fire Corner, and Hell-Blast Corner [was] by the sign-posts erected by the Ypres League'.[124] Another noted the ease of seeing the cemeteries 'right up the road', but 'others in the fields [were] farther away as signboards indicated' and might have otherwise gone unnoticed.[125]

The names inscribed also acted as powerful stimulants to imagination and memory. A pilgrim to the battlefields was much moved by IWGC cemetery signs. At a crossroads with four direction boards, 'each indicating the proximity of a British cemetery ... none strikes me as carrying with it greater pathos than this one – "Blighty Valley Cemetery." Think of all that "Blighty" meant to the men who lie buried there and, if you have any soul at all with you, any heart to beat in sympathy with the sufferings of others, the full pathos of that name, "Blighty Valley Cemetery," will sweep over you.'[126] *The Athenaeum*'s reviewer of *Muirhead's Belgium and the Western Front* was sorrowed by the lack of plans to preserve the trenches or dugouts, but believed the cemeteries would become the true battlefield markers and interpreters, especially if they retained their old trench names.[127] Beckles Willson had been proved right: embedding the Tommies' nomenclature made the signboards far more than simply a marker; they were touchstones of the spirit of the wartime BEF, conjuring up their ghosts in the present. Sign and signified merged as far as Beckles Willson was concerned.

In the salient, the Ypres League and IWGC provided the visitor with much useful local direction, flagging up both cemeteries and other sites of importance and interest. Beyond Ypres and the League's coverage things were far more hit and miss. One visitor found it extremely difficult to find the battle exploits memorials and wrote twice to the IWGC asking why they were not signed in the same way as the cemeteries. In reply an IWGC official politely pointed out its lack of responsibility for this type of memorial and suggested using other information in collaboration with cemetery directions for orientation, from which the memorials could be found. This singularly failed to impress the correspondent, who clearly felt everyone was being treated as if they had the map-reading skills of a good soldier. 'Your letter reads as if your point of view was that we were quite idiotic in not being able to follow what you evidently think is obvious,' she wrote back tartly.[128]

There was one other navigation aid available across the battlefields, but it was a highly specific marker. In 1920 the French sculptor and war veteran Paul Moreau-Vauthier proposed a squat granite obelisk to mark the points at which the German 1918 spring offensives were brought to a halt, and thus the furthest points reached in France by the invader. His designs, on display at the Salon des Artistes Decorateurs, provoked the interest of the Touring Club de

France. Seeing an opportunity to create a network of such memorials as the spine of a battlefield route for its members, the club approached Marshal Pétain to suggest precise sites. He agreed to support the scheme, and the first of the memorials was unveiled at Château Thierry on Armistice Day 1921. By the spring of 1922, the sites had all been agreed and the Touring Club was ready to roll out the programme, funded by its members.

Working with its Belgian equivalent, the club decided to expand into Belgium, with Dixmude adopted as the first site. As monuments designed to mark specific military actions, three variations, based on three pieces of kit, were designed to reflect the particular allied army involved, with helmet, water-bottle, and gas mask respectively representing the Belgian, British, and French forces.[129]

The Belgian Touring Club then raised the idea of further monuments in Belgium with the British Embassy in Brussels, which promptly engaged in discussions with the War Office, the IWGC, and the Office of Works. Sir George Grahame, the British ambassador, was keen to know French reactions to the initial Touring Club de France scheme, and in particular whether France was happy to see stones erected with British symbols, or whether the plan was to restrict it to solely French army actions. Ware felt similarly, and the IWGC remained cool on the idea, concerned that it might be dragged into erection and maintenance expenses. Even when it was made clear that the French Touring Club intended to mark the actions of all allied armies in 1918, the attitude of the IWGC did not alter.[130] Assurance from the two clubs that the scheme was not officially supported by the Belgian and French governments allowed the British government and IWGC to be sympathetic but uncommitted. However, with the two touring clubs determined to press ahead, a British partner was required.

Into this vacuum stepped the Ypres League, which enthusiastically undertook (seemingly at the invitation of the British Embassy in Brussels) to erect seven memorials around the old salient, and devised their English name of 'demarcation stone'.[131] As with the other examples, all were to be sited on roads, and so had the further effect of creating a memory landscape anchored on motor transport. The one that attracted the most attention in Britain by far was that placed at Hellfire Corner on the Menin Road near Ypres. As a good example of the 'Tommification' of the landscape, Hellfire Corner was a place name that had become well known on the home front by the end of the war.[132] Given its familiarity, it was no surprise that the League identified this as the site for its first marker. Formally unveiled in August 1923 as part of the League's reunion weekend, Baron Vink, the local landowner, said the marker would 'remain for ever a vivid token of the gallantry of the British and Belgian troops in their sacred ground of Hell Fire Corner'.[133] It also became a minor curiosity of the war's legacy: in 1928 the Imperial War Museum added a miniature version to its collections, and, along with the League, sold plaster models that were ideal as 'a paperweight or as an ornament'.[134]

As the landscape was gradually restored to something akin to its pre-war condition, wiping out the war's traces, the demarcation stones became more significant. Unsurprisingly, the Ypres League highlighted them in its guidebook, *The Immortal Salient* (1925), flagging them up within the numerous trails around the district outlined in the text. As a correspondent for the *Surrey Mirror* wrote in 1928, a visitor could 'pass along the highway and fail to notice any evidence ... of the war [but] in course of time he will come across what are known as Demarcation Stones' placing the evidence of the war in front of the eyes.'[135] H.A. Taylor noted something similar. Referring to the 9 (Scottish) Division memorial at Point-du-Jour and the nearby demarcation stone, he noted that, but for them, 'it would be impossible for the uninformed visitor even to guess the point, because neither the road nor the fields show any scars of the deep trenches that once seared them'.[136] For another veteran, the Canadian Will R. Bird, demarcation stones served as a sobering reminder of the perilousness of the situation in the spring of 1918. Gazing at the one near Kitchener's Wood, just north of Ypres, he remarked, 'It seems very near to Ypres.'[137] Here was a memorial with deliberate echoes of ancient milestones and boundary markers, designed to be of assistance to a traveller, also acting as a memorial of a historic event. In memorializing that historic event, the demarcation stone was also meant to appeal beyond the intellect and touch the emotions.

*

For the visitor navigating the battlefields involved mental, physical, and spiritual progress, subjecting them to a plethora of emotions and experiences. Battered to pieces while driving along wretched roads or soaked by driving rain while trudging along the crest of a ridge brought about a sense of engagement with the landscape. However, the walker was thought to derive the most from the experience. Immersed in the world of the trenches, the walker, it was believed, was able to detect the ghosts of the past more fully, to empathize and sympathize on a deeper level than those who merely struggled to keep a car moving safely. Place names, particularly those devised by Tommies, soaked into the soul and their meanings became understandable. Battle maps, common from wartime newspapers, assumed a three-dimensional reality. The natural elements of wind, rain, and snow stood in for enemy action. Such an experience all added up to something every true pilgrim knew: the way was as important as the destination.

4

Postcard Scenes: Devastation

And, indeed, the devastated areas of France resemble the sea more than any other natural phenomenon. In traversing them one feels that this is truly 'a land that is lonelier than ruin'.
>> The 'Special Representative' of the *Aberdeen Press and Journal*
>> (January 1919)

Passing through shell-torn country, Passchendaele – now razed to the ground – is reached. All that remains of the church is the mound seen in the background of the photograph.
>> *Illustrated Michelin Guides to the Battle-Fields (1914–1918):*
>> *Ypres and the Battles of Ypres* (1919)

I shall never forget our entry into and the first sight of this ruin. I had read a great deal of the total destruction of many towns by shell fire, I have heard lectures, and have seen large pictures thrown on screens from lantern slides, but no picture, no lantern slide, or pen can describe or make one realise the awful destruction and devastation.
>> J.H. Roberts (February 1920)

Anyone who visited the battlefields of the former Western Front in the twenties and thirties encountered the monstrous trails left by the mighty military machines unleashed during the war. Whether it was through the despoliation of nature, or the obliteration of all traces of human industry and artistry ancient and modern, the visitor was made aware of the titanic encounters fought across north-eastern France and Flanders. For those who came in the five or six years immediately following the Armistice, before reconstruction and restoration had matured, the sight was amazing and provoked a plethora of emotions and responses. Dealing with devastation challenged the mind and senses and, veterans aside, very few were prepared for it.

ENCOUNTERING DEVASTATION

All visitors to the battlefields experienced the jolt of the transition from the rear areas behind the lines – the extensive region in which the army had housed the men and trained them when out of the line, and concentrated its supply dumps, transport hubs, and hospital camps – to the former fighting zones. Those who made the journey immediately after the war using the Channel ports of Boulogne and Calais would have seen plenty of evidence of ongoing military activity in the form of soldiers coming and going, military stores, huts, and much other army paraphernalia. Those who made the longer crossing to Zeebrugge had a slightly different experience, for they arrived in the former German-occupied zone. Moreover, the port was one heavily fortified and scarred by the famous raid of 1918. And for those arriving in the first few years after the war, there was the absorbing sight of the blockships sunk by the Royal Navy in an attempt to block Zeebrugge's canals. The port of Ostend, a little further up the Belgian coast, also displayed many marks of the war and had its own reminders of the 1918 raid. However, at none of these points of entry was it anything like the actual front.

For many visitors heading for Ypres by road from the Belgian ports, it was the approach to Dixmude, very nearly the northernmost point of the Western Front, that made them aware they were crossing into a different world. Visiting in the early autumn of 1921, T.J. Blinkhorn and his friend were sobered by the sight of the smashed town, and the shadow of the war was emphasized when they were suddenly asked to take shelter while ruined buildings were detonated. On returning to the bus and resuming their journey, their descent into destruction's depths became more precipitous: 'From now onward to Ypres it was one continuous sight-seeing of devastated country appalling to look upon and saddening to contemplate.'[1]

Thomas Boal, the *Aberdeen Press and Journal*'s correspondent, also landed at Zeebrugge. Moving on to Bruges, he saw little trace of the war, but at Courtrai things began to change: 'There were many signs of war, including the sight of some British soldiers busy on the task of exhuming our British dead, whose graves are scattered all round, and reinterred in the neat cemeteries.'[2] Passing from 'the still cultivated part of Belgium, a veritable garden' near Ostend to the 'great desert of weeds, unless in the ditches and shell holes, where irises grew' was a shock. But, like many other Scots, Boal was at first surprised, and then delighted, to find the fecundity of thistles across the smashed landscape of the battlefields.[3]

For those heading for the Somme, Amiens was a key transport node, and this meant the first real evidence of the war was the enormous cemetery at Étaples, the largest British war cemetery in France, dramatically sited on the dunes at the edge of the town. Fabian Ware, head of the Army's Directorate of Graves Registration and Enquiries Unit throughout the war, and by 1918 vice-chairman of the Imperial War Graves Commission, was deeply aware of its impact and, to ensure it could be appreciated fully, he even managed to persuade the French railways to slow the trains running past it, allowing travellers to linger over the impressive spectacle.[4]

En route to take up a post with the American YMCA in Paris in the summer of 1919, Alice Knight realized she had crossed a boundary when moving beyond Étaples towards Amiens. In a letter to her parents telling them about her train journey to Paris, she wrote: 'one gentleman knew the country well, so he pointed out all the places of interest – first of all we passed through Etaples and I saw a lot of ruins – then we came to Amiens and saw lots of ruins, trenches, and dugouts galore that place is absolutely in ruins – then we went through Abbeville and all along the Somme', before arriving in Paris in the early evening.[5] On an information-gathering tour that same summer, the travel agent John Frame felt 'the terrible ravages of the conflict begin to overwhelm' him as the train entered the Amiens district.[6] Elizabeth Braithwaite Buckle made the same journey at much the same time as the prelude to finding her son's grave on the Aisne. As the train approached Amiens, she saw the evidence of war in the 'ruined houses, battered, broken woods, then more desolate ruins, then Amiens' itself.[7] 'R.O.P.' gained a similar sense of transition. He saw the damage inflicted on Amiens, 'but it was barely a foretaste of what I was to see next day'.[8]

The first impressions of France gained by actress Efga Myers during her visit to the battlefields in the summer of 1919 was the peace of the rear areas. From the train she saw the rolling countryside south of Boulogne with its scattering of pretty villages surrounded by fields edged with belts of poppies and wildflowers. All was calm, and there was no sign of the extraordinary. It was at St Pol where the first signs of change were sighted in the form of damaged buildings. 'But it is beyond St. Pol, on the straight road to Arras, that the real battle-front begins,' she added. The evidence came in the form of poppies, cornflowers, and marguerites hugging the lines of trenches as well as a sudden drop-off in traffic. Nonetheless, this was still a liminal zone, as she realized on reaching Arras, where 'the sense of utter ruin stabs one'.[9]

In some places the transition to the wasteland was sudden, as if stepping through the door of a finely appointed room into a junkyard; in others it was a gradual transformation of increasing destruction and signs of death. The Canadian journalist John Dafoe was aware of crossing the 'border line between the actual battlefields and the secondary districts of the war – as for instance Amiens and Valenciennes – [where] we saw human life finding its way into normal channels; but over the areas of continued and desperate fighting there was still the abomination of desolation'.[10]

Whatever the path taken, the statistics were numbing. In France and Belgium 3,430,000 hectares of land had been affected by the fighting. In France 4,926 towns and villages and 866,844 houses and farms had to be rebuilt or repaired. The equivalents for Belgium were 242 towns and villages, 100,000 public and private buildings, and 24,000 farms. In France, the combat zone had been home to 4.8 million people in 1914. Although the region was industrialized, some 96% of the land had been dedicated to agriculture, two-thirds of which was arable. By the end of the conflict, the population stood at only 56% of its 1914 total, and in the worst-affected zones it was lower still, at 43%. These people then faced the Herculean task of clearing the battlefields. It started with the need to fill 333 million cubic metres of trenches and remove the 375 million square metres of barbed wire dissecting the land. Among the 1,954 settlements utterly razed or severely damaged were 17,466 schools, mairies, and churches destroyed or in need of significant repair.[11]

The dreadful state of the worst-affected areas was implied by the official label bestowed on them: 'the red zone'. When considering these, the Belgian and French governments had to decide whether any kind of reconstruction was possible. For those areas where reconstruction looked unachievable, there was the option of afforestation as the easiest solution. But even getting that far, taking the very first steps in the process of reconstruction, let alone seeing it through to a successful conclusion, bewildered many. Trying to imagine the totality of destruction was an ability beyond the reach of many too.[12]

Figure 4.1   Temporary houses, Lens, 1920.

Olive Edis noted in her diary the sights she passed on the road to Rheims: 'Mile after mile of shattered, desolate, empty towns and villages we passed. It seemed never-ending. Once more the great avenues of trees, shot and splintered to the level of a few feet from the ground ... when the poor tumble-down villages ceased. One wondered how France could ever pull herself together again.' Making slow progress around Verdun, she found it harder and harder to shake off this sensation. As she watched American soldiers and German prisoners of war filling trenches and removing the barbed wire, she found it hard to believe anything approaching civilization could be conjured from such a mess.[13]

Coming to terms with the sheer expanse of the devastated zone was a common difficulty experienced by visitors. Thomas Boal grasped something of it when passing from Courtrai to Paris on the train. He noted 'for a distance of over forty miles, as far as the eye can see ... there is nothing but devastation – not a tree or shrub can be seen'.[14] A vast, empty country unfolded in front of the visitor, seemingly limitless and revealing nothing. Exploring the Somme in September 1923, Nina Stephenson-Browne saw the extent of the damage and the utter lack of anything close to a recognizable landscape:

Of the villages not a trace remains. Only one hundred and forty-nine wrecked and shattered villages remain of the three hundred formerly situated in this region. A few wooden or iron huts, some only a roof on four posts, the space between filled in with boards or matting, this is all that meets the eye. No crops, no farms, nothing but one vast stretch of arid, shell-torn soil as far as the eye can see. Huge shell-holes, dead stumps of trees, remains of barbed wire entanglements, with here and there a heap of bricks to show what was once a prosperous farmhouse or stately country chateau, or little church, is all that is left of the once fair and prosperous province of Picardy.[15]

Driving down to Berry-au-Bac battlefield on the Aisne front, Elizabeth Braithwaite Buckle was once again brought face to face with the desolation. 'Mile after mile of fertile country left uncultivated for years' meant a sea of weeds for the returning farmers to combat.[16] As the veteran and writer Stephen Graham started tramping towards Ypres he realized he could not see it: 'No, there is nothing on the horizon [where Ypres should be], not a wall, not a wood, only the bare eminence of Kemmel Hill. Before you is a vast fen. Some Flemings are at work on it in shirt sleeves, but not a soul is traversing it. You constantly change direction: there is no going directly. It is impassable.'[17] Around Cambrai, the YMCA worker 'Rover' saw an empty landscape with 'scarce a hut or house [visible] for long dismal miles'; his description of Hooge as a 'howling wilderness' is both economical and arresting.[18]

After stopping in Soissons, where she found the ruins impressive, Alice Knight was taken out into the battlefields surrounding the city. 'We went through several villages which were absolutely reduced to bricks and mortar, and one place called Lafeaux, if it wasn't for a signpost nobody would know that it was ever a little village. The grass and weeds have grown up over the debris, which makes it look just like a wild field,' she wrote in a letter to her parents.[19]

The former habitations marked only by trench signboards were something many British visitors noticed, and stumbling across pre-war settlements that now existed mainly as mere expressions on a map awed and stunned them. Charles Jones and his wife were sobered by seeing nothing 'except for heaps of rubble and debris and the notice boards marking the place where they had been'.[20] 'R.O.P.'

found a signboard inscribed '"Longueval" but ... not even a stone to mark where the town had been'.[21] For Sir Arthur Yapp, a senior figure in the YMCA, finding a noticeboard stating 'This was Villers Carbonnel' marking 'a heap of ruins by the side of the Peronne–Amiens road' was a deeply affecting moment.[22] Neville Chamberlain toured the battlefields in the summer of 1919 and felt exactly the same urge to chronicle this detail of the devastated region. Writing to his sister, he told her of villages 'such as Hooge, Becelaere, Fricourt, Mametz, Longueval etc. [where] there was nothing except a notice board to tell you that a village had been there'.[23]

As all the above chroniclers noted, it was not just a wrecked world, it was also one seemingly free of all signs of normal human life and activity. On leaving the main Flanders towns and arriving in the wilderness of the battlefields, silence rapidly descended. 'One can walk for miles without encountering a single individual – everything is perfectly still,' wrote Dennis Gilmore in the autumn of 1919.[24]

Dafoe visited the site of a former Canadian camp near the front line, which he imagined as once so full of hustle and bustle. 'But on the March [1919] evening when I saw it, it was bleak and cheerless beyond the power of words to express ... Over all desolation and loneliness rested like a pall; everywhere the wreckage of battle, the debris of destruction; everywhere the sense of man's mortality! A grim and melancholy expanse.' Driving between Arras and Cambrai, he experienced 'a profound silence' in the flat, empty plain, which reminded him of Manitoba before the prairie farms began to mushroom.[25] Stephen Graham declared the silence 'unearthly, as if the composed deep sleep of the dead had conquered the ways of the living'.[26] Silence was the major characteristic of the devastated zone according to Olive Edis: 'no fowls, no dogs, no cats, no people'.[27] The silence and emptiness combined with absurdity in certain crazy details. John Frame was brought up against the difference between what life was and what it should be by the sight of a boot-scraper in perfect condition and in its rightful place on a doorstep flagstone, but with no sign of the house whatsoever.[28]

Such a silent and strangely similar barren landscape for mile after long mile tested visitors' descriptive powers. One oft-used metaphor was an apocalyptic image drawn from the Bible: 'the abomination of desolation'; perhaps unsurprisingly given their faith, the two YMCA workers, 'Rover', and Sir Arthur Yapp used this term in their accounts, as did 'A Scottish Chaplain' in an article for the *Scotsman*.[29]

According to Dafoe, the old battlefields were 'still the abomination of desolation ... a man-made desert'. Then Dafoe moved beyond description and towards judgement of the cause, blaming 'the ruthless savagery of man in the sway of his passions'.[30] In condemning the very nature of humanity, Dafoe was very much in the minority, for virtually every other visitor declared the Germans guilty. As 'A Scottish Chaplain' concluded, 'this is what "Kultur" has given the world', using the term applied throughout the war as a shorthand for Germany's alleged inherent barbarity.[31]

A similarly apocalyptic comparison drawn from antiquity made the war's impact seem like a disastrous natural phenomenon: Arras 'is like a Pompeii, a strange jumble of the skeletons of houses and buildings', wrote Efga Myers. Neville Chamberlain also drew upon the fate of the famous Roman city, telling his sister: 'To give you a diary or a detailed account of the tour would take too long so I shall rather endeavour to convey my general impressions. I think the most enduring is the terrific amount of destruction. We ourselves must have passed through 50 to 100 villages in which not a single house remained intact. Over and over again I was reminded of Pompeii.'[32] Monstrous gunfire pouring its own form of brimstone from the skies had created a world of ashes, occasionally marked by the shadows of human existence.

THE MONOTONY OF DEVASTATION

After a while, the ability to find comparisons and to shape metaphors collapsed, and visitors faced precisely the same problem as soldiers had confronted during the war: how to describe the indescribable, how to communicate the incommunicable. Getting to the heart of the matter meant admitting defeat at ever being able to do so. Testimony after testimony fell back on this final position. The shattered world was 'beyond description', 'a broad belt of indescribable havoc', 'quite indescribable', 'scenes of indescribable destruction', while the ruined towns were deemed, with a bit of a poetic flourish, to be 'indescribable masterpieces of explosive destruction'.[33] The final mental and spiritual destination of this harrowing physical journey for soldier and visitor alike was boredom, as it all became so appallingly familiar. At one and the same time, it was unique, remarkable, almost unbelievable, *and* utterly monotonous due to its sheer scale and similarity, punctuated only by the odd significant geographical feature, be it natural or human. As a veteran of the Western Front, T.A. Lowe was aware of this possibility, warning the readers of his guidebook thus: 'I can safely say that the mere sight-seer will probably be disappointed with the devastated zones of France and Belgium.'[34] Neville Chamberlain called the battlefields 'the most dreary sight imaginable' in a letter to his sister, before going on to describe the scene in great detail, comparing it to the bleakness of the Yorkshire moors.

Destruction seemed to gain its own essential, mind-numbing, energy-sapping quality in its endless profusion. After seeing a host of smashed towns from the vantage point of Vimy Ridge, Wilfrid Ewart decided not to press on to Merville as 'ruined towns have only one character – ruination'.[35] Constant exposure to such a mess eroded the spirit; Charles Jones felt it, noting in his diary 'a feeling of depression which deepens with the approach of nightfall'.[36] Neville Chamberlain was similarly affected, writing, 'One can't help feeling a sort of despair oneself when one contemplates such vast areas of desolation.'[37] YMCA worker John Eastwood Hastings expressed an irony of the unchanging landscape: it was precisely because it was so unvarying and indistinguishable that it became so memorable. He told his wife, 'As I was riding against time I dared not stop to have a look round; but the monotony of destruction as I passed through was sufficiently appalling to leave an indelible memory on my mind.'[38] For those immersed in a zone of unmitigated ruin, the sense of release on leaving it behind was noticeable. On completing his circuit and getting back to Nieuport, the exit point of the battlefields for many tours, T.J. Blinkhorn admitted 'it was with a certain amount of relief that our minds were able to turn from the recollections of that terrible conflict'.[39] Captain Mee expressed a similar sentiment, being pleased to leave the Somme battlefields behind him, and, as he neared Doullens, found an untouched country with 'not a brick disturbed, and everything bright and clean'.[40]

A WORLD OF CURIOSITIES

Although a seemingly featureless wilderness, the battlefields were, in fact, one vast, open-air museum of military infrastructure and activity, as the earliest visitors quickly found. Being guided round an extensive German defensive system on the Chemin des Dames,

Alice Knight felt a frisson of apprehension when exposed to the realities of the front line, but, like any good tourist, she was also determined to get a full photographic record of the experience. In another of her highly detailed letters to her parents, she told them of being taken through a dugout that the Germans had occupied for thirty months:

> There were several rooms in it with beds, tables and chairs. I think it must have been the one we used to read about during the war, as I imagine it was well fitted out while the Germans were there. It still contains tons of unexploded ammunition. We were all carrying candles and the men, although they had been told not to touch anything or even put the candles anywhere near it, would persist in picking up different shells, etc.
> It was all most weird and uncanny and believe me I wasn't sorry to get out into the daylight once more. We took more snaps and I'm quite anxious to see them.[41]

Wandering among the genuine trenches allowed those without direct experience to engage their imaginations in an attempt to understand soldiers' front line lives. Finding that the road near Mametz Wood still consisted of duckboards that made it difficult going, L.M. Orton tried to imagine what coming the same way might be like in wet weather. She felt it gave her some insight into the conditions faced by the soldiers, but was aware that it was only partial, for the reality 'must surely beggar description'.[42] In contrast, seeing the evidence at first hand actually lessened John Eastwood Hastings's desire to learn more. A brief encounter with the awful reality was more than enough for him. 'Dugouts, trenches, rusted barbed wire, the various devices of camouflage, the little cemeteries, the shattered concrete watch towers, the derelict vehicles of all sorts combined to humble one's inquisitiveness as to what it must have been like in the months of fighting,' he wrote to his wife.[43]

As a veteran able to use his trained eye, Wilfrid Ewart walked the Somme battlefields interpreting the stages of battle by distinguishing between the different types of wreckage and debris he saw. He came across rifles, helmets, shreds of clothing, gas masks, and scattered personal effects, including photographs. In other places there were great piles of salvage neatly stacked waiting to be picked up.

This allowed him to understand more effectively how the 1916 and 1918 battles had ebbed and flowed.[44]

The twin experiences of ammunition explosions and the ubiquity of human remains were especially jarring, particularly for those with no active military service. Bodies, or at least bits of them, were being discovered, accidentally and intentionally, the whole time. Colonel Goodland of the IWGC was always aware of the effect this had on visitors, which caused him a good deal of consternation. When he came across a Belgian roadwork gang near Ypres in the summer of 1920, on noticing fragments of human remains in the spoil heap thrown up by their workings, he approached the foreman, who told him that they had uncovered several bodies, British and German. When asked how he knew the differences, the foreman revealed a good knowledge of the different uniforms and markings. Goodland recommended a Directorate of Graves Registration and Enquiries (DGRE) unit investigate and, probably motivated by the thought of British visitors coming across the site, he added with great understatement, 'it would seem inadvisable to allow British remains to lie in a little mound at the side of the road', although he did note that the workers had erected a 'little rough cross' over the site.

In the same summer, architect L.H. Sacre, sketching and surveying cemeteries on behalf of the Commission, found the work deeply arduous due not only to the state of roads and tracks but also to the disturbing sight of unburied dead soldiers. On a track near the village of Wytschaete he was hit by a strong smell apparently caused by the corpse of a German soldier, and he believed it highly likely there were more bodies in the surrounding undergrowth. Near Langemarck he noticed an unburied British soldier, or 'possibly more than one', and he was told by local people of further British corpses scattered along the length of the Ypres–Zonnebeke road. What made this alarming to Sacre was the fact that the ground had been declared searched and cleared by DGRE units.[45]

The early autumn ploughing season always brought more bodies to the surface. In September 1924, the IWGC commenced a detailed search of the Moquet Farm district on the Somme when Australian and British boots were discovered along with a quantity of bones. Most of the remains were then buried in a nearby British cemetery, and liaison with the local authorities revealed that seven German soldiers had also been discovered.[46] Incidents such as this were ten a penny throughout the twenties and thirties.

Equally disconcerting were the explosions, which came suddenly, without warning. In most instances these were the deliberate destruction of unexploded ammunition by clearance teams. Occasionally, however, they were the result of some poor unfortunate stumbling across something still live, or an accident during removal. Walking along the Menin Road, Sir Philip Gibbs heard the effect of a controlled explosion. This was a familiar sound to him as a veteran correspondent of the Western Front, but a companion unused to the sounds of warfare was shaken by it.[47] Fellow veteran and writer Stephen Graham was equally blasé. Restless 'in a little bed in a cubicle with a tiny doll's house window' (as he described his room in a Ypres hotel), he heard an enormous explosion. Looking out of the tiny window, he saw a spectacle he was used to, as flames leapt up, lighting the night sky. Realizing it was an accident at an ammunition dump, he went back to bed.[48]

Neville Chamberlain, on the other hand, reacted as Gibbs's companion had. Lacking any kind of military service, he too found the detonation of munitions a rather shocking novelty:

> The country is of course strewn with shells of every size. It is rather disconcerting when you whizz round a corner to find yourself faced with a huge stack of them on a notice board [stating] 'Defense d'approcher.' 'Danger de mort.' At one place we were stopped while they blew up 5 of these stacks. We hid between two houses and when the shells went off the walls shook and the earth trembled with the terrific blast from the concussion.[49]

Visitors discovered a world in which recent history still bore down with great intensity. Trekking around the battlefields north of Ypres in the autumn of 1919, Dennis Gilmore saw a landscape 'hideously scarred, with numerous trenches constructed in all directions'. Everywhere he went on the battlefield showed evidence of the military occupation. He came across munition stores, trench name boards, battery positions, food stores, Red Cross depots, army post offices, orderly rooms, and observation posts. He then caught sight of salvage teams at work, and realized the scale of the task before them. 'It will take a long while before things can be straightened out and the fields turned once more to their legitimate purpose of being ploughed and sown and cultivated in the fashion which the Belgian farmer knows no equal,' he wrote with placid understatement.⁵⁰

ENCOUNTERING THE LIVING

The workers Gilmore saw were just a few of the many thousands toiling to clear up the mess. Scattered across the battlefields in penny packets, these labourers were for many visitors the only signs of life stumbled upon in the wilderness. Some were groups of returned local people devoting themselves to the reconstruction of their homes and livelihoods. Others were soldiers and detachments under military command, including large numbers of Chinese labourers and German prisoners of war. With the war not formally over, British troops remained on the former Western Front ready to supplement the occupation forces in the Rhineland if necessary, but their primary job was to help clear the battlefields of munitions and other dangerous materials, as well as concentrate scattered, isolated graves and unburied corpses into military cemeteries. A great deal of this gruesome and extremely perilous work was undertaken by the Chinese Labour Corps, men who had been recruited by the British government for service in France and Belgium during the war by agreement with the Chinese government. During the war they serviced the war machine by building roads, railways, and camps. Immediately it was over, they started clearing up the mess the armies had made.⁵¹ The German prisoners who assisted them were there in something akin to a hostage role, being held to ensure the new German government complied with the Armistice conditions. The surrealist nature of the battlefields was thus enhanced by these seemingly nomadic parties traversing the otherwise deserted landscapes. Visitors coming across them found a winter kingdom of the dead and its servants.

A quality of the landscape all visitors noticed was the sense of loneliness interspersed with scattered communities, both civilian and military. Sir Philip Gibbs encountered a seemingly marooned group of British soldiers. Isolated and without amenities, they seemed abandoned on their desert island in the sea of devastation. 'I think Robinson Crusoe must have felt less lonely than any man who now dwells on the old battlefields beyond the Somme' surrounded by evidence of untold misery, death, and destruction, he told his readers.⁵²

When the West sisters set off on their battlefield tour in May 1919, they came across a small camp of British soldiers just outside Cambrai. The officer in charge told them they had received no visitor or outsider for six months. The presence of the two sisters brightened the mood in the camp considerably, and 'indeed we were to feel quite public benefactors for coming there at all'. Begged to stay to dinner in the officers' mess, the sisters agreed. The officer in charge promptly telephoned the mess orderly, telling him to 'get a shave and make things look pretty'. Rather disconcertingly, the officer then warned the women that the orderly had been suffering from shellshock and so his personal appearance was not up to scratch. The call was therefore also a way of ensuring a man in an already unstable state was not utterly surprised by guests who were not only civilian but female. Despite these somewhat unpropitious circumstances, 'the dinner party was a great success', as Marjory recorded in her diary, with the orderly managing to beam with pleasure. The West sisters were truly amazed at how they had transformed the atmosphere, 'but one could well understand it in that god-forsaken wilderness, where, as they said, the dreariness of clearing up those lonely battlefields month after month was getting more and more upon their nerves'. Exploring the camp, the sisters found the men had adopted a Belgian boy who had attached himself to them during the war. With nowhere to go at the cessation of hostilities, the boy had stayed with them, and the soldiers had helped rebut requests to repatriate him.[53] When the evening came to an end, the West's hosts phoned through to the British railway transport officer and asked him to reserve two seats for their guests. Their hosts joked that, on finding out that they were female, the railway transport officer had fainted.

For the West sisters this was just one moment in a litany of meetings with the inhabitants of the liminal world. After hitching a lorry ride to Arras, they were intrigued to come across eight British soldiers only too happy to recount their wartime experiences. This 'gave us what will always be the most memorable and unforgettable days of our lives', for the soldiers were of long service and had tales not only of Gallipoli and the Western Front battles, but also of peacetime service in India. Most were proud recipients of gallantry medals and awards, and they were accompanied by their mascot-pet spaniel, who had been with them throughout the war. At an ammunition dump the soldiers took some phosphorous grenades and smoke bombs, explaining carefully how they worked before telling lurid stories about

German conduct during the war. The final excursion the soldiers arranged for the Wests was a tour through the ruined trenches, where the women were very interested to see the trench signboards. 'It was quite pathetic to read some of the quaint names,' Marjory noted, and it made her appreciate the spirit of the soldiers: 'Truly Tommy is a marvel to have any humour left in him at such a time.' One of their guides took them to a well-remembered dugout and he regaled the sisters with his recollections of trench life.[54]

Free, and freely available, expert guiding was an advantage experienced by many of the earliest visitors, as they encountered British soldiers desperate for a break in routine and equally desperate for company. In the case of Lady Norman, a battlefield tour also acted as a family reunion, for at Poperinghe she met her son, who was still serving in the army, and he instantly became the party's guide. Taking them across the Passchendaele battlefield, visiting the ground over which he had fought, and identifying the precise German gun positions his own battery had targeted proved a 'wonderfully interesting day' for Lady Norman, and as Edis noted, 'it was probably quite a unique experience for a Mother to go over the battlefield as she did, tracing so many incidents with the son who had lived and fought through them'.[55]

Alice Knight and 'Wrightie' chummed up with two soldiers for a day in Ypres. 'We spent a jolly fine day with them', she told her parents, 'and it was most interesting because they had both been fighting there. Well we took lots of snaps and promised to change them with one another which we did do.' All told, the two young women 'had a topping time' with their new soldier friends.[56] Only the earliest visitors had such encounters. Once the bulk of the tidying-up was complete and the devastated world put on the path to recovery, such experiences disappeared into memory.

Possibly because of her naivety, Knight proved an intrepid explorer of the battlefields during her time working for the American YMCA, often dragging along the rather more circumspect and cautious 'Wrightie'. One bitterly cold day in November 1919, the two young women set off for Amiens as a base for a day on the Somme. Needing to catch the 5.30 a.m. train for Albert, they were up and out extremely early. 'By jove it was cold when we started out,' she wrote to her parents, but they managed to catch the train, and were helped by two British sergeants who were going part of the way; one was off to Lille and the other to Cambrai to search out his cousin's grave. Arriving at

Albert an hour later, the ladies found the town 'in an awful state'. Frozen and hungry, having had no breakfast due to their very early start, they were pleased to get some hot coffee in an estaminet, but had to make do with dry bread to eat. Like a good tourist, Knight distracted herself from the rather frugal meal by scribbling off some postcards. Then the two set out on foot for Delville Wood in search of a relative's grave. 'It was awfully cold and the sky looked full of snow. We bought some flowers to put on some lonely soldier's grave, and away we went. We walked all along what used to be called "No Man's Land". Nothing but battlefields and trenches the whole way,' Knight was later to tell her parents. Seeing virtually no one as they tramped along, the sheer emptiness of the Somme battlefields became apparent. Offered a lift by some passing British soldiers, the women accepted and were dropped off at Montauban, a village (or what scraps remained of it) close to Delville Wood. And it was here they came into contact with civilization Somme 1919 style:

> Well when we arrived at Montauban to our great joy we found a YMCA hut. I might as well tell you before I go any further that the only thing to tell you that that little place was once a town is a board marked 'Montauban'. We went over to the hut and asked a soldier if we could get something hot. He said we had better go to another little hut where the officer was and he would be sure to do something for us. When we got over there we heard a fire crackling and it sounded just fine to us. We knocked at the door and voice from within said 'come right in'.

They found a YMCA official, who told them to make themselves comfortable and invited the young women to join him and his two female colleagues for breakfast. All three YMCA workers were absolutely delighted to have real guests, as opposed to their usual company of the British soldiers and Chinese engaged in battlefield clearance. In addition, they were full of admiration for Knight and 'Wrightie' at having penetrated so deep into the Somme battlefield: 'They said we were bricks to venture on such an expedition on such a day,' as Knight recorded proudly. And thanks to the hardships of getting so far on so little in the way of rations, she also adored her breakfast: 'I shall never forget it. I never enjoyed a meal so much in all my life … It was the first piece of bacon I've seen since ever I came to France ... Gee! [A nice little Americanism picked up along the way.] How we did tuck in, and how we did enjoy it!', and their mealtime experience was rounded-off by the novelty of 'a Chinaman to wait on us'. Fully reinvigorated, the two women set out again to find the cemetery, with the YMCA officer offering hospitality on their return leg should they need it.

The sense of being unexpected, but most appreciated, guests continued, for their next stop was the local Graves Registration Unit camp. As at the YMCA hut, everyone greeted them enthusiastically and they were offered another meal. Knight recorded insightfully, 'Of course they don't often get any ladies up that way so I suppose we were welcome visitors'. Walking on to Delville Wood, Knight and 'Wrightie' inspected the cemetery, and as well as feeling pity at the number of burials, felt 'sorry for the soldiers who are up in that part of the country digging up bodies from the battlefields and re-burying them. It must be an awful job.' Obviously deeply moved by what she saw, Knight told her parents, 'I want you all to see if you can make a collection for the YMCA Hut for Christmas, so that they can give the boys as good a time as possible.' One of the YMCA women workers 'was awfully delighted and said anything would be appreciated as it was an awful trouble to make ends meet and the Y Hut kept the boys from visiting various estaminets and drinking. Its [sic] really a fine institution and I should love to do some little thing for them.' When they finally got back to Albert station, after another reviving meal with the Montauban YMCA team, they experienced their final bit of inadvertent celebrity status for the day, as a group of officers, enchanted to come across British female company, insisted on treating them to cups of coffee before the train arrived.[57]

As Knight's testimony revealed, visitors were intrigued by the novelty of what they saw, especially the sight of Chinese and other non-European contingents at work. The language used to describe the men of these units revealed all the cultural preconceptions of the time, having heavy overtones of condescension, even while there was admiration for their work or genuine interest in their culture and habits.

Atherton Fleming certainly took on a superior tone in his 1919 guidebook when recounting an anecdote about 'that incomparable creature the "Chink" Labour Corps man'. Bumping into a member of the corps who was busy building himself an oven from recovered German shells while his colleagues were breaking up shell baskets to use as kindling, Fleming sought the British NCO in charge of the detachment, deeply alarmed by the prospect of a terrible accident. After hurrying the men away, the British soldier then let them watch 'and await results', telling Fleming 'quite candidly that he had found the example was much better than precept when dealing with Chinese labour'.[58] The implication of this anecdote, which is presented as a comic interlude at the expense of the Chinese, was that they were not quick-witted enough, and too childlike, to understand abstract principles, and therefore much better instructed through a very obvious practical demonstration.

Just before Alice Knight was served her breakfast by the Chinese assistant working for the YMCA, she had walked across the deserted battlefields, where 'The only living creature that we met for miles and miles was an occasional Chinaman. You see they've got them out there digging up the bodies and reburying them.' At this point, she believed them 'awful looking people but we just walked on and took no notice of them. I think we met three in about eight miles.'[59] Her initial response reveals someone highly suspicious, and perhaps fearful, of those with whom she was totally unfamiliar. For others, they were simply a part of the post-war landscape. A correspondent for the *Aberdeen Press and Journal* made the task, rather than those carrying it out, the point of interest: 'Chinamen and coolies and men of our own labour battalions were passed in groups here and there busy at the gruesome task of gathering in the bodies of the outlying dead into graveyards by the roadside.'[60]

The Chinese were but one part of the cosmopolitanism of the former battlefields, and parties from five continents remained in Flanders for a considerable time after the war. On the Menin Road, Olive Edis encountered many salvage workers, including Indians 'turbaned and shivering in the keen east wind' that swept across the landscape.[61] Near Flers on the Somme, Wilfrid Ewart came across a member of the Indian Labour Corps 'left in this lonely spot to carry on the work of salvage'. Like his fellows outside Ypres, he too was feeling the cold. He had based himself in an old dugout and was keeping himself warm by a small fire.[62] In the Givenchy region, Coop witnessed 'Chinese and other Asiatic labourers' filling in the trenches, while around Cambrai 'Rover' noted 'labour companies of British, Negroes, Siamese, Zouaves and demobilized *poilus*' back in their camps after their daily labours, all looking pretty bored and fed up.[63] Black soldiers, especially those from the American army, were always treated as a great novelty. Olive Edis believed they created a 'picturesque' scene, and described them as 'fine, big, muscular fellows, and their bronze skins were clear and healthy', and near Auxerre, she saw 'a number of delightful negroes of a fine type, probably Americans'.[64]

The German contingent among this polyglot clean-up team often attracted far more pejorative comment from visitors than any other group. Most saw the hideous mess all around them as the sole responsibility of the Germans, and they were often delighted to see them toiling to make amends: the battlefields stressed the magnitude of German guilt. 'They have proved themselves the scum of the earth, more ruthless than their forerunners, the Huns of Atilla, who came to grief fifteen hundred years ago on the plains of Chalons,' wrote an *Aberdeen Press and Journal* correspondent.[65] 'D.E.F.' rejoiced at the 'satisfactory sight' of German prisoners at work near Ypres.[66] Sheffield grandee Sir Alfred Bingham was equally delighted to see groups of Germans at work on the battlefields. To him it was a form of rough and poetic justice.[67] Poetic justice was precisely how Dafoe judged the sight of German prisoner gangs filling trenches and clearing the battlefields. He believed he saw in their eyes men who had realized it 'was the end of their dream of world domination' when they enslaved themselves to 'the homicidal maniac who reigned at Potsdam'. His satisfaction at seeing the Germans labouring at these tasks was partly the result of visiting the ruins of Cambrai, which he declared a 'monument to the malignant spirit of the Hun in defeat'.[68] Having seen the trail of destruction, Neville Chamberlain was unmoved by the miserable conditions in which the German prisoners lived and worked. He told his sister, 'The country swarms with Boches. You see them cleaning up wire, mending roads, watering the streets, feeding the cattle, and being herded into their barbed wire cages in the evening. I felt it was a humiliating spectacle for humanity: they looked like slaves but I felt no sympathy for them.'[69]

Sympathy was equally absent from Alice Knight. She and 'Wrightie' were made distinctly nervous, while also a little exasperated, at finding themselves alone with German prisoners. Suddenly, the men they had seen in propaganda photographs, held in barbed wire enclosures soon after capture, were in front of them. Seemingly unable to suppress details that must have alarmed her poor parents considerably, Knight told them in a letter:

Wandering about in the Trenches we came across a German Prisoners of War camp and all the Prisoners were out for recreation. I was surprised to see them in bunches and no guard with them. They all had something to say to us as we passed them, but not understanding what it was we took no notice. I might say we were the only two girls about and couldn't see a French or an English soldier anywhere near at that moment. Well after a bit we found ourselves in a network of trenches and there were two Prisoners roaming on their own. When they saw us they started to follow us. We began to feel a little bit nervous but just kept straight on and took no notice of them. Well after we got on to the road again we happened to look back and they were waving frantically at us. We just ignored them. I wonder if they felt small! Fancy having the audacity to wave to two English girls, I never heard of such cheek.[70]

Others perceived the German prisoners differently. Seeing German, Belgian, and British temporary memorials on a piece of ground held by the Germans for much of the war, John Frame was moved when his guide told him that the enemy soldiers had carefully preserved all three memorials and not simply their own.[71] Another who saw the humanity and civilized nature of the Germans was 'A Scottish Chaplain', who was enchanted at the sound of a violin being played in a German prisoner-of-war camp deep in the ruins of the Ypres battlefields.[72]

Trying to retain sympathy for the Germans and perceive them as decent and cultured people was challenged not only by the nature of the devastated zone but also by the sight of local people labouring hard to piece together the shattered and scattered fragments of their former lives. Where settlements did exist in a condition beyond noticeboard status only, they were still far from anything resembling normal life. The returning inhabitants, huddled together in ramshackle shanties made from salvaged materials or abandoned military installations, inspired visitors' pity for their appalling living conditions and admiration for their determination in such a desperate situation.[73] In the summer of 1920 Major Charles Fair was driven from Cambrai to Amiens in one long journey and thus crossed the full breadth of the devastated zone in a single trip. Even as a seasoned veteran of the Western Front he found the experience sobering. He saw the awful mess in which returning people were trying to recreate their lives, and the lack of usual infrastructure, schools, churches, places of entertainment, and skeletal transport links. 'One wonders what joy or happiness can ever come into the lives of these people again,' he mused.[74] Visiting the battle zones in the spring of 1920, 'Rover' said 'the havoc of war appalled us' when he and his YMCA party saw the shacks in which people were living.[75]

Travelling down from Feurnes near the Belgian coast, Edis came across people, including many children, who had lived in dugouts throughout the war. In one place they found a family still living in a combination of dugouts and disused military huts. All seemed quite 'cheery [and] they were making the best use of the deserted but quite hospitable little buildings'.[76] The general accompanying Baroness Campbell took her to see his old dugout on the Ypres canal bank. Having identified it, they clambered down and were greeted by some women for whom it was now home and who were busy selling drinks to visitors.[77]

As visitors noticed, people simply found a way to exist regardless of the mess. John Dafoe watched a family unload their belongings from a wagon and carry them off into the debris. At first, he could not understand where they were going or why, but he then realized that they were making their home in the cellar under their ruined house. For Dafoe, the scene was poignant, pathetic, and yet somehow inspiring.[78] Neville Chamberlain worried about the future prospects of such people, commenting to his sister, 'You wonder how anyone is to live if the houses were built for there is nothing for them to do. Factories have been smashed to atoms. Fields are cut up with trenches and shell holes and covered with rusty barbed wire.'[79] This world of people living the life of the old American west on the Western Front lingered for a long time. Despite the disappearance of much of the trench network across the Somme, T.J. Booth, visiting in 1923, saw evidence of the war in a number of villages consisting of nothing but temporary buildings or shacks.[80]

ENCOUNTERING THE WORK OF NATURE

Working hand in hand with humans, often more quickly and more effectively, but far more haphazardly, was nature. As all veterans knew, the power of nature to alter the battlefield was amazing. The moment trenches were abandoned, they began to fall in; the moment spring arrived, no man's land exploded into a riot of colours as flowers bloomed, butterflies floated, and birds swooped and sang. When hostilities finally ceased, nature unleashed its forces unchecked, and the results delighted and confused visitors at one and the same time. Healing was made obvious in the work of nature, and the speed of the transformation was astonishing. A Scottish clergyman saw this effect as early as June 1919, rejoicing in the sound of birdsong and the sight of grass and shrubs colonizing the landscape. 'How swiftly Mother Nature plies her healing art,' he noted.[81] Olive Edis found 'kind nature was already, even in winter, beginning to cover up the wreck with green'. Even on the 'tortured ground' of Verdun, grass covered everything in a lush verdant carpet.[82] Writing in February 1920, the officer-veteran T.A. Lowe commented in his guide, 'Nature is hard at work on the battlefields nursing them back to health and peace.'[83]

Ever a keen observer of nature, Henry Williamson delighted in its effects while watching a summer sunset from the Ypres ramparts. He recorded the flowers, grasses, and small creatures who had made their homes in the embankments and surrounding moat and found consolation and balm in the beauty and serenity of it all: 'How sweet a thing it is to be alive and free in the sunlight among the fair grasses of summer, watching the swallows' wings gleaming blue above the water.'[84] For Stephen Graham there was an element of redemption in the cycle, as the dead were transformed into something palliative: 'There are broken rifles, there are graves. There is all but the blood. But from the blood has risen flowers.'[85] A Kirkintilloch woman felt something similar. One of the sights to leave a lasting impression on her was the banks of forget-me-nots covering Vimy Ridge. She told her local newspaper in August 1919: 'Nature seems to have planted there her own most fitting tribute and reminder of the sacrifices of these brave men.' For her, a female force was at work on behalf of all bereaved women as 'Mother Nature' poured out her profuse powers of beauty on the graves;[86] the IWGC was later to claim a similar role: it planted and tended on behalf of others, meaning they need not trouble themselves with any concerns about the sanctity and upkeep of the graves. By contrast, natural effects could also leave some feeling almost cheated of the anticipated experience. John Eastwood Hastings told his wife, 'Nature veiled something of the grievousness of the scene beneath the mantle of snow and I am told that it is much more impressive when it is unrelieved in that way.'[87]

As well as the celebration of nature's effect there was also a recognition of its 'oddness', not just in disguising the battlefield, but also in its wildness. What visitors found was by no means a bucolic vision showing the beauty of nature, perhaps in harmony with human rural pursuits and endeavour, as in an eighteenth-century landscape painting, but rather a crazy, untamed scar tissue growing rapidly over the wound. 'It is all indescribably wild now,' Stephen Graham wrote, 'Gun trench, Grab Alley, Big Willie, Hohenzollern and the rest, cement-coloured, or yellow with a withered prairie of weeds'.[88] Unlike some others, Efga Myers did not see this as nature victorious, instead it was the merest veneer. Apart from the ubiquitous poppy, 'the flower of battlefields', there was nothing to meet the eye 'but a blackened plain, infinitely dreary, swept of everything – a sheer skeleton of nature, hideous in its appalling devastation'.[89]

For many the strangest aspect of this flimsy skeleton was the effect on the myriad and overlapping shell holes, especially in low-lying Flanders. Here, ironically enough, human action managed to turn the region back into something akin to its the natural, primeval condition. Here was a new Flanders, a new *pays bas* indeed. According to a correspondent for the *Aberdeen Press and Journal*, the Flanders battlefields were a marshland covered in rich, thick grass. It was a weird world made weirder by the constant sound of croaking frogs and sight of huge, innumerable dragonflies flitting about continually. But no matter how strange, this wildlife did somewhat mitigate the intensely dislocating feeling of having 'landed on the moon'.[90] Exploring the battlefields in the early autumn of 1919, Dennis Gilmore found the boggy country around Ypres almost impassable thanks to thick clumps and banks of 'flags and weeds and rushes, wild flowers, [and] red poppies' surrounding the shell-holes.[91] Neville Chamberlain told his sister he could hardly walk a yard across the Ypres battlefields without falling into a shell hole filled with stagnant water, and as far as the eye could see were bulrushes, weeds, and other marsh plants.[92] Another visitor to Ypres similarly walked among a 'luxuriant growth of grass and weeds' in sodden ground 'covered with many shell holes full of water and all kinds of rubbish'.[93]

As a veteran and man of letters, Wilfrid Ewart saw a new natural aesthetic in the scene. He was much moved by the poetic experience of being utterly alone on the battlefields, aside from the presence of the dead, while larks sang 'from an atmosphere of calm unclouded blue'. Nature's reaction to recent human history had distilled into something new and powerful: 'Strange how the horror, the loneliness, the chill, the cruelty of earth can be transfused into beauty by the mellowing sunshine of an autumn day,' he noted in terms highly reminiscent of Charles Sorley's wartime poetry.[94]

The combined effects of human labour and nature could give the impression that reconstruction was progressing with far more rapidity than was the case. Even in the earliest days of battlefield visiting, some were convinced the wounds of war were being wiped clean utterly, making it almost impossible to guess at the tragedies of the recent past. Journalist and writer Frank Heathcote Briant explored the battlefields in the early autumn of 1919. Although he believed the towns and cities would be in a state of reconstruction and repair for many years, he predicted the damage to rural areas would be gone within a year, such was the pace of restoration. Quoting a prefect of the Somme department, he told his readers of the 90,000 acres of trenches filled in and 175,000 acres of farmland put back into use. People were flocking back, he said, and 'many towns and villages have quite a pre-war appearance of prosperity', with Amiens almost back to its pre-war population.

Where the situation was less amenable to speedy restoration, there was always the option of afforestation, Heathcote Briant reported, adding that this was proceeding apace, helping to heal the wounds on the landscape.[95] Writing in the spring of 1921, Arthur Hickman confidently stated that the battlefields were disappearing with such rapidity nothing would be left by the autumn. Barbed wire had been gathered in, tireless work on the fields had restored them to cultivation, and anyone standing on the roads around Ypres at evening time would see a host of people returning home after a day toiling to erase the signs of war. He took this as a sign of great hope, promising as it did the return of prosperity and happiness to the region.[96]

Such was the sense of optimism that in the spring of 1922 the YMCA's *Red Triangle* magazine published an article calling Picardy 'an ideal holiday ground', in which the war and its legacies in the form of the cemeteries and memorials appear as a side detail.[97] Nonetheless, the devastated zone was the elephant in the room, and a year later an article on the Red Triangle touring club's excursion to Picardy was focused almost completely on the former fighting zones and the very obvious signs of the war still visible in the landscape.[98]

And Picardy was a lingering bastion of wartime desolation. The suffering and density of damage to Ypres and its environs exceeded that inflicted on any other sector of the Western Front on which British and imperial forces operated, but the devastation of the Somme, due to its size, stretched much further. Its landscape also experienced two forms of disfigurement and desecration. The first was caused directly by battle in 1916 and again in 1918. In between came the German retreat of 1917, during the course of which everything left standing or still untouched by war was destroyed in systematic and minute detail. Realizing what they had done, and its likely mind-numbing effect, the Germans left a noticeboard in Péronne, 'Nicht agern, Nur wundern [Don't be angry, just wonder].'

After the war, the ability of the region to recover was hindered by what it could not help – its geographical nature, which had also shaped its economic development and importance to France. The area affected by the Somme fighting was mainly rural and agricultural, and while this was important to the French economy, it was not given quite the same priority as getting Lens or the coal mines back into production. Unlike Ypres and its surrounding hinterland, which was on the road to pretty much everywhere important in Belgium and north-east France, and reconstruction could commence no matter how much of a challenge, much of the rural Somme was miles from anywhere, and on the way to nowhere. It was home to dozens of tiny communities cut off from each other by inconvenient chalk crests and ridges (as Will Bird put it in 1931: 'Nowhere else have I seen as many sunken roads. They lead in all directions, and in them you can see nothing of the district about you.'),[99] which further hindered reconstruction. Advancing the forces of restoration across this land evenly and consistently was a distinct challenge, to put it mildly. Then, for British visitors, the lack of easy infrastructure, and the need to swing out to Amiens on the train before shunting back inland to Albert as a preliminary to visiting made travel to the Somme so much harder than the easy hop to Ypres. When all these different elements were added together, the Somme must have seemed like a land caught in the nightmare of war for a very long time; if it was moving towards recovery, it was doing so in the incoherent manner of a sleepwalker wandering along in a daze. The Somme remained the heart and soul of destruction long after its smashed skeleton at Ypres had been re-membered.

THE SLOW PACE OF RECOVERY ON
THE SOMME BATTLEFIELDS

For visitors, the signature of the Somme was stillness. 'There is the incomparable Somme silence, a silence achieved by the tremendous thunderous contrast in history, a silence from the stilled hearts of the dead, a deafness and a muteness,' declared Stephen Graham.[101] Walking across the Somme battlefields in the autumn of 1919, 'R.O.P.' and his guide 'met no living thing, not even a bird'.[102]

Standing on the Butte de Warlencourt, an artificial, ancient mound on the Albert–Bapaume road, Wilfrid Ewart and his companion felt utterly alone. Only the odd passing lorry or car on the nearby road reminded them of life. Walking off in the direction of Flers, even those traces disappeared, and they found 'themselves without sign or sound of human life'.[103] Treading the same paths as Ewart, Sir Philip Gibbs felt a similar sense of loneliness on the Somme battlefields: 'Beyond by Delville Wood and High Wood, Longueval, and Mametz Wood there was an immense solitude,' he told *Daily Mail* readers.[104]

What amazed the Canadian veteran Will Bird was the number of temporary buildings still scattered across the villages and communities of the Somme. Everywhere he went he found ramshackle collections of former military installations and prefabricated huts erected immediately after the war and never replaced. It gave the whole region a sense of gloom; every village, especially in winter, was 'an eyesore to travellers or anyone outside the area'. The junkyard effect was exacerbated by the persistence of scrap metal across the Somme, as people took considerable risks to collect the remnants of ammunition and other abandoned military kit brought to the surface by the weather and the seasons, ploughing, and building. Bird claimed the Somme was one massive heap of scrap, piled up at the side of roads, next to farm buildings, or being retrieved from the ground by scavengers eager to sell it on to dealers.[105] 'You may see them, strung out in groups of three or four, stumbling along with heads bent, eyes glued to the to the ground ... The fund they draw upon is practically inexhaustible,' wrote a veteran watching the metal collectors at work in the mid-thirties, adding, 'It is doubtful whether as yet more than a fraction of the total quantity of metal rained down in the course of the First Battle of the Somme alone can have been gleaned.'[106]

If the effect was still sobering in the thirties, then the years immediately after the war were fearful. Wilfrid Ewart declared the Somme to be the land of the dead in the autumn of 1919.[107] Wandering across the Somme at much the same time, Gibbs saw a world devoid of all signs of life or reconstruction, 'Nothing had been done, or is being done, to restore this devastated region except by the removal of high explosives and the decent burial of discovered dead. No houses as far as I have seen are being built on the site of villages which were blown off the earth.'[108] The following summer, 'Rover' of the YMCA inspected his organization's facilities across the old Western Front; as a result he declared 'the wildest bit' of his visit was the Somme region.[109] A summer later little had altered, 'The impression is one of utter baroness [*sic*] and desolation,' said a *Yorkshire Post* journalist.[110]

As virtually all British visitors to the Somme front arrived in Albert having taken the train from Amiens, its condition acted as a taster of what was to come. All visitors arrived with a visual knowledge of one particular building, which had become iconic through wartime reportage. Albert's basilica was a nineteenth-century affair built in a florid mixture of styles including Byzantium and Venetian Gothic. From it arose a tower topped by a gold-leaf statue of the Virgin holding up the Christ child to the heavens. Hit by a shell in 1915, the statue lurched forward, but was then secured by French military engineers. This bizarre sight of the Holy Mother seemingly about to cast the infant Jesus into the square below became ubiquitous, and it quickly inspired its own myth: when the Virgin finally fell, the war would end. And, of course, lo it came to pass – the British detonated the tower during the March 1918 retreat, and within six months hostilities had ceased. Albert therefore already had a legendary status, one in which ideas of sacrifice, martyrdom, and destruction were intertwined. For Stephen Graham, Albert was the town of 'complete sacrifice', the Golgotha of the battlefields, a town of the dead where 'even its soul ha[d] died' leaving behind the odd revenant seeking solace in the ruins. But even this immolation could be trumped. 'Bapaume lies more abased even than Albert. It is as if its stones had had a soul and been afraid, vibrant with the horror of humanity. The consternation of inanimate matter is expressed in its ruins,' Graham went on to declare, teetering on the verge of Gothic melodrama.[111] Rather more matter of fact, but just as affected by the sight, was Gibbs, who was convinced 'no human hand will ever "restore"

Figure 4.2   Temporary buildings, Bapaume, c. 1920.

the town ... or any of those villages where now some of the people who lived there before the war are raking among the ruins for relics of their old life'.[112]

Standing on the Butte de Warlencourt gave visitors a complete panorama of the battlefield in all its abject debasement. Yapp reported the pathetic, moving, and inspiring sight of the mound topped with its three wooden memorial crosses erected during the war, giving it a resemblance to Calvary, and from its summit 'a scene of awful desolation' opened up.[113] Visible were the fragments of human life scattered across the Somme battlefields in penny packets, living troglodyte lives, existing in a parallel world where the rules and mores of twentieth-century civilized life had no grip. It was little wonder that this was a land of rumours and fears. 'Rover' was told that shots could be heard at night, fired by 'wild men who still live there underground – Britishers, German deserters, French and Australians – who come out to plunder and kill' and who were kept safe by the impenetrable labyrinth of abandoned trenches and dugouts they inhabited.[114] Recovery from this 'primeval jungle of dead weeds, the tripartite heads of brown teasles looking like low-lying spectral regalia of the death-kingdom,' required money, energy, planning, and, above all, time.[115] As such it was a slow process, and the Somme retained its wounds throughout the twenties and thirties. Exploring the battlefields in the summer of 1927, members of the Northumberland Hussars old comrades association said the region of the Ancre valley showed 'more signs of devastation ... than in any other sector of the Western Front' they had visited.[116]

THE LINGERING WITNESSES TO THE EFFECTS OF WAR

A timescale of geological ages seemed in order if the landscape were to recover from the man-made effect on its geography. On the Somme, a particular indicator of the scale of devastation and speed of restoration was the woods and copses, making trees a dominant icon of, and metaphor for, destruction. This allegorical language was another minted and put into free circulation during the conflict. Newspaper reports made each wood of Ypres and the Somme, no matter how insignificant on the map, places of agony and heroism. Chateau Wood, Ploegsteert Wood, Railway Wood, Sanctuary Wood on the Ypres battlefields, and Delville Wood, Gommecourt Wood, High Wood on the Somme: totemic and shamanic in their repetitive incantation across newsprint and in post-war memoirs, they also formed the root of visual interpretations of the war. The Nash brothers made trees the souls and sentinels of their paintings, from Oppy Wood to the Menin Road. War photographers were equally captivated, presenting the public with a world of trees and woods that was haunting in its strangeness, home to the dryads of industrial war. As David Jones said in his enchantment in the woods of the Somme:

> The Queen of the Woods has cut bright boughs of various
>   flowering.
> These knew her influential eyes. Her awarding hands can
>   pluck for each their fragile prize.[117]

For visitors heading towards Ypres, a sign of the transition into the battlefields was the condition of the trees. A party of Manchester and Burnley visitors who made the crossing to Zeebrugge soon after the resumption of normal services in May 1920 realized they had moved into the battle zone at Dixmude when the landscape suddenly revealed evidence of the war, most obviously in the number of blasted trees made all the stranger by the odd one in full bloom.[118]

A YMCA tour in the spring of 1923 saw the remains of trenches at Thiepval, roads still cratered by shell blasts, as well as great mine blasts, and the 'stumps where forests had flourished'.[119] A tourist exploring Houthulst Forest, to the immediate north of Ypres, was haunted by the contrast of what a wood should be and what was in front of him: 'Where at one time grew magnificent trees in their thousands, and a wealth of undergrowth, now only remain acres of tall, inelegant, leafless wood, like a cemetery of scaffold poles, with here and there gaps caused by the effect of big guns.'[120] At the same time, the absence of trees, so expected in the rural areas of the battlefields, helped to confirm the intensity of the holocaust. 'Not a whole tree is in sight', recorded a journalist of the Somme countryside in the summer of 1921, while another veteran standing on Kemmel hill, south of Ypres, gained a 'fine panoramic view' of the battlefield made all the clearer by a landscape 'completely devoid of trees'.[121]

Trees and woods were also perceived as silent chroniclers which had sucked in the experience of battle and were expressing it through their forms. Anthropomorphizing trees as the surrogate bodies of soldiers made them ghostly witnesses clinging on to life or reminders of its dying embers. 'The blasted and withered stumps speak of physical torture, which those who faced death in the wood bore, for human bodies were broken just as the trees,' noted a report in the *Belfast Telegraph*.[122] Stephen Graham was obsessed with these images. For him, the metaphysical symbolism of trees ran its tendrils across the battlefields; trees were martyrs with withered hands 'fixed for years in the momentary awfulness of death'. Graham drew his imagery together in one great thicket when referring to the woods of the Somme:

> The road you traverse to the Somme altar is the road which hundreds of thousands of young men trod, marching to moments of destiny, moments of victory ... they marched from the quiet places of homeland and the empire ... to Britain's quarrel and her mightiest enemy ... Now the desolation of Nature alone suggests what a desolation there was of men. The terrible woods are impressionist pictures of the ruined vitals of great regiments, and you can hold a forest in your mind as you would a skull in your hands and say – *This was a forest. This was an army.*[123]

The ghostly forest armies acted as the gateways to memory for some veterans. They were solid reminders of who these men once were, and of the landscape they knew. Much disappointed, and downright miserable, at finding so little he remembered on his return in 1930, 'J.B.M.' found comfort in a blackened, shattered tree among the new saplings in Thiepval Wood. 'Here at last was an old comrade,' he wrote, 'With what friendly feelings I lingered beside that war-worn veteran.' This allowed him to descend into memories of making the assault on the Schwaben Redoubt, and with that, 'I was no longer alone – I was marching step by step in the ranks of the Deathless Legion in company with my comrades long since "gone west".' On his return home 'J.B.M.' noted 'the greatest memory of my pilgrimage – a blackened stump in Thiepval Wood'.[124] His fellow veteran, William Hyde, felt the lush fields with their blaze of summer flowers provided perfect companions for the dead, ensuring that they would never be lonely. He saw poetic juxtapositions in the then-and-now of the Ypres salient: 'Her fields were red; her fields are green. Her spires were razed; her spires are raised ... again. Her trees were trunks, but praise be, some trunks are ghosts.' [125] Seance was achieved through skeletal trees.

Blasted, shattered, and yet often bolt upright rather than leaning at drunken angles, trees remained impervious to clearance. They were a skein laid across the landscape, a ghostly reminder of the world of the trenches and no man's land. Leuze Wood on the Somme, with its 'tattered and mangled stumps', still gave 'some indication of the fierceness of the fighting in the neighbourhood', noted Coop in his 1920 guidebook.[126] Visiting the Somme in the summer of 1925, G.W.C. Craik said, 'The splintered tree-stumps vista of High Wood ... is practically unaltered and cannot be mistaken.'[127]

Attending the unveiling of the South African memorial at Delville Wood in October 1926, W.A. Michell set off to explore the nearby woods, all of which had been on bitterly contested land. Although at first glance they appeared to be recovering from the wounds of war, it was only an illusion, as on closer inspection he found them 'much as they were when we knew them last'. The new growths of thickets and brambles disguised only thinly 'the naked truth of what once was from the eyes of those who are able to read the story these woods would tell, stories of heroic sacrifice and pertinacious courage'.[128] That same year, eight years on from the Armistice, the rumours of Somme woods clogged with impenetrable barbed-wire thickets intermingled with roots, branches, and undergrowth abounded. Assessing the task of clearing such a mix of human and natural debris as formidable, the veteran H. Channing-Renton was quite prepared to believe they would be left as irredeemable islands amidst the reconstruction.[129] The woods therefore defied the processes of salvage, reconstruction, and clearance, resisting those forces for years after the war.

By the same token, plantations of new saplings and the recovery of established trees could be seen as welcome signs of life, hope, and restoration, of nature and humans working together to wipe out the misery of war. Despite Henry Williamson's seeming dislocation due to reconstruction and restoration, he found the remarkable recuperative power of nature a powerful tonic and stimulant. Standing in Aveluy Wood on the Somme, where the new trees were already twelve feet high, covering the place in a green canopy, surrounded by the sound of birdsong and buzzing bees, he achieved peace and contentment, sparking a string of placid anecdotes, rather than violent or destructive ones, from among his wartime memories.[130] Another pilgrim veteran 'was particularly struck by the aspect of the countryside now that all the woods and hills have been planted with young trees'.[131] Life had found a way after all.

Such outbursts of redemptive energy and observation were rare. The dominant impression made by the state of woods and trees was one of introspection and gloom tinged with stunned awe in the non-veteran. On the Somme, two woods were identified as exemplars of the many sylvan martyrs – High Wood and Delville Wood, 'two of the best-known features of the Somme battlefield', according to R.H. Mottram.[132] Veterans debated the precise standing of each wood based on grim credentials. Graham Seton Hutchison declared High Wood '"The Wood" to those who fought in the First Battle of the Somme ... All ways seemed to lead to High Wood,'[133] while Coop believed Delville Wood was probably more famous, but High Wood was 'indeed, the most sadly impressive of them all'. The reason he gave was intimately associated with the point in time he was writing, 1920, when its shattered stumps were surrounded by only a veneer of undergrowth, meaning the wood was a commanding position from which to survey the battlefield in peacetime.[134] Captain Atherton Fleming, writing just a few months earlier, recommended the wood for exactly the same reason ('if the day be fine and clear, a very good view can be obtained'), and it had the added advantage of being full of 'all the usual battlefield flotsam and jetsam'.[135] Standing on a ridge top – hence its army nickname – High Wood remained a pilgrimage point. At its edge the sightlines remained impressive, and the spot was home to a beautiful, and extensive, cemetery and a number of memorials.

High Wood's only rival as a point of pilgrimage was Delville Wood, and H.A. Taylor had no difficulty establishing his rankings. After considering High Wood in *Good-Bye to the Battlefields*, he noted: 'But the wood of woods here, the vilest of death spots, was Deville Wood.' Although many units fought around, in, and for Delville Wood, the corner of the Empire with which it was most associated was South Africa. The South African brigade, consisting of 121 officers and 3,032 other ranks attached to 9 (Scottish) Division, entered the wood on 14 July 1916. When it withdrew on 20 July, only 143 men capable of parading for roll-call emerged.[136] This almost unbelievable bloodletting instantly transformed the wood into a sanguinary birth rite for the only recently formed Union of South Africa, and it became the focus of its memorial activity.[137] Sir Herbert Baker, an English architect who had worked extensively in South Africa, was approached to design the memorial. Revealing his deep understanding of the space as both a wood and a battleground, an integral part of the memorial plan was to replant it with acorns collected around the original Dutch colonial governor's house outside Cape Town. The process completed a cycle, as oaks taken to South Africa by Europeans were brought back to Europe from South Africa: Europe and South Africa were to meet permanently in a quiet corner of the Somme. The battlefield space was imaginatively recreated through the inscription of the London and Edinburgh street names given to the trenches on stone plinths, creating rides through the newly laid plantations. The battlefield was thus permanently written into the 'natural' landscape. A final flourish was abandoned: Baker wished to crossbreed South African springboks with local muntjac deer. A hybrid wildlife was to skip through the wood happily ever after, but this plan was dropped, doubtless due to the conviction that local farmers would equally happily hunt them to extinction in short order.

For the memorial itself, Baker designed gently curving walls leading to a central pavilion tower (similar to those he used at Tyne Cot Cemetery) topped by a sculpture of Castor and Pollux, the twin sons of Jupiter, clasping hands over the back of a horse. The symbolism was rich. The land and spirit of South Africa itself was the horse; the twin peoples of South Africa, English- and Afrikaans-speakers, were the mythical twins, joyously, vigorously, and powerfully walking with the horse into the future. The inscription on the memorial yelled out the architectural symbolism: 'Their ideal is our legacy; their sacrifice our inspiration.'

The value of endeavour, struggle, and sacrifice was enhanced further by the cross. Baker's version included hints at a 'Voortrekker Cross', which reinforced the Afrikaans presence in the architectural grammar, and his so-called Pilgrims' Cross. The latter design had been offered to the IWGC as a standard feature of the cemeteries but was rejected in favour of Blomfield's Cross of Sacrifice. Baker, therefore, used the cross to allude to the pilgrimages of the medieval world and the cultural heritage of the Dutch in South Africa. (In keeping with the times, the memorial made no specific mention of the indigenous communities of the nation, or of other people of colour who had made it their home, although they did gain representation of sorts at the lavish and powerful unveiling ceremony.)

The memorial was approached by a broad grass avenue creating an axis linked to the IWGC's Delville Wood Cemetery, also designed by Baker. Beautiful, light, and yet solid, finished off with an easily accessible sculptural form, Baker's memorial was appreciated by all who saw it. King's Royal Rifle Corps veteran Gerald Dennis was amazed, describing it as 'glorious' and 'awe-inspiring; its purpose seemed to envelope us'.[138] According to Taylor it was a 'magnificent memorial'; his fellow veteran, W.A. Michell, echoed this sentiment, referring to it as a 'magnificent structure'.[139] Magnificent, but without imposition; allegorical and symbolic, but not cryptic; attention-grabbing, but not demanding – Baker achieved a harmony in the Delville Wood memorial pilgrims could appreciate.

On the Ypres battlefields was a wood with an equally distinctive history and a reputation quite different from the evil or melancholy memories inspired by so many others. Ploegsteert Wood on the southernmost edge of the Ypres front was a place associated with the original BEF and the London Rifle Brigade (LRB), a Territorial Force unit sent out as reinforcements in the winter of 1914. Although it was by no means ever an entirely quiet spot, 'Plugstreet Wood', as the Tommies labelled it, escaped the punishment inflicted on other woods and forests during the conflict. Like Delville

Wood, Plugstreet was rich in London nomenclature. Thanks to the presence of the LRB, and its educated, middle class, white collar officers and men, virtually every corner of the wood was labelled and turned into a randomized version of the Geographia A–Z of the capital. The Strand, Oxford Circus, Piccadilly, Rotten Row, Fleet Street, Somerset House, Bunhill Row, Eel Pie Fort (after an island in the Thames at Twickenham), and London Farm could all be found in and around Plugstreet Wood.[140] Based in the area for a considerable period, the LRB also played an important role in creating the cemeteries in the vicinity, including one to which it gave its own name.

Given this deep association with the LRB, it was fitting that after the war the IWGC gave the task of designing the cemeteries in this area to Wilfred Von Berg, architect and former LRB officer.[141] Naturally enough, LRB veterans made for Plugstreet on all of their pilgrimages. A particularly large contingent went over in the summer of 1927 for the dedication of the LRB Cemetery, which included a small memorial plaque to the unit set in the cemetery shelter. After the ceremony, the men, their families, and friends piled into the wood, filling it with 'familiar faces searching for familiar spots', and easily found these spots were, too: 'The barricades were there. Tourist Avenue, and Tourist Peep, Hunter Avenue and Bunhill Row, overgrown and untrodden as they are now, were nevertheless recognisable.'[142] Perhaps ironically, it was the IWGC's clearance of paths through the thick undergrowth that sprouted up immediately after the war that allowed these remnants of the conflict to become visible once more. It makes it difficult to believe Henry Williamson was entirely correct in describing the wood as 'unrecognizable, the old redoubts and corduroy paths undiscoverable ... Its spirit lives only in memory; with us it will die, so we walk on.'[143] His fellow LRB veterans exploring the wood, seeing the graves of old chums in the cemeteries they had created and helped maintain, seemed able to recover something he could not. Williamson's inability to penetrate the surface appearance also revealed the irony of a veteran who rejoiced in nature's ability to restore and heal the wounds of war while also being disconcerted to find it act, Lethe-like, on the landscape.

Tramping around the wood in 1931, Will Bird stumbled into a realm of contradictions; stuck in its wartime condition, Plugstreet Wood was also a place of tranquillity:

> The willows and elders are almost massed in places, and one has to force his way through them ... There are two other cemeteries in the wood, all beautiful in such a setting. No other place along the front seems so fitting. There is a quiet, a peacefulness among the trees like the calm of a cathedral. Go in among the trees, and rabbits are everywhere [but] watch out for ditches and rotting duckboard walks, and where you see glistening dark water, long pools of it, stop and you will see traces of the old trenches. Bits of corrugated iron, rusted stakes, old trench bays still holding their form are there, with a hundred other relics of occupation. And all is mire, squelching sod, reeking dampness. It seems one vast swamp.[144]

The odd liminal condition of Ploegsteert was encapsulated in a tale recounted to Bird by one of the cemetery gardeners. A British visitor had strayed deep into the wood, got lost, and after much thrashing around trying to get out, fallen into one of the old trenches. He was eventually discovered suffering from exposure. Wartime destruction and wartime dangers were still at work deep in the heart of a quiet wood on the Franco-Belgian border.

*

Encountering devastation, being shocked by it, amazed by it, and often ultimately bored by it was a significant part of battlefield visiting in the years immediately after the war. Although human labour and the forces of nature worked to wipe away the traces of the conflict, the smashed world of the old Western Front was never fully erased in the twenties and thirties. Remnants of trenches and dugouts, clumps of barbed wire, heaps of twisted and rent metal, as well as seemingly dud ammunition were everywhere, and could be found pretty much anywhere. Lance Grocock visited the Somme with his fellow 'Leeds Pals' veterans at Easter 1935. After much comment on just how much everything had changed, removing all evidence of the war, he stumbled across traces of the old German line. The trenches had collapsed, 'but the scars are there and, in the dank undergrowth, an obscene tangle of rusted wire, shattered gunstocks, broken stakes. You walk with care here: it is dangerous ground.'[145]

5

Postcards from Veterans

Travelling over the battlefields, it is possible to realise, perhaps more than during the war, when it seemed the natural order of things, how complete the destruction has been. Everywhere is brand new ...
                     Ex-Corporal Morris, formerly London Rifle Brigade (1923)

We carried on to 'Zouave Villa,' where we almost held our breath. An estaminet stood at the cross roads sacred to us all, and burnt into our memory by livid scars, and we had a bottle of beer in it. I felt uncanny.
                     'Labour R.E.' [Royal Engineers] (1924)

A splendid opportunity to consume innumerable foreign drinks of doubtful origin.
                     'W.E.B.' (1933)

The old battlefields held a mystical appeal for veterans. Despite being sites of horror and misery, they were also the 'land of lost content', the places where life was never more intense, never more exhilarating.[1] Returning to the sites allowed the thrill of the old days to return, but this time without risk, and although their visit might be an expurgated version veterans shared if they were travelling with family and friends who had not served, with their fellows they could recall everything. The land to which they returned was in the process of transformation, and while finding traces of their old world threw down challenges, they were still there, as R.H. Mottram realized when, in 1935:

> I went out to explore a place which has ceased to be what it was, but will never cease to have boundaries recognisable at some places, a land in which the railway is of very little

use, and which must in some corners be visited on foot, and in others cannot be fully seen, for private owners have resumed their rights and rebuilt houses and business premises. Yet I have found enough that is unmistakable to guide me.[2]

And so they went, intent on playing what Siegfried Sassoon called 'the game of ghosts'.[3]

THE MOTIVATIONS FOR REVISITING THE BATTLEFIELDS

In playing such a game veterans (especially when in groups) tended to adopt competing personas. They could be lads having a jolly. Moreover, it was a jolly sanctified by the thought that their dead comrades would smile on it. Jocularity, drinking, and a good sing-song were therefore cast as alternative but heartfelt acts of remembrance and commemoration. Or they could transform into respectful, introspective mourners congregating at a unit memorial or in cemeteries where large numbers of their comrades lay buried together: they became a particular type of pilgrim. Finally, they could act as keen students of military history, investigating the ground over which they fought, seeking to understand better the part they and their comrades played. However, these guises rarely remained stable. Detached observation just as quickly gave way to moments of deep reverie as veterans trudged across a muddy field or up one of the many gentle inclines boasting the label 'ridge'.

In walking, tramping, and in some cases trudging the battlefields, placing wreaths on memorials, and quietly standing by a comrade's grave, veterans also acted as representatives of those who could not make the trip; on their return home, having made their statements, having informed their fellows, they usually concluded by exhorting them to do likewise and undertake a pilgrimage. As only a minority (albeit one of appreciable size) of veterans returned, they felt an almost missionary zeal in this urging on of their old comrades. Visiting the battlefields was a moral duty, an act of homage and gratitude to those who had sacrificed everything. Veteran 'A.O.K.' was convinced of the necessity, but he couched his exhortation in a curious mixture of the poetic and touristic:

> Should everyone make a pilgrimage to Ypres? Of course
> you should go, and pay tribute to our brave comrades
> who lie out there, they saved England, they were a wall
> unto us both by day and night. Go for a trip with the Ypres
> League parties! You will have the time of your life, English
> is spoken everywhere, the Belgian people will give you a
> hearty welcome and it will be an education.[4]

As so many accounts reveal, veterans felt this need to address their fellow ex-servicemen, providing them with an update on the state of the old front line while ruminating on the meaning of a pilgrimage and the value to be gained from it. But, such was the welter of emotions thrown up by a return to the battlefields, recording responses was no easy task, as the former army chaplain the Reverend Neville Talbot admitted. 'I ought, if only for historical reasons, to try to set down some impressions of the reopening and rededication of Talbot House, Poperinghe,' he wrote, although he doubted his ability to define 'more than a fragment of the meaning of that which had brought them together'.[5] Barrages of emotions and thoughts were unleashed by the return to the battlefields.

Motivations for visiting were, therefore, various and overlapping. 'T.M.' said his fellow veterans were inspired to cross the Channel by two particular desires: to see their old billets and dugouts, and to visit the graves of lost comrades.[6] Rex Sargeant, who made an Ypres League pilgrimage in September 1935, agreed. He wanted to visit 'old haunts and to look up the graves of the "men who stayed behind"'.[7] Arthur Behrend heard the siren call of the battlefields while stopping off in Bruges en route to Spain in the winter of 1921: the sound of clogs, the sight of rows of poplars, and the name Roulers on the map began to kindle his excitement. Seeing an Australian soldier walking down the street gave him an irresistible urge to rearrange his schedule. Behrend leapt on a train for Lille, connected to Arras, and then had to get the local stopping train for Achiet-le-Grand on the Somme front before he could make the final connection on to the light railway.[8]

H.A. Taylor had a similar experience. 'It began in a casual fashion, in the course of a holiday in France when the orthodox relaxations of a coast resort began to pall,' he explained in a note at the start of his book, *Good-Bye to the Battlefields*. It unleashed 'a call to old, familiar haunts [which] became more and more insistent'. Then, once he had undertaken the first trip, the fever was upon him. 'One such visit did not, and could not satisfy. Other visits became inevitable,' he added.[9] The gnawing, nagging irritation to return was common among veterans, and after a visit they often felt the need to share the itch with their former comrades. 'There is a lure in the battlefields. Despite, or perhaps because of, the bitter and painful memories which the names of the Somme and of Ypres arouse in the mind, there comes a desire to tread again in peace the paths known so well four short years ago,' admitted a special correspondent for the *Yorkshire Post*.[10]

Writing in the spring of 1927, E.F. Williams told his fellow ex-servicemen that they really had to go back to Ypres. He had heard all the excuses and reasons not to make the trip. Indeed, he himself had used them when others had advised him to make the effort. Finally deciding to go in 1924, he had found it so addictive he had returned on four further occasions. Williams was insistent in his exhortation: '*You must go* [original emphasis]; it's your duty.' And the reason was the debt of remembrance they owed to the dead. Only by visiting the battlefields and the graves of their comrades could the living veteran truly pay tribute to the dead.[11]

'C.N.L.', a veteran of the Green Howards, saw a visit to the battlefields in similar terms. Using stirring rhetoric, he stated it was the veteran's 'duty to go and see those plots of holy ground reclaimed from the withering blast of war where lie the broken bones of many of the Empire's best whose "name liveth for evermore"'.[12]

Writing in 1932 as the fallout from the Wall Street Crash was washing over Britain, J.M. Finn told his fellow veterans that it was worth making every effort to go on a battlefield pilgrimage. He wrote: 'Times are difficult with many of us, but the writer would urge that every ex-serviceman and every relative of the fallen who can possibly manage it should make a very big effort to visit the scenes that meant so much to our Nation and to ourselves.'[13]

The passing of time also seemed to whet the appetite to return. Some wanted to show their families where they had lived, laughed, and fought before it was too late to recall clearly 'the scenes of exploits that are beginning to take on the quality of a fantastic dream'.[14] 'E.C.F.' was 'particularly anxious to visit our old home on the [Vimy] Ridge and to say that sixteen years after I again walked the trenches and tunnels'.[15] 'T.M.', writing up the account of the Whaley Bridge

British Legion tour, was aware of the mystical call of the battlefields, even if he was unable to define it: 'Sixteen years ago we were keen to see Flanders. Four and a half years later many of us said: "Flanders! I never want to see it again." Yet twelve years after the close of hostilities, the desire to revisit becomes too strong to be resisted.'[16]

Major Charles Salvesen, chronicler of the 236 Siege Battery association's trip to Ypres at Whitsun 1935, started his tale by commenting that in May 1918, when the battery finally left the salient after serving in the district for seventeen months continuously, 'not one member wanted to see the "blasted" place again'. And yet, in early 1935 when a plan to return was mooted 'the response was instant and eager'. After a very largely tongue-in-cheek account of what seems to have been a comic riot of a tour, Salvesen finished his account on a far more reflective note: 'The trip was a complete success and can be recommended to all Units who fought in the Salient, who should go and see for themselves what God, Nature and Man can do. Verily, "I shall restore unto you the years the locusts have eaten."'[17] Quoting from the prophet Joel added weight to Salvesen's statement, but he did not explain precisely why veterans should go. He told them that they would see a transformed landscape but did not state what quality or benefit that would impart. As with so many who exhorted others to remember, Salvesen made the commemorative act an end in itself.

A battlefield pilgrimage certainly reinforced the special status of the veteran. Although ex-service clubs and organizations ensured veterans never lost their wartime selves, the most potent way to stir up that identity was through a trip to the battlefields. The descent into, and re-engagement with, their wartime personas started at the moment of departure. Mrs E.A. Smith accompanied her husband on the 1928 British Legion pilgrimage and watched the process in action, commencing with the train journey. At each stop en route for London more and more people joined the pilgrimage special; baskets of food were produced, leading many of the men to comment on the bully beef and biscuit ration packs they once consumed with monotonous regularity. It was the signal for the veteran self to take over. Boundaries broke down as the common experience came to the fore. Men who had come along with their branch members started chatting to others they had never met, the solvent being their military service. Once the much appreciated picnic baskets had been handed round, hot tea followed, though drinking it on a jolting train was a challenge. All the time a sense of jolly camaraderie was being forged, which Mrs Smith believed gave her an insight into the little things 'that kept the boys so brave and optimistic' during the war.[18] According to the *Yorkshire Post*'s special correspondent accompanying the 'Bradford Pals' association pilgrimage, the party divided between the veterans and their families. 'Among the men "Do you remember?" has been the oft-repeated phrase ever since Tilbury was left behind. With every mile memories have crowded and jostled with one another in an unending stream from their lips, and they "are living it all again," while the women say little but think much.'[19]

For Rex Sargeant, the journey into his past started that moment he joined the September 1935 Ypres League pilgrimage at Victoria station. Recalling the scene in wartime in great detail, he remembered the station as 'the gateway to death, and the Road to Golgotha', taking, as it did, men to the channel ports and embarkation for the front. Sargeant was well on the way to his past self before the train left the platform.[20]

As each stage of the journey to the battlefields was completed, the sense that veterans were set apart from the normal ranks of society became more and more apparent, and others reacted to them accordingly. Wearing the cornflower, the remembrance emblem in France and Belgium, the veterans of the 85 Field Ambulance were hurried through customs in Dunkirk as if someone had said 'open sesame'.[21] Their veteran status and willingness to show solidarity with their wartime allies ensured a respectful welcome and preferential status.

The men on the Manchester Regiment tour likewise sped through customs by playing on their status, and as their account implied, their special status filled them with a certain confidence and light-hearted insouciance: simply saying '42 Division' worked with all French officials. On boarding the Paris Express the progression into the old world and their former status continued, as all realized 'the tour had really begun' and 'tongues were now loosening'. With it the last barriers to the past were opened, but this was not perceived as traumatic or disturbing: 'War reminiscences [were] on the lips of most and the party was full of bonhomie.'[22]

As the countryside sped by, the veteran eye adjusted. Drawing upon their stores of knowledge and experience, veterans spotted things others missed. An article in the *Sphere* told of old soldiers suddenly becoming animated, pointing out their former dugouts and trenches while travelling on the train through northern France.²³ The veterans of the 236 Siege Battery acted similarly. As the train moved towards the old fighting zone soon after leaving Hazebrouck, they moved up a garrulous gear. '"Look there's Mont de Cats! They've never rebuilt the Scherpenberge Mill! Good old Kemmel! Do you remember when — ?" and the floodgates were opened.'²⁴ It was such behaviour that caused Henry Benson to compare veterans to Uncle Toby in *Tristram Shandy*, for, like him, 'these gallant fellows take an intense delight in recalling and reconstructing the sense of their former difficulties and dangers in the fields of Flanders'.²⁵

F.J. Lineton and J. West, two veterans of 6th Battalion, King's Shropshire Light Infantry, travelled to the battlefields together in the spring of 1930. Part of their motivation was the desire to reassert and reaffirm their special credentials as veterans. Their account commenced with an apologia-cum-justification that revealed a defensiveness and sensitivity probably caused by the reaction against the war that flowered briefly in the late 1920s and early 1930s. They believed their pilgrimage was to reconnect them with their old ideals and the period in their lives when 'the bright stars of selflessness, supreme sacrifice, and real comradeship stood out perhaps as never before'. Moreover, belief in such values did not mean they loved war and thus were 'not necessarily the friends of Satan'.²⁶ For Lineton and West the only place they could rediscover themselves and find comfort was the place of trauma, which they transformed and ennobled by framing it as a place of sacrifice. They insisted that war stripped men down to their real selves, often hidden in civilian life, and in doing so revealed qualities of endurance and devotion to duty. The reminder of this was most strongly felt at Ypres, 'a holy place', transfigured by the blood spilled to defend it. Their beloved comrades, the dead, and their places of burial and commemoration then exerted a deep power over them: 'Can you wonder that it attracts and fascinates those who were privileged to serve with them and return to peace when the trial was over?'²⁷

This mystical communion with the dead and the landscape in which they lay was a common sensation among veterans. Few felt and expressed it as strongly as the former Machine Gun Corps officer and prolific writer Graham Seton Hutchison. As a man convinced that veterans were intrinsically linked not only to each other but also to the battlefields, Seton Hutchison believed pilgrimage to have immense value, and identified Ypres as the spiritual heart of the battlefields, calling it the 'Sanctuary of Comradeship' in an article for the Ypres League. Drawing upon his earlier writing (and in the process creating copy he would reuse in his later work, *Pilgrimage*), Seton Hutchison explored his ideas on the supposed irrevocable bond between veterans and the landscape of Ypres.

The intensity of the connection was loosely based on a Judeo-Christian sense of redemption through suffering. Something positive could emerge because the battlefield was the altar on which sacrifices were made, and, though horrific at the moment they took place, those sacrifices revealed the nobility of selflessness and comradeship. Utter agony and trauma were experienced in the crucible of the battlefield, forming lifelong relationships between soldiers and with the ground itself. As with Lineton and West, Seton Hutchison let the actual reasons for the war slip into the shadows and escape close analysis. For him, politics was present in an ill-defined way and contained the whiff of fascism in its reliance on 'spiritual' and 'innate' concepts and qualities. Characterizing the post-war world as a place of mean spirits and petty minds, Seton Hutchison offered the veteran an escape through a battlefield pilgrimage. It was the antidote to disillusionment. Once on the sacred ground, the game of ghosts would commence, the old self would be reanimated, and in the memory of suffering the spirit would soar. Seton Hutchison's position validated and vindicated Rupert Brooke's 1914 sonnets. Honour had come back, nobleness did indeed walk again, and the inheritance of the past was delivered:

> There is a comradeship between men who served before Ypres, men drawn from all ranks of life; and as the war years recede the strength of the bond increases. It is intangible yet dynamic, undetermined yet vivid, illogical yet born of nature herself transcending all commonplace emotion and family affection, indeed the love of David and Jonathan. If the clamour of the market place, the bickerings of political factions and the disillusion of Peace have weakened faith in that comradeship, then.... [original punctuation] go back to Ypres. Saturate yourself in its atmosphere, sample again that soil, soaked with the blood of comradeship, and whose shrines are steeped with a spiritual love which passeth all human understanding.[28]

Saturating himself in the game of ghosts by treading the old ways is precisely what E.F. Williams did, and the effect on him was very much in line with Seton Hutchison's predictions. Despite the changes on the Poperinghe–Ypres road, he found it easy to slip into his old self (as Seton Hutchison promised): 'The intervening years fade right away,' he wrote, and the old 'anticipatory sensation as you "wonder what it's like up there this time"' became overwhelming.[29]

INTERPRETING THE GROUND

In making this pilgrimage journey, veterans, naturally enough, felt the need to tread the paths of their particular memories. But, even when alone, this was never a purely individual odyssey, for their recollections were embedded in the collective experience of the units to which they belonged. Recounting his battlefield trip for the journal of his regimental association, the Territorial Force unit the St Pancras Rifles, Major Charles Fair provided a record of his 1920 visit full of details aimed at veterans of 1st Battalion. Fair acted as a witness on their behalf, and as a result, went with a very well-defined idea of what he wanted to see. As his account shows, far and away the most important places were those associated with the battalion's actions in 1915 and 1916, because this was 'the period above all others when our armies were composed of the men who had joined up in the first rush of enthusiasm at the beginning of the war'.[30] In other words, he was seeking to engage with a memory of the battalion when it was still a volunteer unit. Fair wanted to see the real war sites of his real battalion. He wanted the world in which the battalion was a genuine band of brothers before the dilution of that purity through the arrival of conscripts.

Captain H. Waterlow felt similarly after his trip in 1921, and made the focus of his visit the areas around Béthune, where 1st Battalion served in 1915.[31] He provided descriptions of villages in which they had been billeted and the places they had trained as well as the battle sites themselves. On the Somme battlefields he indulged in a micro-history, finding the unit's machine gun positions and their shelters in High Wood. When he got to Arras, he 'managed to identify our old front line' before advancing to the lip of a mine crater giving a great view across the landscape. It took him into a deep reverie: 'It was a splendid moment to be there and I could have stayed for hours.' At 'The Bluff' on the Ypres battlefields a distinct game of ghosts settled on him. Standing near the tunnel entrances he could 'almost hear the voice of R.S.M. Trezona sending off ration parties'. At Loos he sought out 'Bart's Alley', a site he did not actually experience, being away from the battalion at that point, but his role of interpreter on behalf of his community led him there. Then, applying the typical understatement of the war, he said no one who ever served there was 'likely to forget that health resort'.[32]

The ironic tones of trench journals inherent in the term 'health resort' were echoed by C.D. Planck, formerly of 7 Battalion, London Regiment. He recalled common memories of serving at one of the many 'health resorts' of the Ypres salient in a piece about one of his battlefield tours. Planck took a gazetteer approach to his account providing brief, occasionally facetious, entries on a host of locations well known to soldiers. A wonderful run of entries followed, with that for Dominion Camp typifying the style: 'Always was muddy, and still is.' Chateau Segard was being rebuilt with reinforced concrete and extensive cellars: 'What about the next war?' Dickebusch had been 'practically rebuilt. The lake looks lovely, very nice estaminet at Dickebusch end.' The tunnels and craters at the Bluff leading down to Spoilbank were declared 'the best remnant of the Salient' with signs of the old tunnel entrances still visible, plenty of equipment scattered about, 'and shells galore'. By contrast, the only things marking Hellfire Corner, Zouave Wood, Birr Crossroads, and Clapham Junction were the signboards 'and that's all that can be said for them'.[33]

The 'health resorts' referred to by Waterlow and Planck were then analysed as military problems. Walking around High Wood without fear of attracting enemy fire allowed Major Charles Fair to appreciate its full tactical significance, with its sightlines dominating the surrounding countryside.[34] When the Manchester Regiment old comrades associations made their battlefield tour in 1936, they were accompanied by several serving officers and a 'Divisional General' (never named), who provided 'a splendid word picture of events leading up to the situations we were to study during the next three days'. The account provided by the trip's chronicler describes a serious study of military history when on the battlefields. The party was divided into three 'syndicates', each with its own leader; 'The narratives were exceptionally clear, and it was seldom necessary to ask questions.' Everyone followed the 'studies of the actions' with great attention, and 'in many cases, recaptured a tenseness promoted by the vivid word painting of the narratives'. The day ended in Soissons, where they focused on the crossing of the Marne, and were left with it 'firmly imprinted in our minds'. A conference was held after their communal dinner, 'at which the Division General gave us his views on the actions which had formed the subject of a most interesting day's work'.[35]

'H.B.D.W.', an officer-veteran of the Sherwood Foresters, returned to the Aisne battlefield in 1926. Throughout his tour, he flitted between deep, personal memories and a clear-eyed determination to understand how the battle had unfolded. Fulfilling the role of the military historian, he found it 'intensely interesting to walk over the ground traversed during the attack', tramping across fields in order to verify the accuracy of the map produced in the regimental history. But a personal, engaged vision overlapped with this stance of the detached student of the battle when he identified the precise bank 'under which I gave my platoon a breather before making the final charge'. Such a find acted as the link with his fellow veterans: 'Surviving members of No. 1 Platoon will, no doubt, still remember that bank with gratitude, for the pace had been hot,' he noted. His choice of language also revealed his emotion as he described his objective of that day as 'the miserable trench we succeeded in recapturing'. But then he reverted to his historian identity when, after deciding to identify the German positions, he advanced up the slope and was able to assess the value of the defensive position, being 'astonished at the way the German machine guns ... had dominated every inch of our approach ... No wonder our casualties were heavy.' Through an extensive tour of a battlefield, he achieved what 'every soldier who fought in the War has wished ... that he could see the other side of the hill'. This satisfied both intellectual curiosity and an emotional need: 'After twelve years, I could gratify that wish to my heart's content.' But the gratification only brought another kind of longing, bordering on regret, for having advanced a bit further up the hill he realized what a wonderful vantage point they could have secured 'and how different the story of the fighting on the Chemin-des-Dames might have been'.[36] His role as historian, with its detached, critical eye, alternated continually with that of the proud veteran officer as he walked the battlefield interpreting its signs and symbols.

This veteran penchant for interpreting the ground was also something noticed by T.F. Lister, the first chairman of the British Legion, during the 1928 pilgrimage:

To me perhaps the most interesting feature was to see a few people in the centre of a cornfield – which looked as much like the opposite to a battlefield as one could imagine – searching to locate a spot where they had lived below ground or had participated in some sanguinary encounter. These little parties, detached from the main groups were busily engaged refighting old battles, and doubtless, with a more comfortable view of the surrounding scenery, discussing some of the strategy of the war.[37]

'One stood in the old, familiar places, the same and yet not the same, "fought the old battles o'er again",' wrote the former military chaplain the Reverend T.B. Stewart, adding, 'Every turn of the road, every field, every ruin had its own associations. It was an unforgettable experience.'[38]

Moreover, there was the strangeness of strolling at leisure across a landscape where death or injury once greeted anyone foolish enough to show themselves. Wandering around Frémicourt, Arthur Behrend was able to place where his artillery observations posts were once sited, and, being able to stand up in the open, could appreciate the landscape, realizing 'the wide panorama had not changed'.[39] Ferdinand Tuohy deemed it 'amazing to stand on Thiepval' and 'staggering to drive to Passchendaele behind a white horse';[40] simply being able to stroll around and clamber up to great vantage points without danger of shot or shellfire was a novelty of which he could only dream during the war.

Despite the great changes in the landscape with shell holes, dugouts, and trenches filled, veterans could and did gain a deeper understanding of the battlefield and the opportunity fully to appreciate the nature of their wartime missions and objectives, especially if they followed the advice of every true aficionado of the battlefields by doing it on foot. Walking across the Fromelles battlefield, Wilfrid Ewart found the summit of the ridge 'to be more pronounced than one might have expected' and it reminded him of sitting in the trenches opposite and wondering 'whether it would ever fall to oneself to stand beside the red church-tower of that Aubers which peeped evasively above the trees'.[41]

Time and again, it was finding the German positions so much more commanding and better placed than their own that hit veterans now presented with the chance to linger and contemplate. Exploring

'Jerry's Line', the veteran who described himself as 'One of the Party' understood fully how 'one could always look down on Ypres – as he did – and the result of this direct observation is to be found in the many cemeteries, the Menin Gate, and other memorials in the Salient'.[42] For 'A.C.K.', Tyne Cot was powerful not just for the number of graves and its beauty, but also as a way of interpreting the battlefield: 'One has only to stand at this cemetery to appreciate the dominating position held by the German Army for three years.'[43] Passchendaele Ridge impressed veterans with its panoramic views of the battlefields, and fully explained its tactical significance. 'The view from the ridge gives one an excellent idea of the great advantage enjoyed by those who held it, from the point of view of observation,' wrote W.J. Baumgartner, a member of the Ypres League Easter 1928 pilgrimage.[44]

Another veteran, 'T.P.', felt much the same, writing that from the Passchendaele Ridge he gained a fine view across the flat countryside all the way to Ypres. For J.M. Finn, Vimy Ridge produced a similar effect. Looking out across the countryside from this commanding location, 'we got some idea of the remarkable view that our opponents had of our position'.[45] Veteran Lance Grocock covered the unveiling of the Thiepval memorial for the *Leeds Mercury*, joining up with fellow veterans of the 'Leeds Pals'. Unsurprisingly, they headed down to Serre (where Leeds units had attacked on 1 July 1916) to see familiar sights, and it was while exploring the enemy side of the battlefield that Crocock experienced the standard revelation of the veteran: 'Seen from the Serre slopes, our old positions lie before us like an open book. It would appear that the enemy must have been aware of all our movements. He held the crest: we were at his feet.'[46] Indeed, 'the impulse to traverse the heights where the enemy lurked, and to look at the ground which for so long appeared inaccessible to us' was the greatest motivation for visiting the former battlefields, as a *Yorkshire Post* correspondent admitted.[47] Having spent so long trapped in trenches where the only view was the sky above interspersed by the odd, snatched glimpse of no man's land at ground level, it is no wonder veterans positively luxuriated in this freedom to wander and see the same views from a novel perspective.

VETERANS AND THE DEAD

Tramping the ground, following the flow of battle, inevitably led veterans to consider the dead. In doing so they often managed to avoid any kind of reflection on the morality of the cause and the ideals they had fought for. The war was a fact, nothing else. The important thing was what happened to them in particular places as a result of the war, which resulted in a new understanding of military glory. Amazed and awed by the Menin Gate, F.J. Lineton and J. West were struck by the 'memorials to British courage' and 'the absence of any swashbuckling glorification of war' to be found on them. Staring at the lion on the Gate, they believed he would say, '"These are my beloved boys. I guard their blessed memory and this shrine to their courage."' This was a typical veteran conception in which military glory was not rejected but redefined as endurance, stoicism, dedication, and devotion. True courage, true glory were identified in these qualities, and they were worthy of commemoration, remembrance, and respect. Reading the Gate's inscriptions, the men felt 'a glow of pride that we were privileged to help in that defence'. Such sensations infused veteran visits to the innumerable cemeteries and memorials, which acted as the link between landscape, personal memory, collective identity, and dead comrades. The chance to place a wreath, say a few words, and maybe even share a quick joke or two with the silent host before them allowed veterans to believe they had played fair with those the fortunes of war had taken away. Lineton and West made a special trip to Vlamertinghe British Cemetery to see the plot containing comrades' graves. They held a private memorial service for them, during which they claimed to feel the spirits of their former comrades surrounding them. The need to continue this pilgrimage then led the two to Vimy, stopping off en route at Merville Communal Cemetery to visit the grave of the battalion's first fatality.

The 85 Field Ambulance always included Westoutre Churchyard Extension on their tours in order to visit the grave of a comrade, while 'J.J.S.' reported visiting the grave of a popular sergeant still marked by the original cross made by the divisional pioneers.[48]

Stumbling around High Wood, Major Charles Fair came across the graves of many men of his battalion. This cut through any sense of the detached survey of the landscape as a military problem, and 'for me it was I think the most touching moment of the week'.[49] In 1928, Gerald Dennis and his chums made for Flers, where, after gaining permission from the farmer, they walked across the fields before coming 'to the Colonel's grave, all alone in its glory and just where he had fallen on September 15th, 1916'. At that point they 'solemnly placed [their] wreath against its cross'.[50]

Figure 5.1   Veterans and pilgrims at LRB Cemetery, Ploegsteert, 1927.

After working his way across the Aisne battlefield tracing the footsteps of his battalion, 'H.D.W.G.' visited the Vendresse British military cemetery hoping to see the graves of comrades. On finding more than half of the burials were unknown British soldiers, he was left hoping they were, in fact, his old friends.⁵¹ For the Yorkshire contingents on the 1928 British Legion pilgrimage, the tiny village of Fricourt on the Somme was a place all wanted to see, for it 'will always be more than a name to so many Yorkshiremen' due to the large number of men of that county who fell during the fighting for it, as was revealed in the nearby cemetery: 'Almost every one of, roughly, 500 graves … are those either of men of the East or West Yorkshire Regiments. It can, therefore, in truth be called a cemetery for the county of the White Rose.'⁵²

With the battlefields fast being restored to their former appearance, the cemeteries became crucial orientation points. H. Channing-Renton noted of the Somme in 1926, 'The scars of old days are gradually healing up and a few years hence the valleys and hills of the Somme will betray little excepting for the War Cemeteries.'⁵³ A Western Front veteran working as a journalist for the *Yorkshire Evening Post* found that the cemetery at Mametz (presumably the site of Devonshire Trench) left a deep impression on his memory, for here

'I stood at the very corner of the road from which the attack was launched in the grey dawn of that July morning'.[54] Another veteran used the New Zealand memorial at Messines as an orientation symbol, which opened up the scene for him, and in this district he 'was happy again'.[55] Here the cemeteries and memorials became the enduring witness, anchor point, and key to unlocking the landscape and its history, which, thanks to their experiences, the veterans understood more than any other visitor.

During their visits to the cemeteries, veterans were occasionally surprised to find the former enemy wandering around awed by the quality and care of the IWGC's work. Lineton and West commented on the number of German visitors to Tyne Cot Cemetery, where the gardeners told them that German visitors often complained about the poor state of their cemeteries compared with the pristine condition of those built and maintained by the IWGC.[56]

This experience must have made a deep impression on Lineton, for the following year he contributed an article to the *Ypres Times* concerning his meeting with German veterans at Whitsun in 1931. There was nothing planned about the encounter, as he and a friend simply came across a party of German veterans just off the train at Ypres. Lineton felt compelled to walk up to the group leader and offer a friendly greeting, which was at once returned. Conversation was struck up and Lineton was invited to join the group for a commemorative ceremony at Broodseinde German Cemetery. He accepted the offer, finding two other British veterans wishing to join him. At the cemetery, the British veterans stood reverently during the ceremony and at its conclusion all made respectful farewells. Lineton then informed the Germans that they were about to set off for Tyne Cot Cemetery. Soon after arriving, he was amazed to see that the German group had decided to follow their lead. The two groups of veterans explored the cemetery together, and once again Lineton was able to witness the appreciation of German veterans for the beauty of the IWGC cemetery and dedication of its gardeners.

As all stood together at the stone of remembrance, the leader of the German tour called his party to order, and the entire group stood solemnly for two minutes' silence, during which one 'burly Teuton wept'. Lineton imagined the ghosts of British and German soldiers walking hand in hand around the cemetery at this moment and believed the sacred ground of Tyne Cot Cemetery was the perfect site for reconciliation between the two nations. He concluded his *Ypres Times* piece by asking whether the Ypres League's members should reach out and make their Easter and Whitsun pilgrimages a collaborative activity with German veterans' groups. He suggested joint services at the Menin Gate and the Broodseinde German Cemetery as 'a wonderful opportunity for the exchange of experiences and impressions from both sides of No Man's Land, but best of all it would help the cause of peace and for that a million Britishers gave their lives'. In his heartfelt and powerful piece, Lineton at no point blamed the Germans for the war, but instead referred to the 'mistakes and misunderstanding which a few years back had sent two mighty empires flying at each other's throats'.[57]

ENGAGING WITH OTHERS

The curious experience of meeting fellow ex-servicemen whose only difference was their nationality was noted by many other veterans. The men of the 1/7 Northumberland Fusiliers association overlapped with German visitors at Thiepval in the summer of 1928. Both groups chatted amicably, and the Germans were impressed by the playing of the bagpipes, offering a present to the piper.[58] However, such occasions were not always such an easy meeting of minds, intrinsic sympathies, and shared outlooks. When veteran A.W. Keith visited the battlefields in the summer of 1938, he bumped into two men in Langemarck German Cemetery. It transpired that the older man was also a veteran and was showing his son where he had fought. The two veterans then chatted amiably, which was unusual given the fact that the language barrier often made conversation more about physical gestures or facial expressions, but when Keith asked if they understood the meaning of the watercourse stretching round the cemetery, the mood suddenly changed: 'In a voice full of unhidden hatred, he spat out the information that it was symbolic of the Belgians opening the sluices of their canals and preventing the victorious soldiers of the Vaterland reaching the Channel ports!' Somewhat shocked, Keith gently suggested the Belgians were only defending their country from an invader. Ignoring this point, the older man said next time Germany would not fall for a similar trick. Wishing to divert the conversation, Keith took the young man by the arm and, gesturing towards the graves, said the dead should surely teach them the futility of war. It did not work, for the young man was as radical as his father, stating that his own death meant nothing if it ensured the greatness of Germany. Keith decided the best thing was to withdraw, but his fellow veteran insisted on giving him a lift to Houthulst.

On the way they stopped off at a deeply controversial site, the French gas attack memorial at Steenstraat, the inscriptions and design of which firmly accused the Germans of barbaric behaviour. Being in the presence of the Germans must have made the stop a painful one, but Keith tried to work his way through the different viewpoints. This 'gruesome reminder of the horrors of gas war' left a deep impression: 'To us it was German frightfulness; to the Germans, retaliation for the opening of those sluices!' Having just about navigated this tricky situation, Keith walked straight into another delicate issue by mentioning Hitler, which caused the older man to pull at the wheel so tightly they almost span into a ditch. At this point, Keith successfully diffused the situation by joking how his former adversary had not killed him during the war but nearly finished him off in a car accident twenty years later. This caused all to laugh, and when they dropped him off, the father and son insisted on giving him some food to fortify him for his long walk back to Ypres. 'We shake hands, and they wave to me, as I set out ... What would the Fuhrer say to his minions hobnobbing with a Jew from Aberdeen!' Keith added sardonically. The whole experience left him convinced that the British had no understanding of the continental mindset.[59]

The exchange had come about because cemeteries and memorials acted as fixed points in the transformed landscape of the 1930s. They were places where veterans could go and recognize something familiar in the names and regimental badges. In their timelessness, they were the antithesis of the rest of the countryside, in which the speed of transformation could often leave veterans bewildered, disoriented, and even depressed and angry. The irony was that so many veterans did not want to see restoration and reconstruction as healthy. They did not want rebuilding of the pre-war landscape of peace, or even a better version of it in a bright, new world. They wanted the same old trenches in the old front lines, and the same old estaminets in the back areas. War was real in this landscape, and peace the strange interloper for the veteran who could not conceive of the world with a pre-1914 life.[60] In order to rediscover their memories, the veterans actively sought out the places of trauma. As one of the symptoms of psychological trauma is the inability to escape the incidents and things that caused it, then maybe most veterans were suffering some form of psychological disorder when they returned to the battlefields. However, there was a deep irony present, for only when the landscape was in the precisely remembered condition of misery and distress could the veteran feel, oddly enough, happy. What was released by the sight of it on revisiting then appears to have been not so much a coming to terms with traumatic experience as the determination to release a spirit of nostalgia through reacquaintance with their old world. Therefore, if it was an expression of psychological disturbance, it was a very, very odd manifestation of it. The war truly was the best of times and the worst of times for most ex-servicemen.

OWNING OR SHARING THE LANDSCAPE?

One of the major challenges facing veterans was the sense of possession. During the war, when so much of the usual civilian life of the fighting zones was either completely non-existent or massively misplaced and reoriented, it gave the impression that the armies owned the joint. The Western Front was one huge, private world in a strange liminal zone peopled by civilians whose function was reduced to bit-part players supporting the star acts. On returning, the veteran found the original owners and their ways had taken back control, and that seemed odd. 'J.J.S.' asked his fellow veterans whether any of them could ever forget various places around the Ypres salient. In their imagination, he stated, the locations were locked in their wartime condition, and he acknowledged it was difficult for veterans 'to imagine or conceive in any way their present condition'. With these ideas in mind, he decided to return, and his description made the extent of the transformation clear, adding that 'it is hard to realise it was once so desolate'.[61] Desolation despoiled is what they found, and it felt wrong to so many of them.

The strangeness of 'the savage splashes of architectural jazz' in the rebuilt Albert caused Tuohy to wince. For him, the town of cafes and bars was a strange contrast to the 'real' Albert he knew, the one in his memory, the world of ruins.[62] 'J.B.M.' was another to find Albert an awful disappointment. Alighting at the station, he was appalled that people could be drinking and playing cards in the station cafe. Was such activity a near equivalent of dicing at the foot of the Cross for him? The walk out of town and towards the old front lines failed to provide any comfort, as he saw nothing to remind him of what he had once known so well.⁶³ And it always came back to the strange dichotomy – the difference between what it had been and what it was like now.

The Reverend Stewart thought it most strange to be sleeping in a comfortable hotel close by the Menin Gate, having so often marched past the spot and seen so many trudge 'along the red road … on their way to Calvary'.⁶⁴ Another former chaplain felt equally awful at this spiritual and physical dislocation. At Easter 1930, the Reverend Neville Talbot joined a party of former army chaplains on a battlefields pilgrimage led by 'Tubby' Clayton. Talbot was Clayton's closest wartime colleague, and it was after Talbot's brother, killed near Sanctuary Wood, that the Poperinghe Talbot House (or 'Toc H') rest centre was named (see chapter 11). Indeed, it was Talbot who had crawled into no man's land to recover his brother's body and ensure its burial.

Returning to the birthplace of the global Toc H organization he had helped found proved to be a challenge for Talbot. As the notes he made during the trip reveal, Talbot's experience seems to have exerted an almost entirely malign effect on him. It brought no sense of cosy nostalgia, and no sign of cathartic and therapeutic release. Instead, retreading old ground unleashed trauma, and, irony of ironies, it was traumatic because it was no longer traumatic. Peace and normality had replaced war, and in that scenario Talbot was lost. At Talbot House itself he was perplexed and distraught at finding it 'painted and repapered (hideously) and mended by the owner, and not looking the same as it did'. Already rattled and disoriented by finding everything so different, he was particularly upset by the different layout in the garden, desperately trying to reimagine it as it was when Archbishop Lang addressed a huge crush of men just before the opening of the Third Battle of Ypres. Only those who had been there could truly know it, and so the entire tour began to seem 'futile for such as couldn't see it by memory'. But he finally made contact with his old self, and the host of faces he remembered, when ascending the stairs to the old chapel: 'At last we came to the post of the steep staircase (all but ladder) up to the Attic. And when we reached there we got beyond the obliteration of the past by the Belgian owner … There we fell on our knees and kept silence … All this was very tremendous'.

It was but a brief respite for Talbot in a spiritual journey consisting largely of sickening mental jolts occasionally redeemed by places that he recognized, felt comfortable in, or found admirable. The Canadian memorial at St Julien and the wonder of the towering Cross of Sacrifice at Tyne Cot provided redemptive joy for him, but these moments were fleeting and very difficult to record, for they inspired feelings 'too deep and rich for expression – how much is "incomprehensible"'. Unable to define what gave him spiritual enrichment only increased his sense of disorientation and nauseous dizziness. 'Here [looking out from Kemmel hill] the bewilderment begotten of rebuilding was at its height – for the whole countryside is as it was and more so.' Even the new saplings, which he could have taken as symbols of hope and regeneration, instead led to dejection, leaving him compelled to state 'how once and for all must be recorded the practically complete obliteration of all traces of the war throughout the whole Ypres area. It is bewildering.' There was no respite, even at Hooge, close to where his brother fell. What should have been so familiar was so, so different: 'I badly failed in really recognizing the lie of the land as all is so changed, and ... the familiar shape or ground plan of Zouave and Sanctuary Wood is quite unrecognizable. I should like to spend a day there with a good trench map.'[65] Neville was reduced to a desperate man unable to reinterpret and understand the landscape. He was cast adrift by the obliteration of obliteration, the destruction of destruction. No more no man's land. And in those depths of a man made blind in a world he once knew intimately, his account suddenly ends in mid-sentence, the rest of the page blank, the journey into memory halted and never to be recommenced.

Veterans travelling alone, or as the sole ex-serviceman in a party, could feel acutely this sense of being a roaming revenant in a strangely transformed world. A former officer and member of a Church Army tour in April 1922 was disturbed by the new reality of Ypres. Being able to walk to Hill 60 in daylight for the first time confused him. He found it difficult to pick out the route, being so used to doing it by night with great care to avoid enemy observation. Ypres haunted him, and he felt utterly disoriented without the men he had commanded by his side.[66] Arthur Behrend, a former artillery officer, 'felt tired and very lonely' after a day exploring the positions he occupied in the spring of 1918.[67] The problem for him was that the landscape *had* released a flood of memories and conjured up the ghosts of old comrades, but none were now present. His dislocation cannot be explained by simply attributing it to the restoration of the countryside to its pre-war condition. Behrend was subjected to a barrage (the term seems apt for an old gunner) of memories inspired by the ground over which he tramped; it did provide a pathway to his past. His greatest problem seemed to be the lack of companions. He wanted to share the experience and interweave his recollections with those of others.

New farms, rich, fertile fields, neat houses in rebuilt towns and cities, imposing public buildings, all the trappings of normal, everyday life, in fact, destroyed memory. Even worse, the normal activities and pastimes of the local people could be interpreted as distasteful or disrespectful. W.A. Allinson visited Ypres for the first time since the war in the summer of 1926, when the local fair was in full swing. 'Now, instead of the crashing shells,' he wrote, 'we had the din of the hurdy-gurdy and the mechanical organs,' almost as if the wartime noises of destruction were preferable to the actual, cultural life of the city that had been resurrected. The 'real' Ypres was not the ancient city, its inhabitants, their customs, and daily lives, but the one that was brought into existence by the war.

But veteran disorientation and discontent at change were just as likely to evaporate and transform into celebration of the new, unfamiliar world. Despite his many complaints, Allinson himself found the contrast between the wartime scene at Essex Farm and the present one a joy. Standing in the cemetery and gazing at the memorial to his old division, he heard the birds singing, saw local people strolling to church, and felt at ease in the 'harmony and quiet'. Marvelling 'at the wondrous changes' wrought on the countryside since the war, he concluded, 'truly the transformation from war to peace is wonderful'. Once back in Ypres, his doubts returned, however. The hotels, the cafes, the hustle and bustle, and the noise of the fair made him 'wonder if the price we paid to defend Ypres was worth it after all'. Allinson's weathercock emotions were very much dictated by precise location and space. In the peace of the restored countryside, he was able to accept reconstruction and feel at one with his surroundings, especially if the beauty of an IWGC cemetery formed part of the picture, but in Ypres he found it difficult.[68]

Allinson's reactions reveal the complexity of the relationship between veterans and the battlefields. Disappointment and dislocation alternated with acceptance of change, especially when reconstruction was perceived as the fruits of their suffering, a sign of victory. Clambering up Vimy Ridge, Ferdinand Tuohy found the sight of the restored Lens an inspiration. This is what they had fought for.[69] H. Channing-Renton originally saw Bapaume through field glasses while standing in a British trench; he then marched through it after the Germans, having first destroyed everything not already damaged, had withdrawn. To see it restored, proud, and prosperous was something he found admirable.[70] This acceptance also stemmed from the perception that veterans could feel a deeper affinity with the landscape, seeing beyond the surface, and somehow communing with the 'real' world they knew. E.F. Williams thought the new Ypres a 'very attractive town', but confessed that he went back 'only for the memories and scenes associated with the Salient in the War days'. And he found this easy to do, for, despite the extensive rebuilding of the city, it was very quiet in the backstreets, and once among them he could conjure up 'the steady, measured tramp of marching feet – shades of terrible, albeit, wonderful times, bringing back memories both grave and gay'.[71]

For another ex-officer the obsession of non-veteran 'tourists' for places like Vimy Ridge, where it was easy to find traces of the war, was odd. Instead, he placed his fellow veterans in a completely different relationship with the landscape, one rooted in common experience. The ex-serviceman did not require preserved trenches and tunnels to engage him, 'for it is memory rather than sight that he relies upon. It will be a matter of complete indifference to him should level pasture or rolling ploughland confront him where he last saw the entrance to a brigade headquarters dug-outs or tried to wade along a track almost knee-deep in mud, during that spring of '17 – before the affair at Arras.'[72]

For those veterans less able to carry out the mental gymnastics involved in the reconstruction of destruction, the landscape remained liberally scattered with obvious evidence of the war. In fact, almost every veteran testimony bemoaning the bewildering condition of restoration also contained reference after reference to remnants of their world. Just as flares fired by twitchy sentries on dark nights in the trenches illuminated the battlefield for a few moments, veterans stumbled on the ubiquitous evidence of war frozen within the hard lines of a perfectly shot and developed photograph, allowing them good, long glimpses of their past selves.

'A.W.W.' found a world transformed when he took part in the British Legion pilgrimage of 1928. However, as well as experiencing something akin to 'a sense of disappointment at the disappearance' of battlefield features, he also began to realize that 'if one probes deeper, sometimes an old pillbox, where one lived for weeks, may be discovered among the blackberry brambles of Glencorse Wood, or the garden of a new red-roofed farmhouse'.[73]

On his first night in Ypres, immediately following the Last Post ceremony at the Menin Gate, 'A.C.K.' and his chums set off for a walk along the Menin Road. While strolling along they 'tried to imagine the scene as we used to know it and then to bed, tired but satisfied after our first day'. Later in the trip he went up to St Julien and found the pillbox he had sheltered in during the fighting in September 1917. It brought back a flood of memories and details, which he recounted in his write-up for the *Ypres Times*. Far from their being something he wished to suppress or avoid, 'A.O.K.' seemed to find comfort in his memories linked to a particular battlefield feature and was glad 'to hear that the Belgian Government have agreed to the preservation of some 180 of the most interesting concrete shelters, block-houses and dugouts'.[74]

Although Tuohy was highly unimpressed by what he found in the rebuilt towns of Albert and Ypres, once he got out into the old battlefield his temper improved as recollections flooded back: 'Memories, memories! Was ever square mileage so teeming with them?' Here he found remnants of the world he once knew, including the people. At Essex Farm, he stopped 'in a real estaminet' for a beer served by a proper '"Promenade Mademoiselle"!' Hooge proved equally arresting for him, as did the memorial stone at Gheluvelt commemorating the stand of the Worcesters in 1914.[75] Like Tuohy, Frank Hermen found Ypres almost unrecognizable when he returned to the city in the summer of 1928, but on walking out into the countryside 'one is able to find direction more easily'. The Poperinghe road was 'the same as ever, the Hop Store (ex-dressing station) once more being put to its peace-time use'. On heading back in the direction of Ypres from St Jean, it was an olfactory element which took him back in time: 'The old familiar smell of dust and the peculiarly Ypres smell still hangs over these fields and roads. Nothing is so powerful as the sense of smell for plunging one back ten years.'[76]

Many others had similar experiences. 'C.W.P.' reported the huge amount of decaying military equipment and infrastructure still strewn across the countryside in the summer of 1928. 'Passing along anywhere in the war zones, apart from the pill boxes, the traveller sees an amazing collection of the debris of war. Piles of rusting

Figure 5.2 Veterans at Menin Gate, c. 1929.

corrugated iron, some pierced with shrapnel, either lie in heaps or have been used for buildings, barbed wire everywhere, and the British wiring corkscrews for entanglements form posts for boundary lines,' he wrote.[77] Despite the restoration of the farmland around Fleurbaix, for Wilfrid Ewart it was a district 'full of recollections for the [First] Army'. He found an old farmhouse near a crossroads, which he remembered well, and this immediately oriented him, provoking vivid memories of serving in the district in 1915.[78] H. Channing-Renton recognized a house, now entirely rebuilt, the cellar of which he had occupied in July 1916. Walking on he saw the old windmill near Colincamps and came across trees used as observation posts, recognizable by the ladders still nailed in place. In fact, the Somme was a veteran wonderland, with traces of trenches, wire, dud ammunition, and plenty of signs stencilled on walls that once pointed the way to baths, concert halls, and other parts of the BEF's infrastructure across the former Somme front.[79] Old signboards always inspired affection and nostalgia. E.F. Williams found great comfort in seeing the old army sign, 'All traffic this way', on a wall in Poperinghe.[80] These literal signposts to the past dragged veterans back to their old selves regardless of reconstruction's seemingly rapid progress.

No matter how much tidying up occurred, the former battlefields remained a mad collector's box of every and any kind of military equipment, as every veteran found even when making only the most cursory glance down and around. Visiting the Somme with a fellow 'Leeds Pals' veteran at Easter 1935, Lance Grocock saw a landscape little resembling the one he fought over. After much comment on the wiping out of all evidence of the war, he stumbled across traces of the old German line. The trenches had collapsed, but it was very easy to find their traces in the long grass.[81]

'E.W.' wanted to find the sites around Bligny where his battalion had fought in August 1918. The experienced Captain Oswald acted as his guide, and they set off along the long straight road from Amiens to Soissons. Reaching the old front line at Marfeux, they started to move across the battlefield over which he had fought, and then up to his objective of that day, Bligny. Although the countryside was restored, and the silence was broken by the cheery sounds of children laughing, down in the tall grass 'E.W.' found lots of relics and remnants of the fighting. After spending a period of silence in memory of his comrades, he walked back down to Oswald and the car, and then back to Amiens. 'A full day, but one which will linger long in my memory.'[82] Exploring the Arras battlefields in 1932, J.M. Finn happily renewed his association with 'Bois-des-Boeufs (Bully Beef Camp of old), Feuchy Chapel with quarry (site of a cushy dugout) ... Then by Les Fosses Farm. Who remembers the caves under this farm lit by electricity, water tanks below, and an advanced divisional canteen, during the 4th Division days?' he asked. At every turn he found a reminder and did not seem in any way disoriented by the rebuilding and restoration of the landscape.[83]

As the 2 North Midland Brigade Royal Field Artillery association made their motorbus tour of the salient in 1934, a string of comments poured forth as places were recognized: '"This is where we had an O.P. [observation point]"; "Somewhere about here we had a forward gun"; "That's the blinking village Jerry used to shell us from with his 6-inch"; "Here's the very cellar I kipped in"; "This is the road where we had six horses wounded in one afternoon"; were the remarks heard in turn.' The veterans then headed down to Meteren to find the spot where the officers' mess was established, and also where their first officer casualty occurred. At Kemmel, they searched around until they found the first bit of front they covered with their guns and where they established their ammunition column.[84]

Their fellow gunner-veterans of the 236 Siege Battery descended into their old selves with equal alacrity, and with it came the lingo and jargon of their military trade. At Lock 8 on the Ypres–Comines canal 'everyone piled out in great excitement' and they soon found the old battery dugouts. Next it was Brasserie Cross Roads and the estaminet still serving beer. Then it was on to Lock 7, where they remembered receiving orders to fire on 'a target at 11 degrees First charge (Look up your range tables you old 6 in. How. Gunners!)'. Kemmel was next, for the old 'O.P.', after first visiting Klein Vierstraat Cemetery to see the graves of old comrades, and then Swann and Edgar Corner, where they all reminisced on the day Bailleul fell and they 'had to switch 94 degrees right ... That was the day the guns were red-hot'. Busseboom brought back memories of being gas-shelled in April 1918, but ended with 'a glorious burst of firing, completing 28,000 rounds in the consecutive 30 days', which broke the back of the German spring offensive in the region. They were disappointed that the ground was so rough at Ridgewood that they could not find their battery position, but soon after they stopped off for another wander along the canal, which brought them to a well-remembered old dugout, a 'cushy billet' indeed. So deep in reverie did they fall that some decided to return to Lock 8 in order to do a little more exploring before making their own way back to Ypres.[85]

For Rex Sargeant, the whole experience of returning to the battlefields was one of constant slippage into the past where memory was so strong as almost to overwhelm him. On the Poperinghe–Ypres road, he found it easy to visualize the march of troops and the sound of their boots crunching down. Passing an estaminet he frequented regularly, he saw again the young woman who used to serve: 'a cross between Anna Neagle and Helen Twelvetrees' but now aged a little. In fact, as he discovered, she had married a British soldier and there were several 'lovely girls ... the result of an international agreement between England and Belgium'. The jovial nature of these comments was balanced by much darker thoughts and reminiscences. At the Sanctuary Wood trenches he remembered his old pal, Sandy, a 'cabaret on two legs', who 'died bloodily, and his end was terrible'. It caused him to declare Sanctuary Wood 'a gruesome place', which triggered 'the old haunting fear of being buried alive and the Wrath to Come'.[86] This was no veteran stumbling around a reconstructed world desperately seeking remnants of the past. Rather, the past had gripped him by the throat.

VETERAN HUMOUR

While they provided moments of deep introspection, respectful remembrance, and communion with the dead, veterans' tours were also full of fun and humour. Indeed, the humour expressed seemed to be the perfect accompaniment to the tour, for it was usually a resurrection of the wry and self-deprecating sort so often used during the war. Even Rex Sargeant, a veteran seemingly traumatized by his re-exposure to the battlefields, was able to switch moods when he and his chums decided to spend a Sunday night in Lille. They had plenty of fun flirting with local women, and relearned the lessons of the war, as French women apparently had the ability to outflank a mere Tommy with effortless ease: 'The female of the species in this French town is a fast worker and seems to know *all* the answers. Verb. Sap!'[87] The 85 Field Ambulance old comrades association ran regular battlefield visits organized through the Ypres League. Every account provided by the association secretary, Alfred Skinner, was written in a facetious tone. When describing the 1930 tour, the arch comments were unleashed from the start, as he noted the train journey was an added bonus, for 'the L.M.S. [London Midland and Scottish Railway], intent on giving value for money, took us on a circuitous route through the more cultured suburbs of North London'.[88] A year later the train service was still the source of amusement for Skinner, as he wondered whether St Pancras was originally designed as a meeting place for bishops, only for the architect to change his mind, add a few railway lines, and sell it off to the L.M.S.

Skinner's accounts once on the battlefields give the impression of never being more than a few moments away from a smile, giggle, or wisecrack. The comedy started at Dunkirk, where 'Mrs Douane' insisted on sampling some luggage and succeeded in exposing to 'the world the dreadful colour scheme of Capt. De Trafford's slumber-suiting (these bachelors!)'. Skinner himself, 'only possessing two or three hundred cigarettes and half-dozen packs of playing cards', managed to sneak through without problem. Eating lunch in a Lens cafe, the men felt the waitress would make a great 'Nippy' at the Strand Corner House. Although her somewhat mournful nature induced a spirit of reflection, this was deemed a great aid to the digestion. Having partaken in a good lunch, Skinner 'would have given anything (even my Ypres League badge – in handsome enamel, threepence extra) for a long sleep, but our party were too much like camels, who, as you know, can go through the desert for three weeks without sleep (or is it without drink? If it is the latter, then it is not our party I am thinking of), and my siesta had to be postponed'. A good afternoon was then spent touring the battlefield, following which everyone tucked into the teatime cakes with gusto. The spirit of a schoolboy outing free from the usual rules and regulations was expressed that evening when they wandered around Ypres, after a hearty sing-song in their hotel bar, looking for somewhere still open and willing to serve drinks. Nothing was found, but there was one last comic scene to play out, as back at their hotel they found that the key they had been given was for the garage and not the front door, and thus required rousing the hoteliers from their sleep.

The next day the fun continued in the form of an old Belgian man at the station who, offered the opportunity to fill his pipe, took an enormous pinch and then spent much time attempting to stuff the lot into the bowl. This incident was deemed 'quite like old times'.[89] Somewhere among these incidents a battlefield tour of some description took place, but finding its traces in Skinner's account is difficult.

'Yarrl', who appears to have been a former officer, also adopted a comic tone in his account of a battlefield visit. The studied flippancy and insouciance of the old sweat started with the title, 'Wipers, Easter, 1932'. In opting for the Tommy slang term for the city, 'Yarrl' set out his agenda, and continued it immediately with his internal debate over whether he should join the pilgrimage or not. Leading with the arguments in favour, he said he had not been back since 'demob' in 1919, and if he did not join his three chums, he would 'feel all kinds of a fool to miss it', and, finally, it would make a pleasant distraction from the ordinary Easter holidays of 'inevitable golf and other normal distractions'. The arguments against included the uncertainty of the weather at that time of year, the horror of seasickness both ways, dislike of travelling anywhere there was likely to be a crowd, and the difficulty of leaving his wife and children for four days over Easter (this last 'was easily overcome by my wife, who strongly recommended me to accept, and no doubt welcomed heartily my absence!').

Having made up his mind to go to 'Wipers' with his chums, everything then settled into a bit of a jolly adventure. The sea crossing was smooth; their guide, the vastly experienced Captain de Trafford, had a smart, comfortable car; the hotel was excellent and sported 'a lavish table'. Each of his friends was described with a touch of Bairnsfather's style: one was mistaken for a plain clothes police officer thanks to his demeanour and dress; another was an impressive figure whose dignity was enough to ensure Belgian police and porters treated him with utmost respect; the third looked like a theatrical impresario 'on a sly jaunt' but had a round figure thanks to a penchant for the knife and fork in the dining room.

The scene set, 'Yarrl' presents in his account a group of friends determined to recreate their wartime lives in wry humour, high jinks, and larks. 'Troops were billeted by 7.0', which then allowed time to peruse the wine list 'with a great relish'. A bottle was selected on the advice of their epicurean friend, but 'Yarrl' believed it 'to be obviously an early 1932 vintage' rather than the fine one promised. 'Lights Out' was midnight. The next day, 'reveille' was seven o'clock in the morning, and off they set for a circular drive around the salient, led by de Trafford. In the afternoon, they split into pairs and 'Yarrl' went off with 'Feathers' on a five-mile ramble from Ypres. At Oxford Road, 'Feathers' believed he had found the remnants of the trench he and his company had dug in 1917. This was treated as a comic vignette as they debated whether an indentation now eight-inches deep was the remains of a trench. But on agreeing that the company was full of very short men, and that an army marches on its stomach, general agreement was reached on the authenticity of the site. The other two made it no further than the 'Patisserie Belge', arguing their interest was purely academic, rather than gastronomic. On their way back home, they had a quick walk along the seafront at Ostend, 'where no doubt many a foreign feminine heart fluttered at the sight of such British masculinity', and the tea in the cafe was nowhere near Sergeant-Major calibre.

'Yarrl' finished his account with ten conclusions about the trip. Nine were serious, or relatively serious, reflections, while the last one was: 'Why had we waited so long before making the pilgrimage?'[90] From these conclusions, it is possible to realize that 'Yarrl' and his chums had moments of great poignancy, and clearly believed it had all been worthwhile, but they had no qualms about seeing the tour as a bit of fun, too.

The fun element of these tours was often expressed in drink. Knocking a few back (or, it is often implied, more than a few) seems to have been a significant part of any veteran tour. For many members of the 2 London Regiment association, this started on the Channel crossing, where the bar was found in short order. 'Judging from the number of popular airs they sang, I think they found the refreshments to their liking!' as the account of the tour stated. On the last night, some men were so heavily refreshed they found their Ypres hotel door firmly locked. Having no alternative, they bedded down in the bandstand in the main square. 'I am glad to say that it had no ill effects and showed they were ready to go on sentry once more!' the association secretary recorded of the miscreants' condition on the following morning.[91]

When their tour had an enforced stop at Neuve Eglise caused by a temporary breakdown of their charabanc bus, W.J. Baumgartner and his chums all took advantage for the 'purposes of refreshing the inner man'. They were back underway for only a short time before facing the next and 'inevitable delay' at the customs post. Dismounting once again, they quickly 'discovered and consumed strange, mysterious liquids at the adjacent hostelry'.[92]

The men on the Manchester Regiment tour went on to Paris, where the fun really began. Referring to their final night, the tour's chronicler noted with deliberate roguishness, 'over the evening we shall draw a veil, carefully removing it on Tuesday morning when the return journey commenced'.[93]

When the Leyland Motor Company organized its tour to the battlefields in 1929, the vast majority of those who participated were veterans, and, as the souvenir booklet produced to mark the event revealed, all expected to have a hoot. Facetious allusions to drink and flirting with women were made throughout, culminating in the account of the last day, when the party was given six hours in Ostend prior to the return sailing: 'Time to explore, each according to taste and inclination, the cafes, shops, restaurants. Time to put to the test the General Manager's dictum "If tha' sticks to beer tha's orl reet." ... Time to see that indeed "They are often very fair [the women of Belgium]." Time gentleman, please!'[94] Having a grand old time seemed to be the most important objective of the tour, and it seems as if they achieved that handsomely.

Perhaps reinforced by a snifter or two, veterans' spirits soared on wandering around their old patches, and high jinks ensued. On the train back to Dunkirk, one of Major Salvesen's party decided, in a fit of levity, to pull the communication cord and then 'oh, la – la!!'. Among their most precious memories of the trip was the entertainment provided 'by the battery humourist [who] made everyone laugh till they cried' (especially his rendition of the old army song 'Cookhouse Door'). He concluded his excellent turns at Victoria, when they entered a cafe for breakfast, by shouting loudly, 'Does anybody here speak English?'[95]

On the day the Menin Gate memorial was unveiled in July 1927, the crush in the Hotel Splendid and Britannique put such intense stress on the staff that an argument broke out between the chef and waiter. The veterans present promptly saw the funny side, labelling it a re-enactment of the battles of Ypres. According to one veteran, the battle was fought in 'dead earnest' and 'the combat so affected the one and only waitress that she fainted gracefully away'. Desperate for a meal, one of the veterans marched into the kitchen and began bringing out the dishes himself. In carrying out this 'heroic deed' he managed to come through 'without himself having received a single scratch'. Thus, the veterans reduced what might have been regarded as an upsetting and embarrassing incident on a sacred day to a bit of knock-about fun. On the same day, an Old Contemptible sitting in a cafe before the ceremony suddenly shot his head out of the window and began serenading the assembling crowd with a rendition of 'Pack up Your Troubles in Your Old Kit Bag'. 'This raised considerable amusement among the spectators; even the gendarmes could hardly keep a solemn face, and a number of people joined him in the song.'[96] Even on such a serious day as the unveiling of the main imperial memorial on the Ypres battlefields, a veteran managed to wedge in a note of light-heartedness that was also appreciated by his audience.

Some, very few, probed this levity, questioning whether it was, in fact, no more than a veneer. The *Church Times* correspondent accompanying a party of limbless veterans visiting Ypres looked deeper and saw a conscious act of avoiding certain recollections of their war service. These men may well have been grieving for the loss of a physical and spiritual part of themselves, giving them a status akin to pilgrims mourning a loved one. But, instead of expressing pain, they indulged in 'jokes about familiar spots, their miseries being tactfully forgotten'.[97] At the same time, sensitivities over appropriate behaviour on the battlefields meant veteran jollity was classified as the honourable exception. In July 1932, the Ypres League was able to report that its Easter and Whitsun pilgrimages were subject to much better weather than experienced the previous year ensuring 'pleasant times were spent in revisiting old battle haunts'.[98] Having fun and indulging in a bit of nostalgia were therefore seen as acceptable aspects of a pilgrimage, at least if they were conducted under the auspices of the League. By its nature, the League provided respectability underpinned by its spirit of reverent commemoration, and as the fun was inspired by the veterans themselves, it could be divorced from the idea of frivolous tourism.

Avoiding the taint of the tourist was also achieved by the choice of itinerary. F.J. Lineton and J. West advised their fellow veterans to eschew the usual tourist sites when undertaking a pilgrimage to the battlefields, which might even mean carefully controlling who was allowed to join the party. According to the two veterans, those who wanted to see the tourist sites would undermine genuine communion with the landscape and the memories it inspired: 'They will take you away from the particular spots which will thrill you to the marrow, when you put on memory's spectacles and see those old places in war-time guise.'[99] For Lineton and West, the battlefields were a private world and outsiders were not particularly welcome.

Devoting themselves to old haunts and sites of specific interest to their battalion, Lineton and West started their tour by walking up the Menin Road to Railway Wood searching for traces of their trenches. On the following day they went to Vlamertinghe British Cemetery to see comrades' graves. Poperinghe and its environs were then visited for old billets. An essential landmark on the trip was their divisional memorial at Langemarck, which was 'particularly interesting, as our old battalion had a large share in the capturing of this place in 1917'. On their last day, the magnetic appeal of a microgeography of personal remembrance exerted itself and Lineton and West returned to Hooge and Railway Wood. The specificity of their routes and axes of remembrance was made clear in their account of the trip written for the *Ypres Times*, the journal of the Ypres League: 'We had not visited Hill 60, and many other famous spots, because we confined our pilgrimage to places we knew in the war days. We were satisfied that our programme had been fulfilled; nothing remained but to return to England, our minds freshly stored with memories of wonderful, if terrible, days.'[100] Yet even this seemingly 'pure' pilgrimage also contained visits to the 'many famous spots' including the Menin Gate,

Tyne Cot and Essex Farm cemeteries and the Canadian memorial at Hill 62. Veterans were by no means immune to tourist delights.

*

Being back on the old Western Front, regardless of the changes to the landscape, regardless of what they saw, or where they went, veterans were inspired and the experience unlocked memories. W.J. Baumgartner wrote:

> In short, we were a happy company. For those who had known the Salient as a salient, there was the pleasure of coming together again on the scene of their old comradeship; of stirring up half-remembered things in the lumber-room of memory – old, fearsome things, that now had strangely lost their fearsomeness and acquired a new, humorous aspect. Ancient episodes concerning rum-rations and cooks and quartermaster-sergeants were hauled out from the background, and helped to link together the incidents, and to fix their topographical and chronological positions. These veterans were full of the cheery spirit that helped to hold the Salient; so that the uninitiated members of the party listened almost with envy to the old battles being fought again, and the old tales retold.[101]

Therefore, for Baumgartner the pilgrimage did not stir up old memories simply by gathering veterans together. Rather, being on the actual ground was crucially important as the catalyst to memory, and memories were anchored in specific locations and specific times. What he did not make explicit was whether they shared only the memories they felt comfortable expressing, meaning the foregrounding of the light-hearted and comic. But, even if it was merely a selection of carefully edited highlights, the retelling was done in such a way as to inspire envy in those who had not experienced the war. Truly, those who had not been with them must have held their manhoods cheap while any spoke who had survived the test of Flanders fields.

The chance to relive old days and remember old comrades motivated veterans to return to the battlefields, often in large numbers, and often on multiple occasions. Sites of tragedy and great endeavour drew them back, as did places recalled with fondness for their association with relaxation, entertainment, and diversion. Unlike the bereaved, who found it so hard to see the battlefields as anything other than sites of pain, the veteran saw something different. 'What wonderful times they were,' wrote T.J. Booth, and it was his determination to relive them that took him back to the old Western Front.[102]

Returning to remember those wonderful times in the actual locations could be confusing and disorienting. Sitting down in a Vimy restaurant, H.A. Taylor was amazed by the simple act of eating 'a meal as was only dreamed of by the men who fought their way down to the red ruin that was the old Vimy'.[103] But, regardless of their complaints at the dizzying pace of reconstruction, its Janus face made it a world safe for veterans to walk in: the France and Belgium they visited looked both forward and back, never quite escaping the scars inflicted on them by the war, as all veterans found once they stared carefully on the ageing face of battle. As Frank Hermen told his fellow veterans, 'Anyone able to spare a few days of his holiday to visit familiar haunts ... will not regret it.'[104]

6

Postcards from Pilgrims

Yours is a pilgrimage in memory of those who passed this way. You will tread reverently, for it is holy ground. It is the shrine of those who won the right for us all to have a country of our own.

The Pilgrim's Guide to the Ypres Salient (1920)

I can now say that I have seen the last resting place of my loved one, and that is a lot to a mother.

St Barnabas pilgrim (1923)

In their tour many pilgrims will pause and stand sometimes in meditation, reflecting upon the deeds of valour, so many of them untold, and of sacrifice that will never be known, of which every landscape has been the witness and that nearly every village knew.

Graham Seton Hutchison, *Pilgrimage* (1935)

To be considered a pilgrim was the single most important badge of honour a battlefield visitor could attain. A particular status was required to gain the title. A pilgrim was either a veteran or bereaved, or according to the influential remembrance organization the Ypres League, someone with a genuine, rather than sensationalist, interest in the nature of the fighting on the Western Front.[1] But in the wider public forum, the interpretation of pilgrim was often far narrower and referred solely to the bereaved. The popular definition was summed up by a special correspondent accompanying an Edinburgh war graves pilgrimage as a person with 'a craving to see the place where he [the lost loved one] had spent the last moment of his life, or the spot where his body lay'.[2] Henry Benson, a journalist dedicated to telling the stories of pilgrims, agreed, describing them as a 'faithful band ... who have made the long journey, frequently at inconvenience and cost they can ill afford, for the purpose of placing some floral tribute on a tiny patch of greensward which covers the remains of one that they held most dear in life'.[3] He admired such dedication and saw in it a wonderful outcome. Far from being trapped in a cycle of mourning, the bereaved were, in fact, keeping fresh the memory of a lost loved one. 'They are dead only when they are forgotten,' he wrote. For Benson, pilgrimage denied death's victory.[4]

The pilgrim had, and revealed, special qualities. Pilgrims took things slowly, meditatively, immersing themselves in the landscape and memory. It was the antithesis of the tourist approach, as so many proclaimed long and loudly. 'There are two ways to visit the Salient,' a Green Howards' veteran recalled: either as a tourist with a 'hoarse-voiced guide; or as a pilgrim visiting the resting place of a quarter of a million of the Empire's dead. We chose to go as pilgrims.'[5] Here was a sense of moral superiority, indeed, underlined by the decision to undertake exploration on foot, eschewing the comfort of car or bus. 'Tour is, perhaps, scarcely the term to apply to a visit to the areas devastated by over four long years of sorrow and strife,' noted a November 1921 article in the *Belfast Telegraph*, before adding, 'The French have a better word. They call it a pilgrimage.'[6] Pilgrimage, with all its medieval frisson, was believed by many to be the correct term. Ypres League member the Reverend H. Shannon Brisby saw direct parallels between the medieval pilgrimages to the Holy Lands and battlefield visiting. According to Shannon Brisby, both were journeys of the heart to the place where the heart truly resided.[7]

The full force of the religious–spiritual implications was maintained by a *Church Times* correspondent, who described a tour made by a limbless ex-service group as 'in very truth, a pilgrimage as Catholics understand the word; a visit to a sacred shrine to honour before God the memory of comrades who had "passed out of the sight of men by the path of Sacrifice and the gate of Death"'.[8] Pilgrimage maintained the trappings of Victorian medievalism boosted by the war. Allusions to chivalry abounded during the conflict and, as the number of casualties mounted, many relatives turned to ritual and ceremony to help them navigate fear and loss. Protestantism, particularly in England, made a drift towards High Church and Anglo-Catholicism, with processions and ritualistic prayers for the dead.[9] Individual loss was woven into a group activity; the private was made corporate, expressed in public. Battlefield pilgrimages maintained this habit, and maintained the wartime focus on marches and progression, but the battlefield version was very much focused on treading the painful steps of suffering. Pilgrims followed the via dolorosa of their lost loved ones, shared the sufferings while expressing their own pain of loss on the (literal) road to consolation. The battlefields had been elevated to the status of holy places. As such they demanded reverence from people stepping softly and slowly. Pilgrimage was the only way to deal with the battlefields.

As the definition hung on the attitude in the heart and mind of the individual, it was less important whether the person travelled alone or in company, great or small. For those wishing to travel in groups, often because the planning required to undertake the journey was beyond their means or confidence, pilgrimages were arranged with great regularity both by large-scale organizations with national and international reach and by much smaller regional and local bodies. The numbers taken varied greatly according to the size of the organization and the point in the post-war period. After reaching their height in the years immediately following the war, currency exchange issues and the economic downturn of the early thirties saw a decline in the number of pilgrimages, but there was a remarkable recovery in the mid-thirties, and indeed, many pilgrims were on the battlefields of the Western Front as tensions rose between Britain, France, and Nazi Germany in late August 1939.

THE SCALE OF PILGRIMAGE ACTIVITY

Although the grand, large-scale pilgrimages, notably the famous British Legion tour of 1928, gained the most attention, the real heart of pilgrimage activity lay in the huge number of small, penny-packet trips that trickled over to France and Belgium in a constantly flowing stream.

An indication of the expectation of pilgrimage size is provided by the Ypres League tour in the summer of 1924. Having planned to take five hundred, based on the previous year's experiences, the League saw only eighty participants in 1924. Reflecting on the experience, Beckles Willson, the League's Secretary, attributed the fall partly to the rival attraction of the Wembley Empire Exhibition, with its dramatic dioramas and models of the Ypres battlefields on display.[10] In fact, the League had met its high-water mark with the

1924 tour, but thereafter settled into a regular pattern of pilgrimages each consisting of between fifty and one hundred people, and the evidence suggests between twenty-five and sixty people travelled on the tours arranged by smaller, local organizations such as British Legion branches and regimental old comrades associations. The Metropolitan Railway old comrades association took forty-three members on its battlefield tour in 1928; thirty-six went along in 1929, and twenty-two in 1930. In 1930 the Passport Office took over responsibility for the issuing of war graves passes, and on examining the figures, estimated it would be issuing somewhere between 600 and 700 each year.[11] These figures imply that the combined allure of the tenth anniversary of the war's end and the profile of the British Legion pilgrimage made 1928 a bumper year.

However, even in 1932, at the height of the Depression, the Metropolitan Railway old comrades association took eighteen pilgrims, including one child.[12] In 1939, with war once again looming in Europe and in the wake of Hitler's annexation of Czechoslovakia, the Imperial War Graves Commission (IWGC) was able to report that 'there was no falling off in the number of British visitors who signed the book at the cemeteries and memorials, and at Easter and Whitsuntide ceremonies and pilgrimages took place'.[13]

Despite the lingering effect of the economic battering of the early thirties, the Ypres League had at least sixty people on its August Bank Holiday pilgrimage of 1934.[14] Indeed, many made their first visits in the early thirties, and were so inspired they decided to repeat the experience (when the 1/4 East Yorkshire Regiment association made its fifth successive tour in the summer of 1936, it set a new record for numbers[15]). The Ypres League's travel bureau declared 1935 an extremely busy year, which was 'happily out to break all previous records'; between the beginning of March and the end of July nine independent groups used its services, and three more groups were due to go out in the second half of the year, including over sixty people signed up for the League's August Bank Holiday pilgrimage.[16]

On 5 August 1937, the *Yorkshire Post* reported large numbers undertaking a battlefield pilgrimage timed to coincide with the anniversary of Britain's entry into the war; indeed, there were 'increasing numbers as time goes on [of] a flood of battlefield pilgrims'. An IWGC official was quoted as stating that around 100,000 people a year signed the cemetery visitors' books, and in June 1937 alone some 15,700 signatures had been recorded. July and August were reported to be the busiest months, and the IWGC confirmed an upward trend in visiting.[17]

The Great Western Railway boasted not only a veterans' organization, but also one reserved solely for Old Contemptibles and 1914–15 Star holders. In that same summer of 1937, twenty-eight of the latter made a battlefield visit and placed a wreath at the Menin Gate.[18] In the spring of 1938, Henry Benson claimed to have spoken to a London travel agent who had taken 26,000 bookings for trips to Belgium over the Easter period, and using IWGC visitor book signatures as a guide, he calculated somewhere in the region of 600,000 had visited the cemeteries and memorials in the previous year, basing the figure on the rationale that one person in five signed the books.[19] Starting in its annual report for 1926–27, the IWGC published the numbers of signatures in the register books kept at its cemeteries and memorials. The reporting was not always consistent, as on occasion it differentiated between British and other nationalities; sometimes it stated signatures in cemeteries and memorial register books combined and sometimes separately. Adding all such figures together to produce a crude total shows that between 1926 and the spring of 1939 there were some 1.16 million cemetery visits by tours (as opposed to individual visitors, who usually visited multiple cemeteries and memorials on a trip). On the other hand, as tour leaders often signed on behalf of a group, there is something to be said for Benson's method of calculation.[20] Regardless of their precise nature and their calculation methods, what these statistics, and the profusion of newspaper stories, show is the constant passage of British visitors to the battlefields. In addition, the high profile given to the accounts of these tours in local newspapers ensured the awareness of the battlefields remained high in the public imagination.

For some, the evidence of such statistics was suspect. Instead, there was near pleasure in claiming sharp declines in numbers undertaking respectful pilgrimage and battlefield visiting. Advocates of these claims were convinced the decline revealed extreme selfishness and indifference towards the sacrifices demanded by the war years. Ferdinand Tuohy was a committed proponent of this view. Walking the old line from Albert to Ypres in the summer of 1927, he declared it was now largely overlooked and unvisited, while William McMaster, custodian of the Ulster Tower, was quoted as stating, 'only a few dozen come every week now', whereas in the early days 400 people had visited each week.[21] Tuohy returned a year later with the British Legion pilgrimage to mark the tenth anniversary of the war's end. The Legion, a marvel of organization and logistics, successfully transported, billeted, and fed 11,000 pilgrims, and it was estimated a further 25,000 travelled independently for the main ceremony at the Menin Gate.[22] Regardless of these figures, and their implications, Tuohy took most pleasure in unpicking them: the north-western and Midland districts sent the most pilgrims, Wales and Ulster the least, with the Irish Free State being the big surprise in terms of turnout, which he put down in part to the enthusiastic leadership of General Hickie. Yet, after wedging his inquisitorial figurative fingers into the statistics to expose miscreants, even he had to admit the pilgrimage revealed enthusiasm for battlefield visiting was still great.[23]

Outside of the schemes run by the war graves visitation departments of the Church Army, the Salvation Army, the YMCA, and St Barnabas Hostels, the Ypres League was the most important organizer of pilgrimages. Like the Christian charities, it scheduled pilgrimages for Easter and Whitsun (of course, both festivals were focused on triumph over death after agony and sacrifice), as well as a summer trip, usually in late July or early August. Using its network of expert member-guides and contacts across the battlefields, it was able to offer packages at very reasonable rates, usually based on a three- or four-day trip built around well-considered, if slightly breathless, daily itineraries. Always sensitive to the needs of its members, in October 1933 the League launched a staggered payment system for its pilgrimages, giving those on more modest incomes much more flexibility. Within a few months it had proved its worth, with the League declaring that the scheme had been received enthusiastically by members.[24]

Typical of the League's offerings was the Whitsun 1925 pilgrimage, which consisted of 107 people. Captain Parminter of Wipers Auto Services in Ypres met the pilgrims on disembarkation at Ostend and distributed them across Ypres's hotels. He then divided the party into smaller groups, taking each on a tour accompanied by a British veteran acting as guide.[25]

In 1926, the League ran a special pilgrimage for the unveiling of the South African memorial at Delville Wood on the Somme. The trip, to be based in Amiens, was offered in both a three- and four-day form and at second- and third-class travel prices. The itinerary included one day at Delville Wood for the unveiling ceremony, followed by a one-day tour of the Somme battlefields remarkable for the sheer number of sights and stops included before returning to Amiens: Albert, then Aveluy Wood, Pozières, Courcelette, Martinpuich, High Wood, Delville Wood, Longueval, Mametz Wood, Contalmaison, La Boisselle, Bray, Chuignolles, Proyart, and Villers-Bretonneux. Most of the stops were very close to each other, and many were on, or a short way from, the main road linking Amiens to Albert and Bapaume, and thus combined time efficiency with motorbus accessibility.[26] Such considerations show that a pilgrimage was not always dedicated solely to the solemn and respectful business of cemetery and memorial visiting but could include a variety of sites.

Deciding itineraries for pilgrimages was dependent upon the nature of the person or group going, as well as anyone they employed or delegated to undertake the arrangements. Guides often set the programme, and as a result trails were created around the major battlefields providing whistle-stop tours encompassing a range of stops. For example, Captain Parminter devised a standard circular tour from Ypres. On the outward leg he made for Armentières via Kemmel and Bailleul, returning via Messines and the craters.[27] Somme battlefield pilgrimages always seemed particularly intense, regardless of who ran them. Based in Amiens, the Ypres League 1928 Somme pilgrimage drove straight along the road for Albert and Bapaume visiting La Boisselle, the tank memorial at Pozières, Thiepval for the Ulster Tower, and Auchonvillers (presumably to look at the trenches in the Newfoundland Memorial Park).[28]

A year later the Toc H pilgrimage wedged an enormous amount into a very tight schedule: 'The two-days which followed were crowded with wonderful and swiftly changing experiences too full to be recounted in detail,' as the account in the organization's magazine stated. The Santerre district of the battlefield was included and deemed a particularly appropriate term with its roots in 'sainte terre (holy land) or "sang terre" (field of blood)'. They visited Mont St Quentin and Pozières to pay homage to the Australians, as well as the South African memorial at Delville Wood and the Beaumont-Hamel memorial at Newfoundland Park, omitting only the New Zealand and Canadian memorials from their pan-imperial tour of Picardy.[29] Brief mentions of the Ulster Tower, La Boisselle, the Butte de Warlencourt, and the Cambrai memorial were made in the account, adding to the sense of many passing glimpses and hurried returns to the bus. Recognizing the somewhat fleeting coverage offered, the chronicler added: 'This may be a mere catalogue to some readers; it will rouse strong memories in our pilgrims and still stronger in members, now scattered around the world, whose business in more tremendous times took them to those places. These memories are no mere luxury and our pilgrimage no mere "sentimental journey".'[30]

Despite the declaration of pilgrimage credentials, such hectic schedules ensured that the experience of the pilgrim overlapped with that of the tourist regardless of the long and loud protests claiming otherwise.[31] The sheer number of places and stops wedged in to the average pilgrimage must have bewildered the non-veteran lacking understanding of the landscape, especially if their travel arrangements meant they were sleep-deprived after taking the overnight boat. Those on the 1929 Toc H pilgrimage faced this particular problem. Having dozed fitfully on the crossing, they arrived in Poperinghe without a chance to freshen up and prepare for the day. They were then packed into motorbuses for the tour, and once back at the hotels some shared bedrooms to keep costs down. The impact on mind, body, and emotions of these demanding schedules and arrangements must have been marked. As the pilgrimage account admitted, 150 was probably too large a group to manage and keep moving easily.[32]

The blur of places and sites was also experienced on the Ypres League's 1932 Whitsun pilgrimage, as the only regret of the group 'seemed to be that they were unable to remember or record half the places they had visited'.[33] A battlefield pilgrimage tested the physical and mental strength just as surely as a medieval one to a cathedral shrine or holy place.

FUNDING AND ORGANIZING PILGRIMAGES

Many of those undertaking this rewarding but testing experience did so free of charge or at a very heavily subsidized cost. Often organized for large numbers to minimize expense through block-booking rates, the pilgrimages required very careful organization and planning. The Ypres League's free pilgrimage to the Somme in August 1927 entailed taking pilgrims to cemeteries scattered across the length and breadth of the extensive Somme battlefields. Captain Oswald planned his logistics brilliantly, making use of his full team. He scattered his guides across several stations to meet the arriving pilgrims and to drive them to specific cemeteries. At the end of the trip, he presented each of the pilgrims with a map showing every stop they had made, 'which was greatly appreciated by the recipients'.[34] The Reverend George Smissen was deeply appreciative of the free pilgrimage arranged for the dedication of St George's memorial church, Ypres, in 1928. For Smissen, this provision of free pilgrimage for widows, orphans, and families summed up the instinctive, 'exquisite insight' and sympathy of the League for the bereaved. 'That act alone', he wrote, 'would justify the existence of the League.'[35]

St Barnabas Hostels managed to find funds to cover, either partially or in full, the visit of nearly 850 people for its 1923 pilgrimage to mark the official opening and dedication of Lijssenthoek cemetery, and 700 people for the 1927 unveiling of the Menin Gate.[36] When drawing up his plans, the Reverend Mullineux, founder of St Barnabas, maintained a sharp eye for detail. He was well aware of the effects of early starts and long rail and boat journeys (exacerbated by using overnight sailings to maximize time and lower costs) on people, especially if they were elderly or suffering any kind of infirmity. Fully committed to his mission, he cared little about niceties or ruffling feathers. In the week before the unveiling of the Menin Gate, he treated Sir Fabian Ware to a full broadside. Concerned that Ware was too much focused on the dignitaries and not enough on the needs of genuine pilgrims, he pointedly asked him to take 'into account the time it will take 700 people to attend to the calls of Nature'. Returning to the attack in a second letter, Mullineux accused Ware of passing over 'the needs of these aged pilgrims'.[37] Mullineux's pugnacious attitude, not always tactful or open to compromise, ensured his pilgrims were treated with the respect and dignity they deserved.

Having been looked after from start to finish, the pilgrims, many of whom had never been abroad before, were hugely appreciative of the efforts made on their behalf. Lilian Peyton expressed her deep gratitude to the Ypres League for providing her with the chance to see 'our "little bit of England"' and enclosed a photograph of her surviving son at the grave.[38] (The inscription on Private Peyton's grave at Coxyde Cemetery reads, 'Loved and missed by Lily, Baby, Mother and Dad.')

H.W. Allinson, formerly of the West Riding Division, said it was only due to the cheap rates of the Ypres League that he could afford to go on a pilgrimage. Thanks to the tour's 'guide, philosopher and friend, Captain G.E. de Trafford', all went smoothly, and his expertise overcame the delays caused by a mistake in the railway bookings.[39] Many others were indebted to the hugely experienced de Trafford, as Wilfred Hyde realized, telling the League's Secretary, 'I am sure I am voicing the thoughts of others when I say that those, including myself, who were meeting him for almost the first time, left for home feeling they had made a very staunch and a very true friend.'[40] Another veteran added a touch of humour, stating he was truly grateful for de Trafford's wonderful skills with a penknife, and for graciously returning a pair of trousers he had left behind in the hotel. 'Does he always go round to retrieve the property of absent-minded pilgrims, I wonder?' he concluded.[41]

On his return, Rex Sargeant told fellow veterans daunted at the prospect of organizing their own battlefield pilgrimage of the excellent service provided by the Ypres League. According to Sargeant, the sympathy extended to all veterans was because 'they are old soldiers to the last man'. Admitting that he was a slightly reserved person, and had been travelling alone, he had been a little apprehensive about joining a group, but on meeting the party at Victoria station he was immediately put at his ease by Captain de Trafford, whom he judged to be a man 'born to wear a "Sam Browne" [officer's] belt'.[42]

The other organizations could boast similar credentials bearing testimony to the extreme dedication shown to those they escorted and took into their care. 'I am sure it is a great comfort for we parents and wives to have had the privilege of visiting our dear boys' graves, and to meet with such kindness as we have done from the Church Army,' wrote one grieving mother. A widow added her heartfelt thanks: 'I am gratefully thankful to the Church Army and its officers for giving me the honour of treading on the soil of where my dear husband fought and fell.'[43] Mrs c. Briggs of Birgham, Coldstream, was another profoundly grateful for everything the Church Army had done, heaping praise on its work:

> I must say we found everything beyond our expectations. We were well taken care of and looked after. Words cannot express the kindness and patience of the V.A.D. [Voluntary Aid Detachment] Guides: everyone was so kind and tried

to comfort each other. We were all on the same mission. The graves of our dear ones are most beautifully kept. It is a great comfort to think of them so well cared for ... It is very touching to see the mothers, some over 70 years of age. One I was speaking to was 73. She was broken down with grief, but was very brave too. The memorial service at Ypres Town Cemetery was very beautiful, but very sad. We all felt so broken hearted. Some were holding the Flanders Poppies in their hands, which they had gathered in among the ruins of Ypres. One thing which touched most of us [was] when the special train drew into Ypres Station the Belgian and British flags were put out on the flagstaff side by side.[44]

Others had the wealth and time to make their own arrangements and travel independently. As the wife of a major-general, Elizabeth Braithwaite Buckle fell into this category. She set off for France in the spring of 1919, before travel restrictions were lifted, intent on finding the grave of her son. On return she wrote up her experiences and published them partly as a cathartic, public memorial statement, and partly as a stock of good advice for those wishing to do the same. 'Now, this very week, fourteen months after he fell, the restrictions on going to France have been removed and I have been there,' she stated, and after making her memorial statement she felt she could 'be quite business-like and go straight ahead' with her advice. Having poor health, she decided to travel first class, but she believed everything could be undertaken on a much more modest budget. She urged all to get their passports and paperwork sorted out well in advance, to check details with the IWGC, and travel as lightly as possible.[45]

Checking details about the precise place of burial was a crucial piece of advice for all pilgrims, but knowing where to turn for information could be a tricky task for there was a plethora of organizations at work. Sitting alongside the war graves visitation charities in France and Belgium were the official bodies of the Army's Directorate of Graves Registration and Enquiries (DGRE) and the IWGC. In the early period there was the potential for confusion between these two authorities, and much time was devoted to highlighting the correct route for people requiring information as to the graves of their relatives.[46] For the early visitor in France and Belgium the agency on the

Figure 6.1   St Barnabas pilgrim at a grave in Lijssenthoek Cemetery, 1924.

ground, until its final withdrawal in the autumn of 1921, was the DGRE, while the IWGC prioritized the transformation of the cemeteries into permanent sites of commemoration. Of vital importance to visitors was the DGRE's chain of enquiry centres established at Albert, Arras, and Ypres, the nodal points of the main battlefields, which the IWGC gradually took over as the army continued the wind-down of its activities.[47] It was thanks to the clear instructions provided by the Albert office, combined with the assistance of a veteran guide, that 'R.O.P.' made for Longueval on the Somme in order to visit a particular set of graves. Told to walk 1,200 yards north-east of Longueval church, their first challenge was finding any sign of the church. Eventually, after much stumbling around and passing a host of crosses marking the burial sites of unnamed soldiers, the two men found the cemetery and the cluster of Black Watch graves 'R.O.P.' had come to visit.[48] Neville Chamberlain was less successful when he went looking for his cousin's grave in August 1919. Relying solely on the information provided by the regimental chaplain, and not having verified it with the DGRE or IWGC, he 'found the cemetery a little way from the village in rising ground near the road. There were men of his regiment there and an officer killed the same day or the day before, but though we examined every cross there was no trace of his name.'[49]

THE HEART OF THE PILGRIMAGE EXPERIENCE: THE CEMETERIES AND MEMORIALS

As the cemeteries and memorials were the focus of every true pilgrim's visit, the precise condition of the places of commemoration was of great importance. In the first few years after the war, visitors encountered a scene of constant activity, some of it gruesome and harrowing, as bodies were discovered, exhumed, and reburied. Between the Armistice and the early spring of 1920, nearly 190,000 bodies were exhumed from the battlefield and reburied in IWGC cemeteries. At that point it was hoped that the remaining work in France and Belgium would be completed within a matter of months.[50] In fact, it was to last another eighteen months, during which time a further 3,086 bodies were discovered and reinterred.[51] (Between the final searches of the battlefields in 1921 and the spring of 1938, 40,000 further British-imperial bodies were discovered and buried.[52])

As wartime creations, the cemeteries contained only temporary (although often lavish) grave markers, utilitarian fencing and fixtures, and varying degrees of planting. All were subject to a programme of maintenance and improvement as a preliminary to their final architectural and horticultural forms. The IWGC understood that temporary did not mean uncared for.

Therefore, as the first pilgrims arrived, they saw the DGRE and IWGC teams going about their daily business. Indeed, some of the pilgrims were engaged in their own detective work, intent on finding the graves of lost loved ones through their own efforts. Mrs Frances Pumphrey of Hindley Hall, Northumberland, was keen to find out more about her son, who had no known grave. However, her public search was carried out in an almost cryptic manner. In March 1922, she used the letters column of the *Ypres Times* to appeal for information about a convent near Ypres used as a hospital in October 1914. A year later she asked anyone who had served as a stretcher bearer in the 2 Highland Light Infantry or 2 Warwickshire Regiment in that month to get in touch. Two years later she had a specific person she wished to trace, and appealed to him, or anyone who knew him, to write to her. Her final appeal came soon after, when she asked any officer still in possession of the pre-war maps of Ypres marked in thousand-yard squares to make contact.[53] All of this implies that she was gradually building up information about the locations where her son was last seen, and the possible sites of his original burial, with the intention of having the ground searched for signs of his body. Whatever happened, no identifiable trace of the body was ever identified, and her son was commemorated on the Menin Gate memorial to the missing.

The battlefields threw up their own folklore of remarkable stories such as this, in which the moral seemed to be the value of persistence and faith. Henry Benson told of a couple who continually made enquiries about their son's lost burial place. They searched cemeteries in Belgium during the summer of 1920 with no luck but came out again the following year. Seeing an exhumation party at work, they followed them round and observed their work. Then came the breakthrough, as the exhumation party uncovered a body that the woman immediately recognized through her own highly distinctive stitching on the shirt. 'Her relief was unbounded,' wrote Benson, 'for she felt that at last her careworn and anxious mind could be at rest,' and she and her husband then attended the burial in the concentration cemetery.[54]

Seeing the DGRE's Graves Registration Units (GRUs) engaged in their grizzly activities could be disturbing. At Hooge Crater Cemetery, Olive Edis found the site 'a swamp' divided by muddy and greasy duckboards. It was a mass of crosses, some adorned with helmets or even scraps of clothing. Unable to resist peeking into a small hut, she saw stretchers that appeared to contain dead bodies covered by sacking. A label was tied to one reading, 'unknown British soldier, West Yorkshire Regiment, identified by badge'.[55] In the same cemetery, Marjory West saw the exhumation and burial parties at work bringing in the scattered dead. On giving her a lift back into Ypres by a GRU team, the officer in command explained their work and the care they took attempting to identify each man. West understood the importance of these painstaking efforts to those at home: 'It must be a tragic piece of work but a merciful one for the sake of those many who long to visit a grave, all that remains now, or of those whom the terrible word "Missing" has left in an agony of suspense.'[56]

Stephen Graham also managed to gain an insight into the workings of the GRUs after getting a lift, like West, from a GRU team. This one was off to Polygon Wood, just a mile or so away from Hooge Crater Cemetery. Sitting in the lorry, packed with fresh wooden crosses for delivery to Polygon Wood Cemetery, Graham portrayed the scene in Dickensian terms: the men were depicted like 'Old Joe' fascinated by the possessions of the recently deceased Scrooge. Graham told of their excitement and pride at having recently uncovered the body of a brigadier-general, while another team had come across a dugout containing the bodies of three officers '"and they picked up five thousand francs between 'em'" by way of reward from the relatives in gratitude at their identification of the bodies. Discovering that the team often slept on the battlefield, Graham asked one of the men whether they had seen any ghosts. In reply the man smiled. 'He saw none. He felt the presence of none. Imagination did not pull his heart-strings. If it did, he would go mad,' Graham noted.[57]

As every pilgrim soon realized, and as the earliest visitors noticed most starkly, the former battlefield was a world of cemeteries and scattered graves, for prior to the intense concentration work carried out by the GRUs, and with sightlines opening up, their presence in the landscape was both more obvious and numerous: 'Those little cemeteries are everywhere, and the battlefields are still dotted with isolated crosses, and leaning against these here and there one sees the rifle of the dead man or it may be his shrapnel helmet,' Sir Arthur Yapp told readers of the YMCA's magazine.[58] 'Look almost where one will from the road and he will see, here and there, the white cross, or clusters of them, showing where soldiers were buried where they fell,' wrote John W. Dafoe, adding 'the war area is in truth one vast cemetery'.[59]

Those first visitors also experienced the visual effect of the thickly clustered wooden grave markers; the 'forest of crosses', as a former army chaplain described it.[60] The grave markers themselves varied enormously in quality. The DGRE produced its own standard version, but often a man's friends or unit would take responsibility for the design and construction. Fifty-five (West Lancashire) Division produced a particularly fine enamel roundel emblazoned with the red rose and the emblem 'They Win or Die Who Wear the Rose of Lancaster', while pilots often had crosses fashioned from old propellers. However, there was a surface uniformity, which was appreciated by visitors. On seeing the neat rows of uniform crosses in a cemetery, Edis realized the powerful effect created by equality of treatment.[61] Stumbling across many small cemeteries on the Ypres battlefield, Marjory West was much struck by their simplicity. It made her realize, like Edis, that the argument for equality of treatment was a

Figure 6.2 St Barnabas pilgrim at a cemetery awaiting its permanent features, c. 1920.

strong one: 'The very plainness of the crosses and their uniformity makes their great appeal,' she wrote, adding, 'Those people make a great mistake who ask for more pretentious monuments and varied inscriptions.'[62] Such simplicity and uniformity made the discovery of the graves of the distinguished and famous quite a thrill. Alice Knight told her parents, 'We went into the Cemetary [sic] there and happened to notice Raymond Asquith's grave and his brother-in-law, the Hon. E. Tennant. They were both in same row and but one at each end. Oh Mother dear! How it touches one to go into the different cemetarys [sic].'[63]

Sir John Lavery's painting 'The Cemetery, Étaples, 1919' captures the atmosphere immediately after the war, and turns the vast burial ground at Étaples into a world of two separate, but intertwined, domains of gender.

The male domain is represented by the dead lying beneath the uniform wooden crosses. As if to define the transformation from vitality to death, Lavery emphasizes the earthy brown of the crosses. The men had very definitely returned to the soil and clay from which they had been shaped. A few of the crosses are crowned with wreaths, although whether this act was cathartic and helping to keep memory fresh was left a moot point, especially as a pink rose droops down at the bottom on one such wreath rather than remaining pert, vibrant, and crowning the circle. A train rushes by at the far end of the battlefield, white steam billowing in the distance. Was Lavery asking whether the world had already become indifferent and returned to its usual business, or was this the triumph of sacrifice: the men had died precisely for this peaceful normality? Regardless, in including the train line, Lavery identifies something Ware saw as a publicity gift, for he managed to persuade the French railway authorities to slow the speed of trains running past the cemetery, allowing travellers time to linger over the impressive spectacle.[64] Beyond the railway the land is green and lush, in sharp contrast to the sandy soil of the cemetery, and the River Canche flows by in aquamarine blue towards the estuary and the sea. Sand, sea, and grass combine in an allegory of an expeditionary force that had crossed the Channel to aid France and now lay permanently beneath its soil.

The green of the grass at least points to vitality and life among this host of the dead, and, indeed, there is life in the cemetery, and the life in this domain is overwhelmingly feminine. In the middle ground two female gardeners of Queen Mary's Women's Army Auxiliary Corps (WAAC) rest on their spades while chatting. Are they, too, indifferent to their surroundings already? Do they not know it is sacred ground? 'Yes, they do,' seems to be Lavery's answer, for behind them two other WAACs are working near a cross, one down on her haunches, possibly about to lay a wreath. Beyond her another group of mourners is gathered around a grave. They appear to be women in mourning black and one is possibly holding a small girl, dressed in a white pinafore smock, in her arms. To the immediate left of the group Lavery delicately daubed olive paint, seemingly small trees or tall plants, which have an almost spectral quality, as if they are hovering spirits, maintaining a watch over comrades, and perhaps also offering comfort to mourners. Referring to a British war cemetery of the Second World War, the poet Charles Causley described it as a 'tidy wreck of all your wishes'. Lavery conjures up the same idea in paint, and in the process offers an element of ambiguity. Here is a temporary world, the wooden crosses in straight rows, but rather higgledy-piggledy, the sand banked up. This site was temporary and yet being maintained. What would come next? Fabian Ware knew and he was sure it would mend the deepest wounds.

THE GRADUAL MATURING OF THE IWGC'S PLANS

Wishing to show the public that the IWGC's plans for the cemeteries would provide a fitting and noble solution to the commemoration of the dead, Ware was anxious to get the project underway and present completed examples. Three experimental cemeteries, at Le Tréport, Forceville, and Louvencourt, were ready by the spring of 1920 and in September the *Times* carried a highly favourable article written by the poet and co-founder of the Ypres League, Beatrix Brice. According to Brice, the cemeteries were 'filled with an atmosphere that leaves you very humble, that gives you wonderful thoughts', where 'chivalry, knighthood, heroism and self-sacrifice ... are knit-together'. Here Brice called upon chivalric imagery, so ubiquitous in wartime and post-war commemorative culture. In so doing, the dead were symbolically drawn into the warfare of a previous age; they were feudalized at exactly the same time as their democratic equality was emphasized through the cemeteries. But Brice also wrote as a woman addressing other women: 'Your own man has a wonderful grave, the nation has a wonderful monument,' adding that the names inscribed on the headstones were the 'flower of the manhood of our race' forever resting in soil won by 'the valour of her [Britain's] sons'.[65] Mary Macleod Moore was equally impressed, seeing in the architectural and horticultural forms journey's end in the search for consolation. She noted that it was 'possible to see how in time all the cemeteries will look. In each the Cross of Sacrifice towers above the plain white monuments, and the great Stone of Remembrance tells the passer-by the glorious story that "Their name liveth for evermore."'[66]

'A Scottish Chaplain' added his praise for the IWGC's plans in an article for the *Scotsman*. Having visited the battlefields, he assured the bereaved of the beauty and majesty of the IWGC's conception, writing a piece Ware or Rudyard Kipling could have produced as an official statement:

> Let the sorrowful hearts at home be fully persuaded of this, that all that reverent and tender care can devise to do honour to the last resting place of our glorious dead is being done. The cemeteries will be enclosed with a bold plain wall, and each grave will be marked by a memorial stone. A tall massive cross will throw the shadow of its wide arms across the green sward proclaiming the victorious sacrifice and the deathless Hope. And closing in the vista of the central pathway will stand the great Altar-Stone of Remembrance, bearing on the face of it in strong bold lettering the noble and worthy epitaph – THEIR NAME LIVETH FOR EVERMORE – And, there, the birds will sing their sweet requiem, with hope in the heart of it, and the trees will bud, and shed their leaves, while the generations come and go; and the gentle river will glide in long liquid lapses hard by the place of their rest.[67]

It was the start of a stream of unbroken praise, as pilgrim after pilgrim wondered at the marvel of the cemeteries. It was also the start of a massive construction project, and patience was requested from pilgrims crossing the Channel. Reporting on progress in November 1922, Henry Benson sought to raise awareness of the sheer scale of the undertaking, stating the IWGC had 'a colossal task in front of it', before adding, 'probably a decade will have elapsed before its labours are concluded'.[68] In Kipling's words, the building scheme was 'the biggest single bit of work since any of the Pharaohs'.[69]

Although built to a common set of principles, the IWGC's cemeteries were hugely diverse in terms of size, shape, and siting. As the biggest cemetery on the old Western Front, Tyne Cot impressed by its scale and commanding position, but the smaller cemeteries were equally affecting. Alfred Skinner said the modest and plain cemetery at Westoutre produced the 'inevitable lump' in the throat, 'and that feeling of reverence and gratitude for those lying there was as predominant as in the larger and more ornate cemeteries'.[70] Former WAAC Edith Winter, visiting her old base of Rouen along with some fellow veterans in the summer of 1922, found St Sever cemetery to be in beautiful condition, with carefully tended flowers and the white headstones perfect, while 'A Grateful Pilgrim' told *Country Life* 'it was comforting to see how beautiful the [cemeteries] are'.[71] Architecture balanced and enhanced by perfect horticulture was an element impressive to many. Tuohy declared the profusion of plants and flowers stunning, and was particularly moved by Aveluy Wood Cemetery for its neatness, beautiful flowers, and plants in such stark contrast to the still mutilated wood around it.[72] L.M. Orton was equally impressed by the array and quality of the planting in the cemetery where her brother was buried.[73] Visiting in the summer of 1922, Captain Mee described the cemeteries as covered with 'flowers of every description ... and the whole looks more of a picturesque flower garden than a cemetery – a fitting symbol of the flower of English manhood which lies beneath the turf'.[74]

A further, and often intriguing, element for the non-veteran pilgrim was the weird and wonderful collection of cemetery names. The Reverend H. Shannon Brisby was moved, amused, and bemused by them. 'Blighty Valley' struck him as poignant in his evocation of sufferings and yearning for home; some were 'euphonious and even picturesque – such as "Mont Noir" and "Nine Elms," while there are also to be found some that are drab and even ugly. Of the latter I might instance "The Gordon Dump Cemetery" in "Sausage Valley".' Yet, regardless of the name, he found that all were pools of peace, centres of repose and restfulness, and horticulturally and architecturally beautiful.[75] The names also reflected another fact. They put the Imperial into the IWGC. Arthur Hickman cited Montreal, Zouave Valley, Euston, Tigris Lane, and Niagara cemeteries as testimony to 'the world-wide forces which were concentrated in France during the war'.[76] The overall effect of each perfectly blended element – name, architecture, horticulture – left a deep mark on all who saw them. 'There can be no doubt that the Imperial War Graves Commission has performed a great work,' noted a pilgrim in 1925.[77]

Pilgrims found another powerful presence in the cemeteries in the form of the IWGC's personnel, which gave them a direct link to the experience of their lost loved ones. Dedicated to their work and having an immense knowledge of the cemeteries and memorials they maintained, the staff were the IWGC's finest ambassadors, the living embodiment of its ethos, and the greatest interpreters of its estate to pilgrims anxious for knowledge and insight. Veteran and journalist Victor Hyde wrote of 'the army that was left behind' for the *Dundee Evening Telegraph* on 4 August 1931, the anniversary of Britain's entry into the war. The army to which he referred was the staff of the IWGC, and their global role stretching from 'Guatemala ... and half a hundred other out of the way places. Oh, yes, and the Sandwich Islands, too.' For all the men involved it was a 'labour of devotion', seen most clearly when a pilgrim arrived looking for a particular grave: 'A gardener touches his cap, and another father or mother or wife or brother or sister or sweetheart who had abandoned hope finds "some corner of a foreign field that is for ever England"' as the gardener reverently takes them to the right spot.[78] L.M. Orton experienced just this, as the gardener provided an explanation of the cemetery design while taking her to her brother's grave.[79]

The Reverend H. Shannon Brisby was deeply touched by the kindness and consideration shown to him by the IWGC gardening team working in Gordon Dump Cemetery. The head gardener helped him find the particular grave he wished to see and photograph on behalf of the family and then ordered the team to stop work for a moment, allowing Shannon Brisby the chance to take the photograph in a moment of respectful calm. 'Here was a man I felt, who, by his understanding and sympathy, was one link in the connecting chain between the dead and the living. Here was a man who knew exactly what the bereaved at home wanted to know – that their dead were cared for, and who was resolved that no effort was too great to ensure that those graves would be as lovingly and as beautifully kept as if the hands of those nearest and dearest to the deceased had tended them.'[80]

The reverend gentleman expressed a common feeling. Those attending the Ypres League's free pilgrimage to the Somme in the summer of 1927 found it 'impossible to exaggerate the depth of gratitude felt by every pilgrim for the devotion of the gardeners of the Imperial War Graves Commission in their work of tending the cemeteries'.[81] Another group of Ypres League pilgrims, this time at Whitsun 1932, declared the gardeners at the Thiepval and Arras memorials to the missing 'most attentive to our requests for finding the names on the piers, and one felt whilst viewing the various piers with the thousands upon thousands of names carved in stone what a price was paid for victory, but also what a wonderful achievement has been accomplished by the Imperial War Graves Commission'.[82]

For the pilgrim ex-serviceman, the opportunity to chat with IWGC staff offered another opportunity: the chance to mull over the old days and find out a little more about each other. B.S. Townroe spoke to a gardener on the Somme. Although the man had not been back to Britain for six years, and he and his wife sometimes found it a little lonely, 'he would not ask for any more congenial task than for caring for the place where so many of his comrades lay'.[83] Loneliness, particularly in the winter, and particularly in the Somme district with its many, small, scattered, and isolated settlements, was a point commonly made by these veterans to their fellows.[84]

Despite these problems, H.A. Taylor believed the men and their families to be contented and happy, motivated as they were by the importance and honour of their work. He described in summer coming across the jacketless, collarless, sun-tanned gardeners, their cycles resting against the cemetery wall, going about their work with infectious dedication and pride. They required no close supervision, Taylor added, as they were doing it for their pals. They were performing a sacred, and fraternal, task:

> The beauty of the British War Cemeteries is, probably, the dominant memory of every visitor to the old battlefields. No praise can be too high for the skill of the designers, and the work of the gardeners. There is nothing of the soulless Government department about the Imperial War Graves Commission. It was no departmental mind that conceived the scheme of accepting from the relatives of a fallen soldier flower seeds to plant upon his grave, and then to return to the relatives for planting in their own gardens, the first seeds that fall in the autumn from their flowers. War grave and home garden are thus in communion. Truly, there is a heart-beat in the work of the Imperial War Graves Commission! That you learn in the very first cemetery you visit.[85]

It was all in very sharp contrast to the nature of the Belgian, French, and German cemeteries and memorials. Although interested in them, pilgrims could sometimes find it difficult to penetrate the mentalities and cultures expressed through these different memorial forms, and as they never strayed deep into what was once German-held territory, often missed the biggest of the German cemeteries. The 1932 85 Field Ambulance old comrades association pilgrimage to Ypres made stops at the French gas memorial at Steenstraat and the Yser Tower memorial to Flemish soldiers. A tour member, 'M-H', deemed the French memorial 'magnificent', but the austerity of the tower left him unable 'fully to appreciate the architect's meaning as to the design', which was 'so unlike anything to which we are accustomed in our own memorials'. He speculated that 'probably the significance is purely spiritual and represents to the Flemish Belgians their national unity among the other races'.[86] At Poelcappelle, near Ypres, could be found the striking memorial to the French ace pilot Georges Guynemer, which made a deep impression on those who visited it. Taylor was impressed, describing the tall column surmounted by a stork in flight as 'exquisitely carved', while the veterans on the Leyland Motor Company pilgrimage were moved by the 'French idiom and phraseology' of the inscription.[87] Determined to pay their respects to the memory of their French allies, the veterans of the 236 Siege Battery association made a point of visiting Bailleul's 'really magnificent' city memorial.[88]

Far more challenging were the cemeteries of the other combatants, allied and enemy alike. When contrasted with the care and attention lavished on the British cemeteries, these others could seem far from the fitting places of commemoration the veterans felt they should be. Inspired by the beauty of the IWGC cemeteries, a British pilgrim visiting the Somme battlefields was disconcerted by the poor condition of the French cemeteries, the sparse horticultural treatment, and the mounds created by mass burials being starkly protuberant. The German cemetery was different again, being very neat and tidy but totally lacking in wreaths or any other objects brought by mourners and visitors. Overall, in neither the French nor the German cemeteries could anything be found similar to that which 'inspires with courage and hope those who make a pilgrimage to the resting place of the fallen soldiers of the British Empire'.[89]

THE JOURNEY'S END: PILGRIM ENGAGEMENT WITH CEMETERIES AND MEMORIALS

The emotional heart of the pilgrimage was the time spent in the cemetery at the grave of the lost loved one. This was also the moment, above all others, the pilgrim owned entirely. Despite the immense care lavished on pilgrims, perhaps even because of it, they could feel vulnerable and a little insecure. This sense must have been increased if it was an older person, or someone travelling alone, female, or from a humble background. In their desire to relieve pilgrims from all troubles, the war graves visitation charities and services inadvertently induced in their charges a state of impotence. This being realized, the imbalance was corrected with a visit to the cemetery or memorial, as the bereaved pilgrims expressed themselves in their own way and, having seen the place of commemoration, gained in knowledge, insight, and power. They moved from being passive sufferers to pilgrims progressing triumphant after experiencing their lesser Calvary.

The simple act of walking into the cemetery started the essential chapter in the emotional and sensory journey of a pilgrim, as each moved from one realm or sphere into another. The IWGC's architects achieved this sense of translation in every aspect of the design, starting with the great efforts to delineate the sacred space of the cemetery from the quotidian world surrounding them.[90] In the boggy ground of Flanders, the architects were often helped by the ancient roadside fixture of the drainage ditch. These long, straight, sharply cut channels created a barrier, the Flemish Styx, acting as the frontier between the empire of the dead in the cemetery and the world of the living. The act of transformation then commenced in the crossing of the many beautiful cemetery causeways, such as those at Mud Corner, Maple Copse, or Le Trou Aid Post. Deeply incised at the cemetery gate or wall, the name of the 'geometry of sleep' announced the arrival in a new land.[91] Then the olfactory arrived, not in the scent of flowers, but the peculiar and particular smell of Portland and Euville stone combined with brick or rough flint made potent by the sheer dampness of the environment. Next came the colours, especially if spring or summer, with the profusion of plants and flowers like artful paint splashes against the white of the headstones, Cross of Sacrifice, and Stone of Remembrance. At this point the navigation of the cemetery commenced. Whether a carefully laid out example of graves concentrated from the battlefields or the jumbled plots of a battlefield cemetery, the architects created routes through the graves leading eye and foot in a particular direction, reeling in the pilgrim to the locus at which all sightlines converged. There, finally, came the words and symbols inscribed on headstones and the Stone of Remembrance. As with all the architectural work of the IWGC and its sister organizations in the imperial battle exploits projects, the overall effect was one of perfect tensions and contradictions; the cemeteries held the dead and visitors in intimate embrace while also nudging the visitor to step back, step away in order to contemplate the whole, let the eye encompass the full sweep and vista: a white silence of stone and a full palette of colours singing as loudly as the birds; a lotus eater's garden, where time is nothing; and a place of energy where thoughts rush like express locomotives. And all that before *the* name and *the* inscription the pilgrim had come for.

On reaching the graves of particular significance to them, pilgrims must have felt a welter of emotions and thoughts. When Neville Talbot saw his brother's grave in Sanctuary Wood Cemetery, he found it marked by a wooden cross awaiting replacement by the permanent standard headstone. 'I wonder so how much "remains" are contained by it,' he mused, continuing, 'I have not, owing to being in S[outh] A[frica], really been able to follow what happened [to] them since I took the parents to the original site in 1919, which I had marked by the heavy wooden cross (made in Lille and now in All Hallows) after the Armistice in 1918. That cross was the successor to 2 if not 3 erected in '15 and '17, which were blown to bits. The original site of the grave was as it proved at a very specially hell-fire point, fighting surged to and fro over it after '15.'[92] Closure was not achieved by Talbot; rather, standing by the grave seemed to inspire more questions than it solved. He was blinded by memory and unable fully to comprehend, contemplate, and understand what was in front of him. Significantly, it was the lack of the permanent grave marker that seemed to trigger the increasingly fractured and fitful trail of memories.

For Elizabeth Braithwaite Buckle, visiting her son's isolated grave long before a concentration team had moved the body into a cemetery in the early spring of 1919, the progression to the precise point seems to have focused her so intently she lost touch with the world around her. Although her husband, whom she had praised as companion and father, had accompanied her, at this point he became a distant, peripheral, almost invisible, figure reduced to facilitator and guide, rather than someone with an equal emotional stake: 'My husband went on ahead a little way to make sure he was taking me the nearest line, but he has the eye of a hawk for country, and had gone as straight as a pointer. He held up his stick as signal, and in a few minutes I was at the end of my quest.' She did not say '*We* were at the end of *our* quest.' But at the grave she made the transition back to wife and mother: 'There beside his trench, facing the miles of open country, alone with God and with the birds and flowers and butterflies all about his bed – there lay our only son, the joy of our life, the pride of our hearts, but oh! never prouder, *never* prouder, Beloved, than now.' The sense of catharsis was immense. After having dreamt of it for so long, and then wondering whether she would be tethered to the site in grief and never able to leave, she actually found solace and completeness: 'It was not so hard as I had expected. The place was so beautiful, the wide silence so majestic, that I turned away almost with a feeling of content. Had I searched the world I could scarcely have found a spot he would have liked better.'[93] The reason for this was its being a foreign field forever England: 'In its butterflies, flowers and peace, it was exactly like summer days back at home when he was a little boy, and all the joys of nature he held so dear.'

Conditioned by the codes of Victorian mourning, the bereaved pilgrim brought flowers or a wreath to the grave, either from home or supplied by the war graves visitation charity or local florist. When Nellie Burrin visited her brother's grave in 1920 before the cemetery had received its permanent architectural and horticultural form, she was determined to beautify it with flowers. During the long walk to the cemetery, she collected wildflowers, adding them to the wreath she had brought from home. After her initial visit she devised a more ambitious plan. A few days later, having specially purchased a trowel in Amiens, she dug up wildflowers and plants in the ruins of Warfussie, before walking on to the cemetery. Here she carefully planted everything she had collected, scooped up water in an old helmet, and improvised another tool from the cardboard lunchbox she had discarded in a ditch (incidentally, this reveals a casual attitude towards littering, probably encouraged by the sheer amount of debris everywhere). Flowers acted as the leitmotif of Burrin's pilgrimage. She noticed the profusion of cornflowers, daisies, and 'red patches here and there' on the battlefields: 'When we got nearer, we found beautiful red poppies. Even in the trenches here and there poppies can be seen as if to mark the spot where our men had fallen.' And wandering past patches of pansies reminded her 'so very much of some I had sent to me during the war [by her brother] that I felt I must bring a root and plant on the grave'.[94] A floral gift from France made its way back to the sleeping sender.

Flowers became proxy words of grief. During the Ypres League's pilgrimage for the unveiling of the Thiepval memorial in July 1932, the party leader, Captain H.D. Peabody, was deeply moved to witness a League member hand a small posy of flowers to an old lady visiting her son's grave. Peabody recorded her extreme gratitude at this very small, but immensely powerful, gesture. According to Peabody such thoughtful acts justified the League, and created a bond between every veteran and the bereaved, for the ex-serviceman knew 'only too well what these poor mothers, wives and sweethearts were feeling'.[95]

The inscription cards on the wreaths thus compacted into a few lines a lifetime of emotions and memories: 'He will not be lonely now, for Mother has been here'; 'He gave his life for his country, and the King called him home'; 'Let us try to be more worthy of the dear ones who died for us'; 'The pain has been taken away, and a profound peace has entered into my heart'; 'Went the day well? I died and never knew; but ill or well, England, I died for you.'[96] The messages on wreaths moved two journalists from the *Belfast Telegraph* deeply, and as so often, it was those from bereaved women that affected them most. 'More touching still was a little wreath – "From Mother" – the only little bit of colour' on a chilly winter's day.[97]

Other small, and deeply personal, rituals were carried out by relatives. When Mrs Benbow arrived at her son's grave in Duhallow Cemetery, Ypres, she asked the gardener to explain how the bodies were placed in the plot, particularly wishing to know which end the head lay. On being told, she fell to the ground and kissed the grass.[98]

Henry Benson described the pilgrims on the 1925 Ulster and Scottish pilgrimage as including 'many old folk [who] looked almost too fragile to cross the village street, and not one of them had ever crossed the sea before'. Practically all came from the working and poorer classes and would never have been able to attend without the support of the Scottish Red Cross, the Edinburgh War Graves Committee, and St Barnabas Hostels. Among the group was a seventy-year-old man from Aberdeen who insisted on walking the three miles from Ypres to the cemetery where his son was buried. Asked why he did not want to take the car provided, he said he had to walk the same route as his son. An old lady from Londonderry, who had been partially paralysed by the shock of losing her husband, kissed the wooden cross marking his grave, saying she was sure Christ would one day reunite her with her husband.[99]

Another elderly woman, using the services of the Salvation Army war graves visitation department, returned to her hostel after visiting her son's grave in the pouring rain. When the officer in charge offered to have her shoes cleaned, the woman refused and wrapped them up carefully in paper so as to take home soil from the grave.[100]

Seeing the precise place of death or fatal injury was also important. The Ypres League August Bank Holiday pilgrimage of 1931 to the Tyne Cot memorial included a woman who saw the names of her two sons inscribed there for the first time, as did a bereaved father of a fallen son. Thanks to having the expert guide Captain de Trafford with them, the man was also able to see the precise spot where his son was killed. 'In much the same manner was each individual pilgrimage – exquisitely sad to the pilgrim – satisfied.' Here someone saw not only the space where death had occurred but also the site of its commemoration. The place of pain, agony, and obliteration was pointed out to the bereaved father as well as the redemptive place of fitting remembrance.[101]

THE SPIRITUAL VALUE OF PILGRIMAGE AND THE SEARCH FOR CONSOLATION

The message that consolation could be found in the cemeteries and battlefields was spread both by pilgrims and the organizations that arranged war graves visits. 'There is a tragic consolation in these soldiers' cemeteries of France. They leave an impression on the mind that is ineffaceable,' declared an article in the *Belfast Telegraph*, while Henry Benson wrote of finding a 'peace and silence which, like the Egyptian darkness of old, could almost be felt at the Noeux-les-Mines Cemetery'.[102]

Benson went on to describe the great benefits a war graves pilgrimage imparted to the bereaved: 'These visits to the graves bring to them [the bereaved] consolation that can be obtained in no other way. "I have seen where my son lies, and I have watered with my tears the kindly earth that covers him," a Newfoundland woman confessed to me a few days ago. "The burden of my grief has been removed, and I am returning home comforted and supremely happy."'[103]

Henry Beckles Willson, Secretary of the Ypres League, believed the pilgrimages of 1923 were a cause of deep satisfaction and proof of their value, for among the parties were many 'who had postponed their visit for years, fearing to face a pain overwhelming in its intensity'. But 'contrary to expectations they found instead a blessed peace and consolation in the knowledge that their loved ones lay at rest amid beautiful and carefully tended surroundings'.[104] Pilgrimages created a sense of group identity, shared concerns, priorities, and community. 'The party returned to Amiens ... with the satisfaction of having been able at last to see with their own eyes the resting-places of their loved ones and lay their own personal offerings of flowers as a token of love and remembrance,' noted an Ypres League pilgrim in 1927, deeming it 'to have been a delightful, comforting and instructive experience'.[105] Beckles Willson wrote of the League's 1934 August Bank Holiday pilgrimage that 'it was touching to watch the personal interest shown by the other pilgrims of the party and their anxiety to share in some small way the sorrow endured by the bereaved relatives whenever a particular grave was visited'.[106] As this 1934 pilgrimage revealed, the emotional power of a pilgrimage and the opportunities they presented for grief to be expressed remained undimmed despite the passing of the years.

For the Church Army, the Salvation Army, the YMCA, and St Barnabas Hostels the overriding desire was consolation of the bereaved. By placing this at the heart of the pilgrimage experience, the paramount role of the guide was to make spiritual sense of the journey, and therefore 'one of our evangelists, missionaries, or a clergyman' acted as escort on all Church Army pilgrimages.[107] As a result, a Christian message of redemption, comfort, and even glory through suffering was stressed. And it was one that seemed to meet the emotional demands of pilgrims. Frank Hewitt concluded his account of a 1923 battlefield pilgrimage organized by the YMCA with a ringing endorsement: 'Such a very happy time my wife and I had with your people in France last September that we are eagerly anticipating a renewal of our experience.'[108] A bereaved mother informed the Church Army: 'I am sure it is a great comfort for we parents and wives to have had the privilege of visiting our dear boys' graves.' Another grieving mother wrote, 'I thank you again and also give thanks to God for answering my prayer to see my son's grave.' A widow was 'gratefully thankful ... for giving me the honour of treading on the soil of where my dear husband fought and fell'. For a bereaved couple, the kindness of the two Church Army guides was overwhelming, especially 'the card you have sent [which] is so comforting when my heart aches for our boys'.[109]

In the many press accounts of pilgrimages, local and national, the quintessential grieving pilgrim receiving spiritual benefit through seeing the grave was a woman, either a widow or an elderly bereaved mother. Women were shorn almost entirely of the active roles they had played during the conflict. They were no longer munitions and agricultural workers or tram conductors. Instead, and aside from the women veterans and those engaged in the war graves visitation services of the charities, women were portrayed as passive sufferers. The men had fought and died. Women had stayed at home, wept, and longed for the moment when they could see the grave of their man. The *Church Army Gazette*'s front cover illustration for its 1923 Armistice edition encapsulated this dominant motif. It depicts an elderly lady on her knees weeping in front of a row of wooden war graves crosses. By her side stand her grandchild and her daughter-in-law, who is pointing up to a vision of Christ on the Cross. In the immediate background is an IWGC cemetery with its Cross of Sacrifice and headstones and the silhouettes of a male and a female visitor seemingly looking at a grave, while on the horizon there are beams of light from a rising sun.[110]

Although teetering on the verge of cliché, the story of suffering, mourning, and redemption shown in the illustration was one verified by experience. While visiting a cemetery near Corbie on the Somme, H. Channing-Renton saw the pitiful sight of an old woman breaking down in uncontrollable tears at a grave. She had lost two sons in the war, and her husband had died a year earlier, leaving her alone, and yet she had managed to make the journey from her home in Scotland. Seeing her rally and express her satisfaction at having visited the grave, Channing-Renton finally understood the appalling anxieties faced on the home front by people concerned about the plight of their loved ones.[111] Reporting on the Salvation Army pilgrimage of 'a large party of women' from northern Scotland, the *Scotsman* told the pathetic story of a woman who spent hours looking for the grave of her son. When, at last, she gave up the search she stopped before an unknown grave. 'With tears in her eyes she said, "This may be my boy," and kneeling, she reverently placed a wreath of heather, carried from home, to the foot of the cross.'[112] Just as they were leaving the cemetery, one of the party came up to say she had found the grave. The woman then went to her son's grave, checking the inscription very carefully several times to reassure herself; 'The wreath, however, she left where she had laid it on the grave of the unknown.'

Mrs Petrie of Aberdeen finally got to see her husband's grave in August 1939. She told her local newspaper she had married a few days before war broke out in the summer of 1914. Her husband had then volunteered and was killed in July 1917. 'I waited 22 years for it ... All those years I had thought of it. Sometimes I felt I couldn't go to the Belgian war cemetery and at others I had a great desire to see the grave. The other day my great chance came along, and I seized it. I don't regret I went.'[113]

Starting in 1920, a Leeds woman made an annual pilgrimage to her son's grave near Arras. The experience may have been consoling, but it also caused concern, for 'the saddest part of the visit is the homecoming, and the thought – shall I ever see the place again?' as she told a reporter. At first she had to take wreaths of artificial flowers, as nothing grew nearby, but she started making her own as soon as the wildflowers returned. As the trip was her annual holiday, her presence must have become familiar in the small villages around Arras, and the sight of an ageing, lone woman provoked local sympathy. On one of her earliest visits, a local carter offered her a lift to the cemetery. He accompanied her to the grave and returned regularly to place fresh flowers. In 1922, another young man offered to take her, and she was surprised to find he was English.[114] It was, in fact, Philip Vyner, a veteran and battlefield guide, who became the Ypres League's representative in Arras. The ability of such stories to provoke the interest, pity, and wonder of readers made them a staple of press coverage.

But such tales were not inventions, as realized by Mrs E.A. Smith, even if the press devised stock ways in which to tell them. A member of the 1928 British Legion pilgrimage, Smith witnessed something that amazed her. Slumping down on the ground at Beaucourt station on the Somme after a long and tiring day, she saw another woman, who had lost touch with the main party, wandering towards them. 'She was no longer young, and she looked very fatigued,' wrote Smith in her journal, 'but I admired her pluck for she had kept going until the tramping of the day was accomplished. It seemed as if some Divine Power must have been given to these men and women who sort of felt, "Well, they stuck it, and so must we," and so on they trudged.' Someone then pointed out the awful fact that the old lady had lost five sons.

A few days later at the Menin Gate, Smith noted an amazing act of serendipity. There was an elderly widow who had lost her only son at Ypres, his body never discovered. The woman then found herself standing by the panel containing his name, 'and when one tried to realize there are 55,000 names on those panels of men whose burial place were never found it seemed an incident that was more than passing strange. She could have touched it with her left hand as she stood. Her grief must be left to the imagination. Mothers will realize better than anybody how she felt but she made no scene.'

As the wife of a veteran, Smith also had the opportunity of watching her husband's reactions as he and his comrades navigated the battlefields. For most of the time the veterans were jolly and enthusiastically searched out old haunts. However, the commemorative service at the Menin Gate brought forth a very different reaction, which Smith noted carefully in her detailed, handwritten account of the pilgrimage. Recalling the Archbishop of York's service, she noted he ended by referring to those inscribed on the memorial panels: 'There were 6 words only and they must have conveyed unbounded relief and almost joy to stricken hearts. They dropped from his lips slowly and distinctly "But it is well with them."' Following this moving statement, they moved straight into the final hymn, 'well those of us who could. Men openly confessed that they could not sing. Some squared their shoulders and did their best, others stood with bowed heads and made no secret of their tears. The women just did the best they could.' [115] The emotional fireworks of a pilgrimage were spectacular, indeed.

On completing their own acts of remembrance and commemoration those fortunate enough to embark on a pilgrimage took great care to visit graves and sacred sites on behalf of others, whether friends, neighbours, or local community. Visiting the cemetery at Étaples in 1919, Charles Jones and his wife found the grave of their friend and neighbour Mr Smallwood's son.[116] During its summer 1930 pilgrimage, the Whaley Bridge British Legion tour visited twenty-one cemeteries and memorials where relatives or comrades were buried or inscribed on memorial panels.[117] In the autumn of 1923, a reader of the *Berwickshire News* contributed a letter published under the headline, 'A Visit to War Graves by a Berwick Man'. He described the condition of the cemeteries, as well as providing a photograph of a plot containing the graves of local men.[118]

For grieving widows and mothers there was the opportunity of linking themselves to the masculine-military brotherhood of their lost loved one. When Mary Matthews went to France to see the grave of her husband, she also visited the graves of other men of the battalion and then listed them all in a letter to the regimental magazine.[119] The travel agent John Frame took care to find and visit the grave of the son of a close friend and colleague. 'After a considerable amount of searching, we eventually came across poor Deas's grave,' he wrote, and having done so he could provide the comforting message that it was marked by 'a beautiful stone ... suitably inscribed; the little plot in which his remains reposed was beautifully laid out with flowers'.[120]

The pilgrim's desire to reassure others about the wonderful work of the IWGC was an extremely common one, and pilgrims combined it with a desire to act as a surrogate mourner and visitor. 'E.C.F.' made it a point to inform those who had not or could not visit the cemeteries and battlefields: 'For the benefit of relatives who have not visited the cemeteries in France and Belgium, I can state without fear of contradiction that there are no better kept cemeteries, and great credit is due to the work of the Imperial War Graves Commission,' he wrote in the *Ypres Times*.[121] Describing St Pierre Cemetery at Amiens at great length, emphasizing its beauty and the degree of care taken by the gardening staff to ensure its upkeep for readers of her local newspaper, L.M. Orton stated, 'If any parents read these words, I hope they will feel comforted by the knowledge that their loved one is lying reverently and lovingly placed.'[122]

William Miell felt a strong duty to inform others about the power of a war graves pilgrimage after he and his wife made a visit to their son's grave. On their return he wrote an account for his local newspaper before circulating it to several regional journals. His motivation was the hope that it might 'be a solace to the heart or many a sorrowful one mourning the loss of a dear one in France'.[123] Miell's comments had the added authority imparted by his position as a grieving parent and not a well-meaning outsider.[124] The letter had the effect Miell wished, for it provoked a flurry of correspondence from people asking how to arrange visits, while others were anxious to know whether those deemed missing were likely to be traced. Given this response, Miell felt compelled to provide further information. He recommended those interested in making a visit contact the YMCA; the IWGC was the organization for more information on the missing, and he assured everyone that the Commission was meticulous in its investigations. In this letter he provided a particularly detailed description of the cemetery at Terlincthun, its grandeur, beauty, and calm. The wonder of the architectural and horticultural features produced 'a certainty in one's mind that the end of all does not cease with this life', he added. He concluded by returning to his original motivation for writing, to provide some comfort to the bereaved: 'Trusting my feeble letter will produce a brightness in some still gloomy heart and point the way to Hallowed Spots, although across the water "yet for ever England".'[125]

The final element in a pilgrimage was the battlefield tour and the completion of the occasionally hectic itineraries devised by organizers and guides. Perhaps to counterbalance their swift progressions across the battlefields, pilgrimages organized by particular groups usually slowed the pace when visiting sites of specific interest and importance to their collective identities. For Toc H this meant a communion service using the Stone of Remembrance (usually in the Ypres Reservoir Cemetery) as an altar, a visit to Talbot House in Poperinghe, and a visit to Gilbert Talbot's grave at Sanctuary Wood Cemetery, where the pilgrims laid their wreath at the Cross of Sacrifice rather than at Talbot's grave 'for so he would surely have wished it done'.

The veterans were used as guides and interpreters of the landscape, bringing the war years to life for those with no direct knowledge of the battlefields. Although nostalgic, cheery, and light-hearted tales were integral to the veteran testimony, this did not mean they shied away from the appalling realities. In 1926, former army chaplain the Reverend Pat Leonard told the tale of 'Reggie', who 'was seventeen and half' in March 1916. Recounting how 'Reggie' ordered his platoon to take a crater lip, Pat told how the young officer was hit twice, but continued to advance despite his batman, 'who loved him like a son', urging him to seek aid. Pat had then helped carry the severely wounded 'Reggie' to the Bedford House casualty clearing station, but soon after arrival his 'spirit was released'. 'Such was the tale, told on a still, bright Sunday afternoon at St. Eloi, as the Pilgrims sat, quieter than many a Sunday School class, and received their lessons from the Elder Brethren through Pat's lips,' as the pilgrimage's chronicler put it.[126] Here was the reality of war given a particular interpretation. The death of a very young man was freely told but translated into a tale of Christian redemption by a veteran army padre for his mixed audience of ex-servicemen, families, and friends. It was a true pilgrimage tale of a spiritual blessing and enlightenment achieved through the process of travel.

A year later Clayton did something similar, combining his faith and his recollections as a veteran in an address to the pilgrims at Tyne Cot Cemetery. Gathering them around the Cross of Sacrifice, where doubtless a few took the opportunity to sit down on the deeply cut stone steps surrounding it, Clayton described in detail the condition of the 1917 battlefield, trying 'to make them realize the nature of the ground upon which they were looking down'. Another veteran then took up the tale, giving them a vignette of one night in the salient. The account in the Toc H magazine noted, 'No hour during the Pilgrimage spoke so uncompromisingly to some of the pilgrims as the one they spent in the sunshine at Tyne Cot.'[127]

Clayton's pilgrimages were an emotional kaleidoscope dominated by his personality and determination to mix light with shade, fun with deep, meditative reflection, and, he hoped, spiritual nourishment and growth. One moment it meant listening to harrowing stories in Tyne Cot, and the next it was marching back into Ypres cheerily singing the 'absurd old songs ... along the old roads where other men had marched: the Pilgrims' happy foot-soreness was a tiny type of the immense weariness of which the gallantry of those others had made so light.' As they passed through the villages, the local people opened their doors and shutters to hear '*Tipperary* and *Rogerum* again after all these years'. In showing an obvious desire to gain a deeper relationship with the battlefields and ponder its meanings, Clayton ensured the moments of levity could never be confused with the perceived shallowness of the mere tourist. The boundaries between the pilgrim and tourist were carefully policed by Clayton and felt most strongly in the passionate admonition against souvenir hunting: the unthinking tourist might '"souvenir" rusty bayonets, if so minded, but also warily, for the scratch of them is poison. Not every battered "tin hat" is empty, and here you may come upon the rubber ground-sheet in which a British soldier was once buried, a rag of khaki cloth and bare shoulder-blade sticking out of the naked ground – Tom, or Dick, or Harry, "Missing" in the list one day, and still missing from some fireside at home.'[128] In such comments the contrast between the casual tourist blundering through the landscape and the true pilgrim was made most sharply.

Because, it was believed, the true pilgrim immersed themselves in the battlefield, they also learned from it, coming to understand its secrets and former character. A member of the 1928 British Legion pilgrimage, Mrs E.A. Smith, was awed by what she saw at the Newfoundland Memorial Park on the Somme. Standing in the remains of the trenches made her realize, for the first time, the true nature of the fighting: 'Never in my wildest imagination had I imagined anything so positively devilish or fiendish as that which I saw,' she wrote of the ubiquitous remnants of barbed wire strewn everywhere.

Once back at the caretaker's lodge, she stood on the veranda and stared across the park trying 'to visualise the scene as it would have been in the late autumn of 1916 ... A sketch of land where sunshine seemed unknown, a pitiless rain driving in sheets as it would do on that ridge, a vast expanse of mud, shell holes which would have been filled with water, and men fighting for their lives in this seething morass, amidst the fury of the guns, gas, barbed wire and every other fiendish invention. I have no idea whether I stood seconds or minutes, but I came to myself again on hearing a small group of men discussing the position.'[129]

In studying and imagining the battlefields so intently, and thus showing the qualities of the true pilgrim, a broader set of meanings could be percolated from them. For some, the sight of the dead was a call to reconciliation among the nations and lasting peace, as King George V himself had said during his pilgrimage of 1922. A special correspondent of the *Church Times* believed the war cemeteries and battlefields were a perfect place for all to consider the scale of the sacrifice and whether the commitment to peace was a strong as it should be.[130] Thirteen years later, with the international scene looking bleak indeed, the *Daily Express* correspondent, Harold Pemberton, took a two-day motor tour along the Western Front. He said the journey was 'difficult to write about sanely because of the depth of emotion stirred by the vastness of the burial grounds and the richness of the seed they contain'. Reflecting his newspaper's commitment to disarmament and peace, Pemberton believed the 'cemetery-strewn countryside' taught the lesson of the 'futility [and] the madness of the wastage of life'.[131] In 1938, the chronicler of Sheffield's annual battlefield pilgrimage expressed a similar sentiment: 'I feel the pilgrimage re-established our profound gratitude to those who made the supreme sacrifice, and I am quite sure that nothing can do more to give effect to their great desire to ensure there should be no more war.'[132]

Domestic concerns could also be addressed through the medium of pilgrimage. In the autumn of 1919, an editorial in the *Berwickshire News* delivered a socially conservative message to counter unrest and discontent:

> We have returned home from a visit to the devastated areas
> and the thousands of graves of the boys out yonder. We
> have met fathers over there visiting the graves of their only

boys – other parents out searching for a trace of something of their missing ones. We stood the other day near Ypres 'mid the wreck and ruin, and in the wilderness we saw a wooden cross bearing the inscription 'To A Brave British Soldier' – somebody's boy, buried there, unknown, without identification. A strong stout-hearted Britisher looked at it – and said to us 'This is too sad for words,' and broke down in tears. Is it for this we have this miserable upheaval at Home? One father, a fine type of Scotsman, whose only boy sleeps in Flanders, told us that the night before his son was killed in a desperate battle with the Germans, the lad told his batman that if anything happened he would find a letter in his pocket addressed to his father and mother. The letter was found on the dead body, and sent Home. It said, 'If anything happens to me you will know that I fell doing my Duty for the Homeland.' And To-day at Home we have 'The Strike.'[133]

Writing seven years later, soon after the collapse of the General Strike in 1926, the Reverend T.B. Stewart saw in Tyne Cot Cemetery a lesson for all. The uniform headstones making all equal before the Cross of Sacrifice was a call to social unity and the way to ensure the men did not die in vain.[134]

PILGRIMAGES FOR YOUTH

These wider meanings, and the values underpinning them, were thought to be particularly appropriate to children, teenagers, and younger people. A hybrid status was ascribed to these who remembered the war only hazily or not at all. They were not quite full pilgrim, but certainly not the equivalent of Thomas Cook's tourists being sped round the highlights of horror and then home again; they formed their own special subcategory of visitor. The role imposed on younger people, particularly those travelling with their schools or as part of a youth group, was complex. The tour was often thought to be educational, but that was a term carrying many meanings. On the simplest level, it was literally seeing for themselves the places where the war had been fought and through that experience gaining a better understanding of recent history. At the same time, it had a spiritual and moral element. By seeing the battlefields, youth would understand what had been sacrificed for them. According to this outlook knowledge would buttress patriotism, internationalism, and a commitment to peace and pacific ways of dealing with global disputes. None of these tenets or qualities were mutually exclusive and they were often presented in overlapping, intermingled ways. Equally, often younger people were simply exhorted to remember while on the battlefield, with no one making it quite sure what was to be remembered or why. Remembrance and commemorative acts, in and of themselves, were the important things. The rite and ritual were both sign and signified.

Army cadets made regular visits, during which they were exposed to a plethora of messages. The eminent war correspondent Sir Philip Gibbs covered the 1925 tour, deeming it a wholly laudable activity. A cult of forgetting the war had sprung up, he claimed, though noting that this was based on the understandable desire to pass over its miseries. Fearful of boring their children, veterans were not telling their tales, adding to the silence. Given this scenario, the best way to enlighten young people was the battlefield tour. The value of a battlefield tour was great. For the future soldier, gaining an understanding of military operations was educative. However, such visits were much more important than this simple utilitarian demand. The real significance lay in the spiritual enhancement offered. Boys would be taught the value of peace and the desolation of war: seeing its results in the cemeteries, they would be able to heed the warning. Gibbs saw no contradiction in the battlefield tour preparing the young soldier for war while also drilling him in the cause of peace. A highlight of the tour was the unveiling of the 18th Division memorial on the Menin Road, conducted by General Sir Ivor Maxse, a former commander of the division. Maxse addressed the cadets, banging home his message as forcefully as his famed wartime training exercises: the war was fought for liberty and peace and 'it was up to them to take proper advantage of these valuable possessions'. At its conclusion, the boys were taken across the battlefield by expert guides who interpreted the landscape, recounting each and every engagement.[135] Somehow these contradictory bugle calls were accepted as a cohesive message and moral.

The particular rhetorical framework, with its high diction and highly stratified gender roles, was also expressed in the account of a Boy Scout troop visit to Ypres in 1928, seemingly written by one of the adult members of the party. During the conflict, stated the trip's recorder, 'glorious deeds [were] achieved by their kinsmen', who had resolutely held back the invader 'in order that our women-folk, our aged and infants might dwell in peace and security at home'. He then went on to outline a relationship between a group of bubbly, excited young lads with no direct knowledge of the war and the battlefields dotted with memorials, cemeteries, and other reminders of the conflict. Linking the two elements together was their scoutmaster, a veteran, who annotated the landscape through reference to his own memories. The chronicler concluded the account by projecting his own thoughts on to the boys. While saying it was 'impossible to describe each boy's thoughts', he was nonetheless convinced that at the Menin Gate all visualized the flow of soldiers marching to or from the battlefield. This emotional moment was then rounded off by four of their number striding forward to sound the Last Post. The moral of the pilgrimage was to lead lives worthy of such sacrifice and always remember the dead.[136]

Major J.M. West provided an account of the 1934 British public schools' tour. His tone made it sound as if a slightly more exotic Officers' Training Corps (OTC) camp was unfolding, in which the difference between being a soldier and playing at one was blurred. At Vimy Ridge, 'it was amusing to observe how the whole party seemed to disappear beneath the earth's surface and it was only with much persuasion and blowing of whistles that everybody was finally recovered'. West himself seemed to drop into a liminal space, for he noted that after dinner 'an atmosphere singularly reminiscent of the old comradeship of the war' descended. West lived out a highly simplified version of his war experiences with and through the boys under his charge. And, like the chronicler of the 1928 Scout tour, he took the same option of providing a script for their thoughts and emotions: 'What were the younger generation making of it? This perhaps was in their minds – "Here we are on the spot where our elders fought, struggled and died during four long weary years and what is the result of it all?

At home, money troubles and unemployment. Out here, painfully new villages, cemetery after cemetery, endless memorials with their vast lists of "Missing" such as the Menin Gate, Tyne Cot and Thiepval – the tragedy, stupidity and futility of it all.' He then answered all the questions of doubt and disillusionment. He said it was the job of men like him to teach the boys the wondrous stories of those who had fought and died for their freedom, and if they began to grasp that, 'the tour was indeed not made in vain, and further, if our efforts have enabled them to understand that they must work for an honourable peace so that such catastrophe can never be repeated, we may have added another stone to the Peace Structure which this country is striving so earnestly to build'.[137] West here turned veterans into those who held faith with the dead, as John McCrae demanded in his famous poem, and it was youth that needed careful guidance away from the understandable, but nonetheless mistaken, path of disillusionment and doubt about the war, and a crucial method for doing this was the battlefield tour. At the same time, West failed to appreciate two fault lines in his rhetoric. First, that his earlier tone had implied war was a bit of a lark, which young men could readily appreciate if taken back to the original sites. Second, the reason for the war was stripped down to a simple, post hoc justification: it was fought in order to teach humanity the lesson that war was a redundant tool in human interaction. This was a commonly used formula in the twenties and thirties and effectively defined the conflict as an enormous ethical experiment and demonstration put on for the edification of humanity.

The contradictions in the way some veterans presented the war was noted by 'One of the Youngsters' in a letter to *Britannia*, a news, current affairs, arts, and culture magazine aimed at a male readership. The writer had recently returned from a battlefield pilgrimage undertaken to understand the conflict he remembered only vaguely as a very small child, and had stayed at a hotel run by a British ex-officer, who also acted as his guide. The correspondent was somewhat dismayed to find 'the old "Hymn of Hate" being sung lustily' by the veteran. Such displays of 'the old hatreds' made youth's mission to ensure peace infinitely more difficult, he concluded.[138]

These strange concoctions of adventure, sombre remembrance, patriotic zeal, and consideration of war's utility and costs were distilled in a very male atmosphere, but this did not preclude the expression of genuine, personal emotion. Thirty-three members of the Queen's College, Taunton, OTC visited the battlefields in the summer of 1923. At Beaumont-Hamel they walked the battlefield, taking careful note of 'the wonderfully accurate work accomplished by British artillery' on the remnants of the German trenches. But the mood then changed. 'The "Age of Pilgrimage" is not past,' noted the account of the tour, before continuing, 'What more touching scene of the tour than that of the Cadet kneeling before one of the wooden crosses in Hunter's Cemetery at Beaumont-Hamel and placing on the grave of an only brother a wreath of wild flowers plucked, perhaps, from the very ground where that British youth had given his life to rid the world of Prussian militarism, and to vindicate the principles that Christianity holds most dear?'[139] The pain and anguish caused by war was accepted, but just as quickly wrapped in the redemptive tones of Christian sacrifice.

Groups of younger women and girls, such as Girl Guide troops, certainly visited the battlefields, but it would often be a component of a trip to France or Belgium, rather than the central focus.[140] Instead, young women and girls often went as part of a family pilgrimage joining parents, elder (sometimes widowed) sisters and brothers. An intriguing case was Mary Dodge, who joined the Ypres League 1922 pilgrimage along with her mother and fiancé. Their precise reason for going is never made clear in her account, and there was certainly no desire to see a particular grave. Part of the interest may have been to see where her brother, Bill, had fought. Bill was either unable, unwilling, or uninterested enough to join them, but he did send a letter to his sister suppling detailed advice on how to find a trench he remembered well. Drawing solely on his own, wartime, understanding of the landscape, he told her that she could find the spot where he was wounded by walking up 'the track' to Hell Fire Corner, at which point she would find a trench, which 'you can't miss'. In addition, if she crossed the market square directly opposite the Cloth Hall ruins, she would find a cellar covered in corrugated iron, where he was billeted for much of 1917.

Mary reported back enthusiastically in a letter of unwitting comedy and naivety, while also revealing someone determined to understand the landscape. She started by revealing an interesting set of priorities, making sure Bill knew the food was nowhere near the quality of her favourite haunt, the Lyons Corner House on Oxford Street. Fortunately, this did not dim her desire to see the battlefields and carry out Bill's request to find his old trench. Mother, Mary, and fiancé Henry dutifully walked out to Hell Fire Corner: 'Such a quiet, cosy spot. I'd love to spend a whole day there,' she reported utterly without irony. Believing themselves in the right neck of the woods they were delighted to identify his trench, but a local man told them it was an entirely fresh excavation for the new drains.

Equal disappointment was met in the market square. Following Bill's instructions, they thought they had found his old billet, but unfortunately there was now 'a five-storey building on the top of it, and the people wouldn't let us have a look in'. Henry, who presumably was not old enough to have served, nonetheless appointed himself guide, often shattering the illusions of mother and daughter. On being shocked at seeing 'two buildings only half there … Henry said they were half-built, not half-destroyed', which was obviously a great disappointment to them. Henry returned to the role of party-pooper at Ypres station. Greatly excited to see the platforms swarming with soldiers, the two women felt they had been given a glimpse of the war, 'only Henry said he thought they were porters etc., not soldiers'. Perhaps his most crushing intervention came on the ramparts. Staring at a trench, both women 'imagined you standing there in 1917 defending it, only Henry came up and said the Germans didn't get within several miles of the wall'. Deploying sense and insight, Mary struck back with the waspish comment, 'Still, somebody must have stood in those trenches sometime.' Perhaps riled by Henry's stream of somewhat passive-aggressive interventions, Bill's letter back remarked ruefully: 'You should have had somebody that knew the place when you went out. Henry doesn't know anything about Wipers … I'll show it to you myself some day.'[141] Here a mother and daughter went out as dutiful pilgrims. The only difference between them and so many others was that they tried to piece together the experiences of a living, rather than dead, relative.

ENGAGING WITH LOCAL PEOPLE

Circulating around and weaving in and out of the lives of pilgrims while they trod the battlefields and lingered in cemeteries were the people of north-eastern France and Flanders. For all the complaints about the supposedly sharp practices inflicted on pilgrims by exploitative locals, there were as many expressions of gratitude for their help and hospitality, as well as admiration for their fortitude in repairing and restoring their communities. In the summer of 1922, Captain Mee traversed the still deeply scarred and torn Somme battlefields. Seeing an old couple living in a shanty of old boards crudely nailed together, he was touched by their 'cheery and kindly' responses to his enquiries about the neighbourhood.[142] He may well have mentioned his veteran status, which usually helped elicit a warm response. The mayor of Dixmude greeted the many veterans on the Leyland Motor Company tour with great enthusiasm, stressing the gratitude of the citizens towards their British friends who had fought so hard for their freedom.[143] When the men of the 1/7 Northumberland Fusiliers association stumbled across a wedding procession in Notre Dame de Lorette, they stopped to take photographs. Warmly welcomed by the party, they were encouraged to return the next day. On doing so they were feted and 'entertained like royalty'.[144]

Pleasure at their presence was something many pilgrims realized after even the most cursory of chats with French and Belgian people. Every year Sheffield's delegation was greeted warmly by the villages the Yorkshire city had so generously supported back to life through the British League of Help scheme, under which British communities adopted a devastated village, town, or city in France.[145] According to the *Sheffield Daily Telegraph*, the link was a very intimate and local one, which might not be understood by the wider community. 'Whatever sophisticated Frenchmen in front of cafes on Parisian boulevards may say about the English, these simple, shy peasants, who have seen their homes smashed and fields devastated in the cruel times of war, are very grateful to Britain for her generosity,' it stated.[146] A structure and hierarchy of relationships was outlined in this construction. It implied that the people of Sheffield, and by extension, Britain, were more magnanimous and certainly more sincere and genuine than the Parisian stereotype allowed: on the battlefields the real souls of France and Britain met.

Local reverence for the British cemeteries and memorials also reassured and inspired pilgrims. A member of the 85 Field Ambulance association pilgrimage was surprised at just how many Belgian families visited the IWGC cemeteries at the weekend. Although it was treated as an afternoon's outing, it was one in which nothing but the deepest respect was shown.[147] The Reverend T.B. Stewart Thomson witnessed a scene of even greater emotional intensity in a war cemetery. An old British lady placing flowers on her son's grave was noticed by an equally elderly French woman. The French woman (who had also lost a son) told the British woman she had been present on the day of his internment in the cemetery and recounted the story of the funeral service. 'Her statement was translated and the two bereaved parents wept in each other's arms. Not a word was or could be spoken; but the universal language of sorrow and love sufficed,' Stewart Thomson recorded.[148]

Engaging with local people was, then, the path to deeper understandings and a true education. Western Front veteran 'J.B.M.' spent much of his time feeling slighted and ignored by the French and Belgian people he met on his summer 1930 tour. Then an old lady held up a metaphorical mirror that shocked him. She said she well remembered British soldiers, and had seen lots of British tourists, but she found it so odd that Britain was now so indifferent to France and seemed much more interested in being friends with the Germans who had caused such misery to 'poor beautiful France that has suffered so much'.[149] He tried to explain to her that Britain would never forget what it had done in the war, and pointed to the cemeteries as evidence, to which her reply was that Britain was more interested in trade and money than anything else. 'J.B.M' was clearly nonplussed by this reversal of opinions and roles: for the British, the usual response was that rapacious locals saw nothing in the memory of the war bar making money from gullible and pathetic British visitors. Here, a Belgian woman delivered the opposite message: they were the true guardians of the meaning and memory of the war, a memory abandoned by Britain as inconvenient to its current conditions. The pilgrim was given another way to consider the phrase 'Their Name Liveth for Evermore'.

*

Even though they often did exactly the same things as battlefield tourists, the pilgrim remained a figure apart. Whether any were left unmoved, angered, frustrated, disconsolate, or merely hollow following a pilgrimage is less clear. Some were disappointed at what they perceived to be the indifference of others, particularly locals, and stated it. As for those who might have felt a sense of alienation by the entire experience, it is impossible to tell. What is obvious from the testimonies of so many is that they found pilgrimage rewarding. Painful and difficult it might have been, but the trial was worth it. Tied to the soil of France and Belgium through the blood and body, present or missing, of a loved one, the pilgrim moved through the landscape with the simple, overriding focus of seeing the grave or site of commemoration. Everything else was the frame and backdrop for the primary act of personal remembrance. Cemeteries, memorials, and flowers were the stage and props for the act, and most pilgrims found that stage, created with such genius and care by the

IWGC, powerful and fitting to a degree they never thought possible. Grief, consolation, even pride and joy, flowed from those moments of meeting between pilgrim and the grave or inscribed name. For those who organized the pilgrimages, the culminating moment of union between the living and the dead was worth every effort and they committed themselves to the cause of the pilgrim with tireless zeal and energy. Redemption and hope were found in pilgrimage, they argued with utter sincerity. Pilgrimage gave the bereaved the chance to live out the exhortation inscribed on a headstone by one mourning family: 'Tread softly our hero sleeps here.'

7

Postcards from Tourists

Madame, please,
You are requested kindly not to touch
Or take away the Company's property
As souvenirs; you'll find we have on sale
A large variety, all guaranteed ...
The *path*, sir, *please*
The ground which was secured at great expense
The Company keeps absolutely untouched,
And in that dug-out (genuine) we provide
Refreshments at a reasonable rate.

From Philip Johnstone, 'High Wood' (1918)

Beyond Hooge the road passes through the 'Tank Cemetery,' where 14 tanks lie embedded, just as they were put out of action.

Muirhead's Belgium and the Western Front (1920)

Make no mistake about it, this war area is the most fascinating corner of Europe ... It provides a succession of thrills to anyone who has an ounce of imagination. It will inculcate a store of memories which will abide long after the more transient charms of natural scenic beauty have passed from the mind.

Bernard Newman, *Cycling in France (Northern)* (1936)

Among the motivations for visiting the battlefields was sheer curiosity (morbid or otherwise). For those many millions whose war experience never entailed any kind of uniformed service, let alone proximity to the actual front lines, the battlefields were both a constant motif in their lives and a deeply intriguing and mysterious presence. People read about the battlefields continually, they saw fragments of them on cinema screens, and they flocked in huge numbers to see photographic exhibitions depicting scenes from the trenches and the landscapes of the Western Front, all of which provoked the desire to see the real thing. However, with so many of the bereaved demanding the right to be prioritized as visitors, friction was caused almost immediately by reports of 'tourists', or worse still, 'sightseers', flocking to the battlefields. All visitors to the battlefields, therefore, had to expect questioning of their motivations for going and their behaviour while in France and Belgium to be discussed in great detail.

A categorization scheme for visitors based on a hierarchy of moral and spiritual worth rapidly crystallized into labelling people either as a pilgrim (good) or tourist (bad). The people of the devastated areas were dragged into this debate. As they returned to their homes, they realized that the influx of visitors had provided a new commercial opportunity ready for exploitation. Unfortunately for these people desperately seeking to rebuild their lives and reacting to a market largely created by the British, this also meant that they, in turn, could be labelled, and rather crudely, too, as either decent sympathizers or wicked exploiters and shysters shamelessly making money from death and misery. The public debate over the nature of the battlefield visitor, and those who provided the facilities for them, after reaching its height in the first five years or so after the war, lingered throughout the twenties and thirties. The people visiting the battlefields were deemed to be on holy ground, and so their behaviour had to reflect that fact. Whether tramping the marshy, desolate no man's land or the perfect green swards of an Imperial War Graves Commission (IWGC) cemetery, the visitor was expected to tread carefully.

In anticipation of the restoration of normal travel conditions, the spring of 1919 saw intense speculation as to how the battlefields would be presented. According to the journal *Answers*, companies were being 'formed to "exploit" the battle zone'. Allegedly, Ypres was to be left in ruins but Arras restored, except for its cathedral, which was to remain in its battered condition as a memorial. The Butte de Warlencourt, a fiercely contested ancient, man-made mound on the Somme battlefield, was to be fenced in, ensuring the preservation of 'that tragic spot', complete with its dugouts, trenches, and shell holes. It was even rumoured that a company had been formed to buy up half a million shell cases, intent on converting them into souvenirs. Clearly expecting a deluge of cheap and nasty diversions, the *Answers*'s columnist archly noted that nothing would trouble the '"joy-riders"' other than the inconvenience of crowded hotels.[1]

The esteemed war correspondent Sir Philip Gibbs waded in, reporting in September 1919 French national tourist office plans for 'ten huge hotels in the form of wooden barracks at different "key points" of the battlefields for the accommodation of the invasion of tourists which is anticipated next spring'. He believed that fleets of motorcars were being organized to help take people across the devastated regions, and some FF30 million was going to be invested. The idea of such activity made him 'shudder selfishly at the thought of those tourist mobs'.[2] *Country Life* also speculated on the birth of a new kind of tourism, and like *Answers*, claimed the Butte de Warlencourt was to be preserved, as were the ruins of Bapaume, and the Thiepval Chateau. Unlike Gibbs or the *Answers* correspondent, the *Country Life* writer saw such a decision as spiritually enriching: 'As they now stand, only changed and softened a little by kindly nature, they will make lighter the task of the historian and help generations of Englishmen yet unborn to realise something of what their fathers did.'[3]

As the speculation intensified, all sorts of fantastic schemes were reported, often based on the flimsiest of evidence or rumours. In the spring of 1920, it was reported that a company based at Le Bourget airfield was about to launch aerial tours of the battlefields. Two flight itineraries were to be offered, the story alleged; the first would cover Lille, Arras, Cambrai, St Quentin, and the Somme battlefields, while the second was for Château-Thierry to Rheims, including a pass over Soissons and the Chemin des Dames. Americans were expected to be pioneers of such tourist trails across the battlefields, and the commander of America's Expeditionary Force, General Pershing, advocated such visits to his compatriots.

In the whirl of excitement aroused by the prospect, the only possible downside was the lack of facilities. Interviewed in the weekly journal *Opinion*, the director of the French Trans-Atlantic Company stated the number of Americans desirous of seeing the battlefields far exceeded the infrastructure available. Victor Cambon, a senior French engineer, told the same publication that every American he met wanted to visit the scenes of fighting, 'but he warns Americans are exacting over comforts in the hotels they visit, and believes that there is not sufficient hotel accommodation in France at the present time to care for even the first contingent of tourists'.[4] Another report told of Franco-American collaboration to ensure a smooth tourist trail across the Western Front, with railways hurriedly repaired, temporary hotels established, and suitable guides identified.[5] F. Heathcote Briant, Western Front veteran and journalist, wrote of an American syndicate scouting out sites around Ypres and Arras for 'monster hotels'.[6] According to the *Aberdeen Daily Journal*'s correspondent, Americans would be the catalyst for changes to the visitor infrastructure across the former battlefields. 'We believe that by next spring [1920], when the Americans will come over in their hordes, better arrangements will have matured; in the meantime, for comfort and completeness, Cook's hold the field.'[7]

American soldiers and veterans picked up on these stories, and a magazine for recently demobilized American soldiers gently lampooned the likely accommodation of the first tourists. Referring to the crude shelter available in the devastated region, it joked, 'Dad will have a first-floor chamber de luxe with all kinds of exposures and thorough ventilation, and Mother's room, en suite, will be adorned with furniture of the Jean Doughboy period.' The piece went on to explain the actual situation: it was the jest realized. Lacking an alternative, French hoteliers were using abandoned YMCA, Red Cross, and other huts, many of which were constructed of iron sheeting, as a temporary measure. These were to be divided into separate cubicles to make 'as many bedrooms as possible ... They will not be luxurious, but they will be serviceable,' it added.[8]

THE EMERGENCE OF A TOURIST INDUSTRY

With hotels being improvised across the battlefields, a commercially viable tourist industry was up and running within a remarkably short space of time, and a crucial element was the battlefield guidebook. Famously, Michelin had got in on the act during the war itself, producing its first guides in 1917. As soon as the war ended, the company expanded the list, and all the major battlefields were covered by 1920. The books were a curious mixture of describing what was once there and what had replaced it. This shrewd move allowed the company to draw upon its existing catalogue for text on famous buildings or cities with slight modification to suit the current, highly peculiar, conditions.

Typifying this style was the guide to Ypres, which included a detailed description of the Cloth Hall with careful interpretation of its architectural features, but placed everything in the past tense, and prefaced it with the statement, 'Nothing now remains but a heap of ruins.'[9] In its own way Michelin was simply adapting its established business practices for a new market, and they were interpreted in that manner, as part of a continuum. *Country Life*'s review of Michelin's new guide to the Somme was written in exactly the same style as it used for other tourist guidebooks, praising its useful notes, well-selected itineraries, valuable illustrations, and handy size.[10] The emergence of this hybrid approach from the big players meant they could capitalize on their credentials. In its review of *Muirhead's Belgium and the Western Front* in November 1920, *Athenaeum* magazine praised the work for its impartiality and exclusion of advertisements. This editorial independence, honesty, and integrity freed the book from 'a very grave' danger, for 'we ourselves could frame the most misleading and magnificent advertisements did we own a sandbag cabaret in the old Bund dug-outs at Zillebeke Lake, or a cellar hotel near the Thiepval château', it told its readers with a hint of arch humour.[11] In reviewing such a publication in such a manner, *The Athenaeum* also revealed the broad parameters of discussion about battlefield tourism. Here was acceptance of the need for authoritative travel guides while also believing it was perfectly acceptable to discuss them in a light-hearted manner.

At the same time, the emergence of the battlefield guide allowed their authors, and reviewers, to make statements on the intended readership and how they should go about any tour of the devastated regions. When reviewing Atherton Fleming's *How to See the Battlefields* in the autumn of 1919, the *Graphic* saw it as part of a natural process. 'The inevitable battlefield tours have begun' said its reviewer, and given this inevitability, Fleming's 'excellent book should be a part of every pilgrim's kit'.[12] The *Tatler*'s comments were of a similar nature. Battlefield tourism was bound to happen, and therefore so was the battlefield guidebook: 'Now that "visiting the Front" will be undertaken by tourists rather in the same way as they "do" Paris, guidebooks of Northern France and Flanders are inevitable,' it stated. As with the *Graphic*, the *Tatler*'s reviewer was very impressed by Fleming's book, and found the 'little humorous touches' made it 'delightful to read'. Beyond the reader intent on using it as a practical tool, an armchair audience was also guaranteed. These were readers with not 'the faintest chance of visiting the battlefields until they have been, as it were, cleaned and garnished ready for the first army of Cooks and Cookesses who are already preparing their hideous equipment'.[13] Having acknowledged the likely demand for battlefield tours *de luxe*, the *Tatler*'s reviewer concluded on a regretful note. If they were transformed from their 'raw' condition into a tourist show, the battlefields would lose their moral and spiritual integrity and effect. Such views may have been received sympathetically by many, but they also revealed a typically British perspective, for it in no way countenanced the idea that the battlefields might be transformed by the original inhabitants wishing to rebuild their world.

Showing an awareness of different perspectives, as well as a very keen eye on the market, the veteran writer on all things French, Sommerville Story, quickly produced a range of booklet guides relating to each major battlefield. Taking great care to identify his target audience, he stated:

> I am not addressing these remarks to those who have a special reason, as many of us have, for visiting some specific spot or spots where they or their friends have spent months in dugouts or billets or the scene of any particular engagement, but to the man or woman who has not too much time to spend and desires to acquaint himself or herself at first hand – as all should do – with the varied scenes of the terrible struggle for right and freedom.

Story thus acknowledged the status of a superior form of visitor without in any way invalidating or condemning the motivations of those wishing to go for other reasons. Having made his distinction, Story felt able to address the tourist, and he did so with skill and subtlety. Stripping such a visitor of any residual sense of guilt at the manner in which they wished to explore the battlefields, he sympathized with their position and assured them of his ability to interpret their needs: 'We may start with heroic ideas of doing it all thoroughly, but the courage of most of us will ooze away when we behold these long stretches of bad road and the monotonous miles of trenches and dugouts, interspersed with ruined villages – all extremely interesting, though often, it must be confessed, lacking in variety. Furthermore, it is all very difficult to follow by one self [*sic*] or to recognise landmarks unaided.'[14] No one need be ashamed of their desire to see the battlefields, or their objection to roughing it, was Story's message.

With tourists piling over to the battlefields long before the official travel restrictions were lifted, and many of them wishing to proceed according to the Story model, apt behaviour was a subject of discussion from the start. In March 1919, the *Daily Express* railed against those desiring a frisson of morbid pleasure from the experience. Appalled by the thought that people might find a macabre thrill in seeing exhumation teams at work, it warned, 'If the battlefields must be – as no doubt they must – a subject for sightseeing, the actual horrors of war cannot be made the object of ghoulish survey.'[15]

Within another couple of months the *Graphic* was able to produce a sketch that almost allegorized the debate: it depicted large numbers of people picking their way over a devastated landscape, with each person made a representative of the different motivations for visiting. The bereaved were identified in a woman and her child clad in mourning black approaching a row of ragged wooden crosses. At another cross, a man and woman are on their knees staring at the inscription. Near them is a man standing in a shell hole studying a map, who is possibly a veteran revisiting old ground; another man could also fit that description, for he is perfectly equipped for the task of clambering around a battlefield, shod in good boots and carrying a sturdy rucksack and binoculars. Others are clearly tourists, wandering along the trenches and poking around for things of interest.

The image was labelled 'The new invasion of the battlefields', and catalogued the different visitors. 'Some go to the battlefields for the purpose of picking up souvenirs, others in order that they may better understand the nature of the conflict which at long last ended with the victory of Right over Might. Pathetic is the sight of the mother and the widow walking sadly in the cemeteries searching for the graves of their loved ones, or mournfully surveying the scene where the brave warriors fell.'[16] All types and conditions of people were on the battlefield, according to the *Graphic*. And, it was argued by some, among those types were some who simply did not deport themselves in a fit and proper manner.

By the autumn of 1919 newspapers were reporting acts of sacrilege. The *Manchester Guardian* told of women removing helmets placed on the crosses of soldiers' graves and 'that parties of tourists have been met boasting of their trophies'. Such behaviour seemed to prove there was 'no law of decency governing the tourist who is also a confirmed souvenir-hunter'. They were deemed Jekyll and

Hyde characters, who were most likely 'excellent citizens' at home, but 'when they are souvenir-hunting they are entirely different people'.[17] As women were expected to be bereaved mothers, wives, fiancées, and sisters, they were also expected to behave according to the highest standards of decorum and respect. The article implied that in collecting souvenirs such women had shamed their sex and their nation.

In November 1919, the issue of visitor behaviour was put on the agenda of the IWGC's monthly meeting. The commissioners were rattled by a series of stories and letters in the press complaining about the behaviour and activities of visitors. Typical was Frank Heathcote Briant's impassioned piece for the *Daily Mirror* in which he reported motor-camping parties working their way round the Somme battlefields. Full of anger, bitterness, and anguish, he wrote, 'I cannot bear to think of the Devil's Mile that runs from Shrapnel Corner to where Zillebeke once stood as the happy rendezvous of char-a-bancs; of laughing crowds at Mouquet Farm; or morbid sightseers among the unknown graves at Thiepval.'[18]

Seemingly inspired by this story, the *Nottingham Evening Post* carried a front-page story claiming that some people were camping on or near grave sites, although no eyewitness statements were quoted to substantiate the claim.[19]

Quizzed by the commissioners as to whether there was any hard evidence to back these statements, Ware said that, other than stories of some picnicking in Delville Wood, there was very little else. Reports back from Graves Registration Unit and IWGC teams on the ground noted 'just mere joy-riding and sight-seeing' with only the odd helmet stolen from a grave as a souvenir. Ware turned to Major Thomas Nangle, the Newfoundland representative who was in almost constant transit up and down the old front line overseeing burial concentration work, for his expert opinion. Nangle stated that he had seen nothing of concern, even when on the previous Sunday some 7,000 people were in Ypres. (It was very probably the first Armistice Day anniversary that drew people to Ypres in such numbers in deep autumn.) Given these comments, Ware thought the press coverage was much exaggerated, but he was nonetheless concerned. Seeking to cool temperatures and provide advice on correct behaviour, he believed a press statement was a useful way forward. Ware was fortunate in having Rudyard Kipling as a member of the IWGC, and he promptly volunteered to write a letter for circulation through the Press Association. As a bereaved father who had heard stories of picnicking on the battlefields, 'which looked to me very much like levity on the part of visitors', he was keen to take on the job.[20] Producing a piece of great delicacy and subtlety by avoiding direct confrontation and accusation, Kipling outlined the differences between the pilgrim and the visitor, explaining the feelings of the bereaved and, through this perspective, gently appealing for considerate and respectful behaviour:

> A number of people from all parts of the world are visiting, and there is every sign that an immense number may be expected next year to visit, the French and Flanders fronts and the cemeteries behind. A proportion of these visitors will be relatives and next-of-kin to the dead, whose pilgrimages will be made with heavy hearts, but very many others will be drawn by curiosity and a natural interest in historical ground. But that ground, it should be remembered, is also holy – consecrated in every part by the freely-offered lives of men, and for that reason not to be overrun with levity.
>
> It is inevitable that the handling of such multitudes of sightseers must be managed on ordinary tourist lines, so it rests with the individual tourist to have respect for the spirit that lies upon all that land of desolation and to walk through it with reverence. It is said that there is a tendency among some visitors to forget this obligation. Nothing would be gained by giving specific instances of what, after all, is more in the nature of unthinking carelessness than any intentional disrespect; but the Imperial War Graves Commission have asked me to express our most earnest hope that all who visit the battle areas will bear in mind that, at every step, they are in the presence of those dead through the merits of whose sacrifice they enjoy their present life and whatever measure of freedom is theirs to-day.[21]

As was intended, Kipling's letter was immediately picked up by the press, giving his comments extensive circulation.[22]

Regardless of the degree of evidence, there was a perception of inappropriate behaviour, which many found troubling. As a result,

Figure 7.1 Tourist excesses, Thiepval, c. 1920.

a distinct division between the tourist and the pilgrim was opening up, and as it did so emotions heightened. The veteran W.A. Allinson delineated clearly between tourist and pilgrim. Referring to Ypres, he believed the tourist was incapable of truly engaging with the city, especially once rebuilding commenced. Only those with a deeper, emotional attachment to Ypres could actually get in touch with its true nature and self: 'Ypres conveys nothing to the tourists who come, gaze up at the Cloth Hall, and pass on. Widows and orphans come, and weep over the grave of the loved one – *they* [original emphasis] realise.'[23] E.F. Williams agreed. He wrote of watching tourists swarm around in Ypres's central square being shown around the Cloth Hall ruins by 'a raucous-voiced guide' pointing out the obvious and adding nothing of value, while battlefield tours clattered out through the Menin Gate in motorbuses, 'and when they get back home they think they have seen Ypres and the Salient, and perhaps begin to wonder what all the fuss was about'.

According to Williams, such tours represented an utter failure to engage with the battlefields because it allowed no time to contemplate the scale of the sacrifice. 'This is decidedly not the way to visit the glorious resting-place of a quarter of a million of the Empire's dead,' he wrote, adding, 'for what do they know of Ypres, who only the Grand Place know?'[24] One veteran even believed a group of visitors to be on the tipsy side, which very much angered him.[25] Even when the visitors were highly motivated and intent on gaining insight from the experience, the mode of expressing it could imply something of the sheer wonder at being in such a strange place, which had the potential to create misunderstandings with others. Neville Chamberlain told his sisters that the Somme battlefields are 'rather horrible, but desperately interesting', and regarded it as 'thrilling to see the actual spots about which one had so often read'.[26] The battlefields were a place of weird wonder, and visitors could not help saying so, even if this might seem offensive or inappropriate to others, especially veterans.

Ferdinand Tuohy, a veteran and writer, was at his happiest when bad weather drove tourists inside, leaving him and his chums to tramp along in communion with the landscape made sacred once again by its loneliness. By contrast, when the sun was out, it was a crush of visitors down from Ostend packed into charabancs displaying their 'rubbernecks (cameras, giggles, guides and all)' as they craned round to glimpse the sites while zooming along to the next cafe.[27] He was reminded of a prophetic poem he had come across in 1915:

> And when o'er ruins we know so well,
> Night in Flanders has come to dwell,
> The Ypres Astoria will ring,
> With furores from the Ragtime King.[28]

Even worse for some was the idea of making the battlefields but a small component in a holiday tour. Reducing *the* destination to a mere stop was no way to appreciate a sacred site, according to this argument. And yet that is precisely what happened, and with equal rapidity. In the summer of 1921, the *Graphic* reviewed northern France as a holiday destination. Just one among the many distractions the region offered was the battlefields, easily accessible by the fully restored northern French railway system.[29] A year later, *Country Life* carried an advert for Ostend. Highlights of holidaying in the seaside resort included polo, yachting, sea-bathing, tennis, golf, fencing, clay pigeon shooting, and the concert hall, as well as viewing the summer residence of the Belgian royal family. All of this added up to a wonderful package, as the advert underlined: 'Go to Ostend for happiness, for gaiety, for health, where all the pleasures of peace are to be found and where may be seen nearby the ravages of war and the battle sites of some of the greatest and most glorious engagements in British military history.' Moreover, these were accessible through a short journey on the daily excursions provided by the road, rail, and tramway services.[30]

Firmly placing the battlefields into a wider package, the *Aberdeen Daily Journal*'s correspondent praised Thomas Cook for realizing 'that too much battlefield day after day (and they are long days) would pall upon the visitor, and so the visits ... are interspersed with visits to the quaint, historic towns of Bruges and Ghent, as well as to Brussels and Antwerp'.[31] However, such diversions were by no means the preserve of the tourist; even those who would be considered pilgrims managed to wedge in alternative delights. While the main focus of Nellie Burrin's trip was to visit her brother's grave, something she did twice in three days, she made a day in Paris an integral part of her itinerary.[32] Veterans, too, were susceptible to the call of other attractions. The men of the 1/7 Northumberland

Fusiliers association provided one free day on their four-day tour, which some used for further battlefield exploration although others went off to Paris or Lille.[33]

DELINEATING BETWEEN TOURIST AND PILGRIM

For those convinced of watertight compartments dividing tourists, and their behaviour, from respectful pilgrims, there was the easy target of the Americans. Americans were rich, vulgar, unthinking, and utterly untroubled by even the slightest scintilla of self-knowledge, according to the definition so beloved by the British. Americans were seen as egregious exponents of all that was most reprehensible about battlefield tourism, and with them the cliché of the tourist grew up. They were caricatured as displaying the extremes of insensitive behaviour: obsessively ticking off sites with little attempt to understand their significance, snapping them with equal fervour, buying up tasteless souvenirs with alacrity. The stereotype of such tourists allowed everyone off the hook, and even more so when the loud, uncouth behaviour was firmly identified with Americans. American visitors were most definitely tourists and most definitely not pilgrims.

From the moment hostilities ceased, expectations of an American deluge were high, and they increased with the ending of travel restrictions. Meeting in November 1919, the IWGC's commissioners discussed the issue in an atmosphere of some alarm and anxiety. French concerns over the large numbers visiting the devastated zone, and the pressure this was putting on food supplies and accommodation, were noted. It was a situation expected to worsen considerably with the rumoured arrival of one and a half million Americans from the following spring.[34] The prophesy appeared to come true. The newspaper for American forces on the Rhine reported the hills around Verdun as 'literally black with thousands of visitors, a majority of whom are Americans'.[35] For the American military this influx added up to quite an embarrassment. 'There are millionaires and their wives sparkling in their gems that the profits of war have brought', noted an article in the *Watch on the Rhine*. Dismissed with almost equal disgust were the congressmen wearing scraps of military uniform, as well as 'Percy', the 'erstwhile conscientious objector'. With withering contempt, the piece sneered at their activities on the battlefields. Such visitors did nothing but look for souvenirs and curious subjects for their cameras, while imagining themselves as gallant soldiers leading their men to victory, it was contended. In sharp contrast was the pathetic sight of the 'sad-faced, silver-haired mothers who knew only the losses of war, and who make the pilgrimage of devotion to little wooden crosses in the wheatfields'.[36]

With the Americans established as the least worthy visitors, the opportunity to have a chuckle, or morally superior tut, at their ignorance was often gleefully accepted by the British. While accompanying a large Scottish pilgrimage in the summer of 1926, the Reverend T.B. Stewart mocked the naivety of American tourists. He told of one amazed that the Germans had managed to destroy the Cloth Hall but nothing else around it during the course of their bombardments. The tourist had, of course, totally misunderstood the pace of reconstruction in the city and the difficulty of rebuilding something as large as the Cloth Hall, which was why it alone remained in its wartime condition.[37]

Given his generally dyspeptic nature, it is little surprise that Henry Williamson took very great pleasure in making acid comments about Americans. Among his anecdotes was one concerning an exchange with an American tourist who wandered over while he was examining the German howitzer on display in Ypres' station square. The man asked what it was, and so Williamson provided an explanation and directed him to a better example on the other side of the square. At this suggestion, the tourist said, 'This one will do … I've only got to see Hill 60, the holes at Messines and the Bloodchapel at Bruges, and the Death Trench, then I'm through,' before hurrying away. That Americans came simply to canter through the sights with no intellectual or spiritual engagement, was Williamson's unmistakeable message.[38]

While such clichés made British visitors feel terribly superior over seemingly jolly dim and uncouth Americans, they ignored just how many Britons saw no problem in treating a battlefield tour as 'a bit of a lark'. In the mid-thirties, W.S. Sanderson, a prominent local businessman and councillor in Morpeth, started organizing excursions for his fellow locals. To the carnivals, races, and other days out, he added a Whitsun battlefield tour, which was a great success. The atmosphere of the trip was described in the coverage provided by the local paper. The headlines for the 1938 tour read, 'Ald. W.S. Sanderson's Whitsuntide Tour/Prospects of an Enjoyable Outing', before going on to advertise his credentials 'as an organiser of events which provide pleasure for others'. After that, the praise kept coming. Sanderson was 'without peer in the North of England ... And undoubtedly, as the pioneer of the week-end Tours of the Battlefields Alderman Sanderson stands alone, and those who journeyed by train and boat have very vivid recollections of those very enjoyable Whitsuntide outings'.[39] Just in case anyone was concerned about the atmosphere of the excursion, the *Morpeth Herald* was able to state that 'the trip from the north-east has established a reputation for cheery good fellowship', and a 'battlefields visit [w]as an enjoyable and memorable way of spending the Whitsuntide holiday'.[40] If you fancied a laugh, then a Sanderson trip to the battlefields was clearly the thing. As the Sanderson tours show, crude stereotyping of different nationalities and different types of visitors did not reflect the complexity of the situation or the ambiguities of status.

By the 1930s battlefield tourism was so ubiquitous in British culture that it had matured into a subject suitable for comic treatment. Laurence Kirk produced the short story 'Charabang! And This, Ladies and Gentlemen, Is the Strange History of Mr Crummy!' for *Nash's Pall Mall Magazine.* The story followed the experiences of Mr and Mrs Crummy on a 'Pinker's Historical Tours' eight-day excursion to the battlefields of Flanders and other famous Belgian sites. Captain Lancing led the tour on behalf of Pinker's; indeed, 'all Mr Pinker's battlefield tours had a conductor with Army rank'. Obviously, the veteran status was regarded as the seal of authenticity but, it is explained, Lancing had earned his officer rank while serving in the Labour Corps at Le Havre and had never been near the firing line. The tour quickly descends into farce as Mr Crummy succumbs to lumbago and the travellers indulge in ceaseless gossip, argue among themselves, get ticked-off by Lancing, and visit dull and bogus places. During the course of the visit none of them actually see a battlefield or visit a cemetery or war memorial, but they don't appear to notice, being far too intent on their group affairs. On returning to London, Mr Crummy tells Captain Lancing, 'I hope I haven't caused any inconvenience,' which causes the guide merely to look 'towards his Maker'.[41] Mr Crummy and his party were most definitely not pilgrims.

THE BLURRED LINE BETWEEN TOURIST AND PILGRIM BEHAVIOURS

In reality, virtually all visitors (then and now) alternated between tourist and pilgrim because it was quite simply impossible to do otherwise, as, regardless of how they viewed themselves, or were perceived by others, they at least shared a similar basic set of needs: eating, drinking, and accommodation. Although the detail of precisely what was eaten and drunk and where the person slept differed hugely according to wealth and precise intention in visiting, the essential similarity remained. When it came to food options, the guidebooks made suggestions and offered advice. With this advice came a set of implicit assumptions. The person really committed to seeing the battlefields would be prepared to rough it that bit more, whereas the sightseer might want something slightly different. Either way, planning was important, especially in the early days when the region was still in recovery mode.

Lowe's advice was extremely spartan. The 'inner man', as he called it, could be sustained on sandwiches and something to drink, but it might also be wise to make arrangements for a 'light midday meal'.[42] Coop's guide meanwhile was a little more sybaritic, but started by hinting at possible dangers, noting 'accidents and contingencies may arise when a supply of food is most welcome, and an emergency ration, therefore, should invariably be carried'. He recommended making arrangements with the hotel for a packed lunch, but then provided a series of cautions seemingly based on his own experiences. One hotel provided a wonderful lunch, but no crockery or cutlery, and another a quantity of eggs, 'which were intended to be hard-boiled but had just missed'. It was therefore advisable to inspect the provisions pack carefully before setting off. In addition, he believed it wise to carry a Thermos flask, a tin-opener, and a corkscrew to 'prevent subsequent exasperation'. 'Little details like these are apt to make the difference to the comfort of a day's journey in the battle area,' he added.[43]

Baroness Campbell's experiences revealed both the wisdom and sentiments behind such advice. During the course of her trip, she had the daily task of collecting food supplies for each expedition across the battlefields. In Soissons she could not source butter, which apparently had been out of stock in the city for a fortnight. Finding suitable places to eat once out in the field proved a challenge.

On one occasion she and her party ate lunch in a ruined house on the Somme, and at Albert they opened their luncheon baskets while perched on the ruins of the basilica.[44] It was a scene visualized by the *Illustrated London News* under the headline 'Picnics on the Old Front: Motor Tours of the Battlefields', captioned: 'With an old ammunition box as table: a party of sightseers [lunches] in a camouflaged shelter on the former battle front in France.'[45] The sketch provides some fascinating insights into how early battlefield visiting was conceived. The party consists of four people, at least three of whom are well-dressed women, while the fourth may well be a male and is possibly in a subordinate position as chauffeur or hired driver. A car sits in the road with three of the four taking lunch in the makeshift shelter with its remnants of camouflage netting still stretched over it. A tall bottle sits in the middle of the ammunition box, implying wine is being taken with the meal. The fourth woman, dressed in fashionable coat and hat, stands in the foreground staring across the landscape; she is placed next to a fallen shell-blasted tree, which was as much of a visual shorthand for the front as the trenches. Class status and wealth are obvious, and there are the clear overtones of the tourist rather than the pilgrim.

Tourist behaviour and status were revealed through other traits, among which was the desire to photograph everything. The guidebooks recognized this, and indeed some advised the visitor to create a photographic record. Both Coop and Lowe agreed that a camera was a necessity, and in the process exposed just how important the photograph had become as both memento and authentic relic of the experience.[46] Atherton Fleming concurred with his fellow guidebook writers, stating: 'I suppose the majority of tourists will want a photographic record of the trip. It is, indeed, a great pity that the British troops were not allowed to use cameras during the war, as no amount of imagination can picture some of the places or the conditions under which our troops worked and fought.'[47] Alice Knight lived out this advice to the full. Virtually every letter she sent to her parents recorded details of more snaps added to her collection. 'Believe me I've spent a young fortune in photography since I've been over here; but I feel they're the best souvenir anyone could have,' she told them in November 1919.[48]

Indulging in the desire to record in this way required some planning in the earliest days due to the conditions in the devastated zone. Both Coop and Fleming urged visitors to stock up on photographic plates and film before leaving, as they were likely to be in short supply across the former battlefields.[49] Within a year supply was much better: in May 1920 an article in the *Dundee Courier* told its readers that there were no restrictions on photography across the former battle zones, and film could be obtained from all the 'usual places'.[50] On the downside, enthusiastic pursuit of photographs made it easier to label someone a tourist. Dennis Gilmore saw parties of tourists, 'mostly from Canada and the USA ... armed with Kodaks ... groping amongst the trenches and debris and snapping photos of any typical bit of ruins to be seen'.[51]

SITES OF MEMORY, SITES OF WONDER: GUNS, TANKS, CRATERS, PILLBOXES, AND TRENCHES

When it came to snapping the Wonders of the Western Front, the photographer was offered an enormous number of subjects. For those making the crossing to Zeebrugge or Ostend, the opportunities started on arrival. Raided by the Royal Navy in April and May 1918 (escapades that were lionized out of all proportion to their actual military worth by the British media), the two ports were already legends in the minds of British visitors. Arrival then tallied with expectation, at least for the first few years after the war, as the ruined blockships and German coastal defence installations were clearly visible. Ostend was deemed 'herself again' by a correspondent writing for the *Aberdeen Daily Journal* in the autumn of 1919, although it was possible to see the bulk of the *Vindictive*, 'seaweed hanging from her side – a grim and battered monument of British naval valour'.[52] In August 1920, the *Dundee Evening Telegraph* told its readers that visitors to Belgium could easily access the Zeebrugge mole, on which the raiding parties had landed, describing the port 'as the extreme northern spur of the battle-line, and ... familiar to all Britishers as the scene of the most daring and original exploit of the war'.[53] The Anglo-Belgian Union, a group of influential British and Belgian bankers, traders, and politicians, who were keen to ensure future co-operation, took the lead in erecting the main memorial to the raid. The memorial, designed by the Belgian sculptor M.A. Dupont and architect M.J. Smolderen, and consisting of a tall obelisk decorated with bronze reliefs, topped by a statue of St George slaying the dragon, was unveiled on 23 April 1925 by Albert I, King of the Belgians.

Another place of interest was the Zeebrugge museum, opened in 1923 by the Belgian antiquary and honorary colonel G.H. Stinglhamber. Stinglhamber amassed a fascinating collection of artefacts and documents. He also created a replica of the German naval officers' club in Bruges, located in the basement of the main hotel, complete with elaborate wall paintings and souvenirs. The *Daily Mail* described it as a 'thrilling' site, and the *Dover Express* was delighted to find a veteran among the museum's guides who managed to bring everything to life, losing 'none of its effectiveness from the naval language employed or the pride with which it was told'.[54] The equivalent attraction in Ostend was the panorama of the Yser–Ypres front painted by Alfred Bastien on huge canvases stretching around a specially built hall. Seeing them in 1932 before catching the boat home, 'T.P.' was deeply impressed by 'the wonderful panorama of the battle of the Yser, 1914. I think everyone who visits Ostend on one of these trips ought to see this, for it is a most realistic exhibition.'[55]

When it came to sheer impact, the biggest attractions along the coast were the fortifications and huge guns, and these were places most visitors stopped on their way either to or from the main battlefields. As well as the scale and extent of the emplacements, what impressed many were the facts and figures about the guns. 'West Londoner' saw the great German gun at Moere and was told of its ability to shell Dunkirk, over thirty miles away.[56] A Manchester and Burnley party was impressed by the Leugenboom gun and the wonder of having Dunkirk within its range was similarly recorded, a fact also noted by T.J. Blinkhorn after his visit in the late summer of 1921.[57] Blinkhorn then went one stage further, for he was able to provide the readers of the account he produced for his local newspaper with details of another artillery position, but one slightly off the usual track, 'not included in the battle-field tours, and consequently not visited by many tourists'. Identifying the site as the major German gun battery at Bredene on the coast, he supplied a great deal of detail as to how to find it, and what remained to be explored. So impressed by this location was he that he dedicated a disproportionate amount of his account to the description. It is difficult to understand why he rated this particular site so highly given its relatively minor military significance. It may simply have been the novelty of finding somewhere relatively overlooked, the impressive remains of an extraordinary piece of military kit, and perhaps even the cachet it gave him as an authority on the battlefields.[58] Binkhorn was setting himself up as the expert tourist.

The other major gun positions were on the Somme, at Chaulnes and Chuignes. Nina Stephenson-Browne stopped to see the Chuignes gun, which impressed her greatly, on her way from Amiens to the Somme battlefields.[59] L.M. Orton, taken on a tour by a former British officer, was equally astounded at the sight of the fifteen-inch German gun used to shell Amiens.[60]

For veterans, with their expert knowledge and experience, the batteries were especially interesting sites. The Chuignes gun was visited by the 1/5 Gloucester Regiment association party during its trip to the Somme in September 1925, and they found it a fascinating site.[61] B.S. Townroe was equally pleased his French driver took him to see the position, carefully recording the precise capabilities and details of the gun. He also noted that tourists had 'amused themselves by writing their names over every part which they could reach'.[62] For Frank Hermen, a veteran from Diss in Norfolk, the coastal batteries were well worth the entrance fee of two francs collected by the one-legged Belgian ex-serviceman custodian. Herman was intrigued by the enormous scale of the positions, complete with the mountings in wells the size of gasometers, which had filled with water.[63]

As most soldiers never saw such artillery pieces in action due to their relative rarity and their carefully concealed positions a long way behind the front lines, they were as amazed as tourist sightseers. However, their insights meant they were able to interpret what they found, whereas for the tourist it was a place to get a good snap, and, of course, scribble a name, as Townroe saw.

Another irresistible draw for cameras was ruined tanks. Having become stars in their own right during the course of the conflict, tanks continued to have an almost mesmeric effect on battlefield visitors. For those who had 'banked on the tank' during wartime propaganda drives for war bonds, there was the chance to see this somewhat novel technology on the very battlefields where they had served. Just as wartime travel restrictions were being lifted in July 1919, the *Dundee Courier* carried a story under the headline 'Tours to Battlefields of Ypres: Visit to the "Cemetery of the Tanks"'.[64] Referring to some 500 passengers embarking on the renewed Dover–Ostend service, the story noted that many were bound for the Ypres battlefields, and in particular were off to see 'the "Cemetery of the Tanks" where many of these armoured monsters lie partially sunk in the Flanders mud'.[65] The Ypres battlefields had two locations labelled as either 'Cemetery of the Tanks' or 'Tank Cemetery'. One was near the village of Poelcappelle, but the more famous of the two was on a bend in the Menin road near Hooge, christened 'Clapham Junction' by the troops. Both locations marked points where tanks had been deployed in large numbers but had come to grief in the appalling conditions of the Third Battle of Ypres in 1917. One of the earliest guides to the Ypres battlefields, Toc H's *The Pilgrim's Guide to the Ypres Salient* of 1920, called attention to this 'Tank Cemetery', where the country was 'strewn with the wreckage of over a dozen tanks', as did Coop, who highlighted the 'dismantled tanks' that could be seen along the Menin Road.[66] Other guidebooks made it a fixture, with *Muirhead's Belgium and the Western Front* of 1920 flagging it up, as did Ward, Lock, and Company's 1921 *Handbook to Belgium and the Battlefields*.[67]

Exposure to the marshy ground around Ypres often caused amazement that tanks had ever been deployed in such terrain. A correspondent for the *Edinburgh Evening News* ran into a veteran who was revisiting old sites on the battlefield. They saw 'dozens of Tanks are still lying about half-buried in the fields. Whoever gave the order for the Tank attack over this ground must honestly believe Tanks could swim.'[68] A similar sense of bafflement was expressed by T.J. Blinkhorn. Stopping at Hooge to inspect the array of ruined tanks, 'the nature of the ground, almost at sea level, swampy and sodden by mist and rain during the time of the fight, rendered these machines of war almost helpless and an easy task for destruction by the enemy'.[69] J.H. Roberts, a member of the British Red Cross, toured the front in early 1920 and wrote up his experiences for his local newspaper. He too felt the need to explain the specifics of the Tank Cemetery at Hooge. The soft ground caused many of the machines to sink, making them targets for German shells, he noted, before adding, 'Most of them, or parts of them, remain there to this day.'[70] According to Lieutenant-Colonel T.A. Lowe, the remains of 'these heroic monsters' at Hooge were a profound reminder of the 'unhappy battle' of 'the tank disaster of August, 1917'.[71]

Tanks, through their power and weight, left a literal impression on the battlefields, which visitors could clearly see many years after the end of the war. 'Tanks with broken bodies half-buried in the mud' could be seen outside Amiens, stated Anna Bowman Dodd in her guidebook aimed at the US market, *Up the Seine to the Battlefields*.[72] Near Pozières on the Somme was another cluster. Such was the gathering of smashed machines at this point, it was rumoured that the French government might preserve the spot as the 'Tanks Churchyard'.[73] (Indeed, the village became home to the Tank Corps memorial.) However, the local authorities on the Somme front seemed keen to clear away this kind of debris in the desire to restore the farmland as quickly as possible.

Some of the great metal monsters had already been removed by the time Atherton Fleming's guide was published in the autumn of 1919. Making them lost features of the battlefield, he remarked of Pozières, 'On both sides of [the road] could be seen, until a short time ago, one or two derelict tanks of the earliest type – relics of the original tank attack in front of Courcelette and Martinpuich.'[74]

Writing for the *Kirkintilloch Herald* in September 1919, the month in which Fleming's book was published, the Reverend H. Y. Reyburn found himself drawn to the tanks, and in the process subliminally made them a proxy for human bodies dismembered by modern weapons. Having seen 'no fewer than five tanks within fifty yards of each other' near Passchendaele, he noted one 'with a great hole in its side, another with the whole internal machinery torn out, another cut in two almost as clean as if it had been done with a knife'.[75] Tanks also provided their own distinctive trails across the battlefields. A Ballymena visitor was able to navigate the old no man's land by following the deep print of tank tracks to the old German line just outside Cambrai.[76]

Tanks were also a prevalent photographic image: dramatic in scale and shape, made even more arresting in their abandoned condition, and often reduced to a state of startling grotesqueness through the profusion of twisted, buckled metal and caterpillar tracks caught stiff and alert, almost as if suffering a mechanical form of rigor mortis. The prolific and prestigious Ypres photographer Antony produced numerous postcards and privately commissioned photographic images, as did the postcard company Nels of Brussels. In 1921 *Country Life* produced *Ypres to Verdun: A Collection of Photographs of the War Areas in France and Flanders*. The images were taken by the experienced photographer Sir Alexander Kennedy, who served with the Royal Engineers during the conflict. In 1919 and 1920 he travelled the old Western Front recording the state of the battlefields. Among his images of the salient were two of ruined tanks along the Menin Road: they had become a vital visual shorthand for the intensity of destruction in the salient.[77] And every visitor with a vest-pocket Kodak ensured they got a good snap of their group gathered round one of the wrecks, including a 1919 Thomas Cook party, which judged two wrecked tanks a 'fit subject for our cameras'.[78]

Due to their sheer size and mass of metal, the tanks took a long time to either be removed or disintegrate entirely. The Ypres League's 1925 guide, *The Immortal Salient*, pointed out 'the remains of tanks' near Poelcappelle.[79] Three years later, a group of Yorkshire visitors went to the same spot and declared 'the remains of British tanks may still be observed in the tank cemetery'.[80] Nonetheless, exposure to weather, the work of scrap metal dealers, and clearance for reconstruction purposes saw the number gradually fade away. Much, much more difficult to erase from the landscape were the concrete pillboxes. Removing these took extreme measures, and so they were among the last remaining elements of military infrastructure. As H.A. Taylor noted in 1927, 'The road from Poelcapelle to Langemarck is fruitful in reminders of the intense fighting that went on hereabouts in August and September 1917. The fields are still sown with "pill-boxes" large and small which defy destruction, or which have not been destroyed because demolition would damage the land so considerably as to make their removal an imprudent undertaking.'[81] Often large, often in prominent places, these concrete monsters offered another great photo opportunity.

Pillboxes were ubiquitous reminders of the war, especially around Ypres, where so many were constructed. 'There are a lot of old pillboxes left standing in the middle of cornfields,' noted the schoolgirl Jean Simmonds, who spent her 1930 summer holiday staying in Ypres with her uncle, an IWGC gardener.[82]

Surveying the landscape from the edge of Kitchener's Wood to the north of Ypres, Will Bird counted thirty-nine of them.[83] As large, solid constructions, pillboxes could also be put to other uses. Bird found a farmer using one as a pigsty; another veteran commented on the number converted into stables and tool sheds.[84] Having rebuilt his chateau, Baron de Winck of Hooge near Ypres integrated the pillboxes on his land into an ornamental grove for his new garden.[85]

There was one further function for pillboxes, as the famed war correspondent Sir William Beach Thomas found on his return to Ypres in May 1920. Delighted to see the people of the region determined to recover from the desolation of war, he nonetheless noticed that some were 'living in shacks and even dug-outs and "pill-boxes"'.[86] W.E. Wilford, a member of Leicester council, visited the battlefields as part of a wider tour of France and Belgium. Commenting on the extreme shortage of housing in the devastated zones of Belgium, he stated: 'The situation is such that some people are actually inhabiting the pill-boxes left by the Germans.'[87] Members of the Ypres League were informed of a similar story by one their number living in Ypres. Published in the April 1922 issue of the *Ypres Times*, the article told of a resident of St Julien, just to the north of Ypres, who had returned after the war to find his house destroyed, with a pillbox built over the site: 'Not seeing the means to erect his house at present, [he] fitted up, and is now living in, the pillbox.'[88] Among the most famous was a large pillbox at Festubert, which was occupied by a French family from January 1919 until the 1970s, when the last remaining member, an old lady who had lived there all her life, died.[89]

As the other battlefield features disappeared, pillboxes became the enduring reminders of the war. For 'J.E.H.', recounting a trip to Belgium in the summer of 1935, 'Even now, no matter where the traveller goes in Belgium, the war cannot be forgotten. Its signs are everywhere ... upon still scarred fields where concrete pillboxes still rear their ugly shapes.'[90] Pillboxes provided the sole signs of continuity and recognition for veterans among the Ypres League's 1927 Menin Gate pilgrimage.[91] It was only once they started seeing pillboxes from the train window that the veterans on the Ypres League's 1935 August Bank Holiday pilgrimage realized they were crossing the battlefield they once knew so well.[92] Pillboxes thus marked out the path of memory for veterans. G.W.C. Craik declared them 'the most vivid reminder of war on the whole Western Front', especially the cluster on the slopes up to Passchendaele.[93] 'A.W.W.' found a world transformed when he took part in the British Legion pilgrimage of 1928, but was reassured when he stumbled across a pillbox he knew well.[94] Those on the 2 London Regiment association tour of Easter 1932 were given a free day to explore as they chose. Many got on the tram to Hooge, from where they walked up to Glencorse Wood intent on exploring the ground they attacked in August 1917. Examining the pillboxes, the association secretary 'at once recognised the one that made Colonel Kellett a casualty on the morning of our attack'.[95]

Ugly to look at, utterly miserable to occupy, and often contested savagely, pillboxes somehow became sacred objects after the war.

Determining a single reason for this transformation in status is difficult. Possibly it was due to their sheer resistance to easy removal, which then made them remnants of a lost world, or it may have been their bloody histories causing them to become a focus point for memories. It was probably a bit of both. Taylor tried to understand their appeal. Examining the remains of one damaged, but by no means obliterated, by a botched post-war attempt to blow it up, he saw something curious. 'What remained, perched up on a little eminence, reminded one of nothing so much as a miniature cromlech – one of those rude tables of stone associated with the sanguinary religious rites of the Celts.'[96]

The pillbox as altar and memorial exerted a strong hold on veterans. During the war, 46 Division placed a memorial cross on the roof of a German pillbox its men had captured. At the war's end, the permanent memorial was situated close to the original, which was still standing precisely where it was first put.[97] The engineers of 34 Division placed their memorial next to a pillbox at Broenbeke, near Langemarck in the Ypres salient.[98] For the war memorial committee of 9 (Scottish) Division, the German pillbox on the Arras–Gavrelle road at Point-du-Jour was the perfect orientation mark for its memorial.[99] At Messines, the architect Samuel Hurst Seager preserved two pillboxes in his design for the New Zealand memorial. The twin German pillboxes at Pozières, one embedded in the foundations of a windmill and the other in a house (and nicknamed Gibraltar), were integrated into the Australian memorial landscape in this small Somme village as the ruins of the windmill site were acquired, and the 1 (Australian) Division memorial was erected directly opposite the Gibraltar position.[100]

The significance of pillboxes was also recognized by the IWGC. At Derry House Cemetery No. 2, south of Ypres, the architect W.H. Cowlishaw built the cemetery wall over the roof the pillbox, which straddled the boundary. Far more famously, pillboxes formed integral parts of the largest IWGC cemetery in the world, Tyne Cot, on the slopes in front of Passchendaele village. These German bunkers were captured by 3 (Australian) Division. The largest was then used as an advanced dressing station, and the ground around it for the burial of the dead.

After the war, the cemetery expanded massively as huge numbers of bodies exhumed from across the Ypres battlefields were brought to Tyne Cot for reinterment. Commanding spectacular views across the salient, the cemetery quickly grew in fame and status, which was emphasized further by the visit of King George V on his battlefield pilgrimage of 1922. During his inspection it is alleged that he suggested the retention of the pillboxes, especially the central one as an observation platform.[101] The king's association with the Tyne Cot Cross of Sacrifice then became a standard component in any kind of comment about the cemetery, and according to the precise source, he either suggested the idea or the exposure of the original concrete within the overall design.[102] The task of designing the IWGC cemetery fell to Sir Herbert Baker with the assistance of J.R. Truelove, and it seems likely that it was Truelove who undertook the task of transforming the Tyne Cot pillbox into a plinth for the Cross of Sacrifice.[103] By encasing the pillbox in Euville stone and forming an ascending pyramid with steps, Truelove created a viewing platform from which to study the battlefield, and with it linked the cemetery space to the battlefield space. The visitor could stand in a battlefield transformed into the beauty of an IWGC cemetery while looking across the wider battlefield. To ensure the visitor knew precisely what was underneath, a section of the pillbox was deliberately left exposed, revealing its original concrete and framed with a bronze wreath. Within the cemetery two other pillboxes were left exposed in the lower plots, this time with little alteration other than the banking up of soil around them as part of the landscaping process. Taylor found this architectural translation poignant and fitting: 'Thus the observant can see the white symbol of Christ [the Cross of Sacrifice] towers over a structure which once spoke of death and bestiality,' while the other two pillboxes were softened 'with beautiful rambler roses gradually spreading themselves all over the concrete'.[104]

The strength of the emotional investment made in pillboxes was expressed by many other veterans. An Australian ex-serviceman exploring Pozières in 1924 found the windmill in 'exactly the same shape as when I last saw it after the recapture of the main German line in front of Pozières, except that its base was then covered with the bodies of dead Germans and Australians'.[105] He then walked on to the Gibraltar position, and was disgusted at the way this relic had been treated. 'Despite the adjacent memorial of the First Division and the warning notice of the French authorities, the villagers of Pozières had entirely disregarded any sentiment attaching to the pill box. The inside appeared to be used as a village latrine, and the place was indescribably filthy.'[106] This veteran seemed to believe the structure was akin to a tomb that had been desecrated. Concern over the condition of pillboxes grew during the late twenties and early thirties as reconstruction matured and landowners finally had the time, energy, and money to concentrate on their removal. Supported by 'Tubby' Clayton, the Ypres League and British Legion combined in a campaign to preserve examples across the Ypres battlefields. After discussion with the Belgian authorities, in the autumn of 1932 it was announced that some 180 pillboxes and blockhouses were to be preserved as a memorial to Field Marshal Lord Plumer. The decision was regarded as significant because 'the "pill-boxes" and "blockhouses" are the sole reminders of the scenes of great deeds of valour, fellowship, and service witnessed by them during that period'.[107]

Their status was then enshrined in a specially prepared gazetteer, *The Pill-Boxes of Flanders*, written by Colonel E.G.L. Thurlow. Thurlow's approach to the subject leapt between strictly utilitarian military prose, as he described construction techniques, uses, and detailed instruction on how to find them, and high diction as the structures were venerated as tombs and sites of heroic endeavour and endurance. Through their preservation, Thurlow argued, a valuable moral quality would be enshrined in perpetuity, as the pillboxes would 'enable the younger and coming generations to visualise something of the great spirit of sacrifice shown by their elder brethren'.[108] The foreword by 'Tim' Harington, General Plumer's senior staff officer, also insisted on the sacred nature of the pillboxes. Like Thurlow, he instructed the younger generation to see them as places where 'glorious fellows' fought 'against terrific odds' for 'your liberty'. When visiting a pillbox, they should try to think of the men who 'died for us'. It placed an onus on the living to lead good and worthy lives, and to ensure that the horror was never repeated.

This was a long way from the standpoint of the mere sightseer who viewed pillboxes as similar to ruined tanks. They were things tourists could clamber over and merrily snap as weird and wonderful remnants of the war. Using a different gaze based on their wartime experiences, veterans could invest them with deeper meaning, regarding them as symbols of the horror and mess of the battlefield over which they and their comrades, alive or dead, triumphed.

As well as playing a leading role in the preservation of Ypres' pillboxes, Clayton was also the prime campaigner for the battlefield's remaining craters. During the course of the war each side had undertaken an immense amount of mining activity with the intention of destroying enemy positions through the use of huge weights of explosive. As a result, the Western Front was not only pitted from end to end with shell holes but was also marked by crater fields caused by the detonation of mines.

For the British, the craters of the Somme and Ypres were the most famous. Although it was not actually the biggest of the detonations, the La Boisselle crater on the Somme took the laurels as far as visitors were concerned. This was largely due to sheer practicalities. A short way along a track leading from the main Albert–Bapaume road, the crater was easily accessible and on the main tourist route across the Somme battlefields. 'Some of the more agile members' of the Ypres League free pilgrimage of 1927 could not resist clambering down to the bottom of the massive crater, where a photograph was taken showing the main party on the lip and the intrepid team at the bottom.[109] Nellie Burrin also 'made sure to take a photograph'. Although it came out very well, she was a bit disappointed that it did 'not show anything near what a big hole it is', as she recorded in her diary.[110] For the veteran G.W.C. Craik, the crater never lost its ability to amaze: he said it was 'as awe-inspiring as ever' when he returned in the summer of 1925.[111]

On the Ypres battlefields, the most significant craters were those south of the city, blown during the battle of Messines in June 1917. As with the pillboxes, by the late twenties there were growing concerns over the gradual disappearance of this extensive belt of craters, and in September 1929 'Tubby' Clayton wrote to the *Times* urging the preservation of the St Eloi crater south of Ypres. For him it was not just a reminder of the battlefield, but a place of contemplation and the grave of many soldiers. With 'scarcely a spot near Ypres … now left undisturbed, away from motor-horns and manufactured souvenirs. Here is a pool of peace, where man's wrath is God's praise,' he argued. A few days later the *Times* published a striking photograph of the water-filled crater, which added to the power of Clayton's appeal, and a letter of support sent by a veteran Royal Engineer who had worked on the tunnelling operations around Ypres during the war was soon published by the paper.

The letter spurred activity, although not directed at the St Eloi crater. Instead, Lord Wakefield, a friend and admirer of Clayton, announced a plan to purchase Lone Tree crater at Spanbroekmolen,

Figure 7.2 Servicing the needs of visitors: The Café de la Grande Mine, La Boisselle, c. 1920.

and secure it as a 'Pool of Peace'.[112] For H.A. Taylor, the Lone Tree crater was a place of beauty and calm, while also serving as an excellent position from which to view the surrounding countryside. Peaceful meditation went hand in hand with the opportunity to understand the ebb and flow of the battle of the Messines Ridge.[113]

Closely associated with the wonder factor of craters were the remains of the trenches. Every visitor wanted to see the most dominant feature of the battlefields, the one they had read about continually in the newspapers and seen on cinema screens and in photographic exhibitions. Trenches were also the reminder of battle to disappear most quickly as reconstruction gathered pace and nature got to work. Obliterating all traces of them was much harder, as the deep incisions across the landscape were obvious throughout the twenties and thirties, especially in the winter months.

Where deliberate decisions were taken to preserve trenches, they became great attractions for visitors. At Vimy Ridge and the Newfoundland Memorial Park on the Somme, the trenches were integral parts of the overall memorial scheme. For Newfoundland, the trenches in front of Beaumont-Hamel represented a site of trauma and hallowed memory at the same time. On 1 July 1916, its men had advanced into no man's land only to be cut down within minutes, causing mass mourning in this oldest corner of the British Empire.[114] Determined to commemorate the sacrifices of the Newfoundland Regiment, the unit's padre, Father Thomas Nangle, raised funds and purchased the battlefield soon after the war's end, employing the Dutch-born landscape architect Rudolph H.K. Cochius to help him plan the memorial scheme. Cochius decided to freeze the battlefield as it was that July morning, preserving the British and German front line trenches and orienting the visitor through the unit's attempted line of advance. The horror and reality of war was kept, but nature was used in a very deliberate manner to redeem, as Cochius imported 35,000 trees, mainly Scots Pines, to frame the perimeters of the site, creating sylvan walks, while grass softened no man's land and the trenches, ensuring a wild meadow in spring and summer.[115]

Cochius created a rich commemorative landscape in which visitors could immerse themselves in a liminal space between the past and present, allowing for contemplation of war's realities and the nature of sacrifice. It was also the ideal spot for the tourist-in-a-hurry to see real trenches and collect a few postcards. 'Plenty to be seen here,' wrote a member of the Ypres League's 1932 pilgrimage with great enthusiasm, continuing seamlessly, 'old trenches intact with all kinds of war equipment, rifles, grenades, tin helmets, old rum jars, machine gun positions, etc. Log cabin, where various souvenirs, war maps, photographs, etc., may be seen, and a visitors' book which you may sign,' which completed the catalogue of its distractions.[116] Will Bird told his readers, 'Fo not leave that area without going to the Newfoundland Park. It is a splendid place, with a log cabin at which you may buy postcard views of all the Somme, and with a most interesting caretaker in charge.'[117] Neither happy customer got round to mentioning the two IWGC cemeteries included in the park. The trenches and the souvenirs seemed to be the headline attractions.

The trenches at Vimy Ridge (see chapter 9) and Newfoundland Memorial Park were part of commemorative sites, but others were in private hands and were run for profit. The Ypres battlefields were home to the two most famous, and lucrative, sites of Sanctuary Wood and Hill 60.

At Sanctuary Wood the returning owner very quickly realized the commercial potential of the trenches running through his land.

Assisted by the new road built to allow access to the Canadian memorial on Hill 62, tourists came in droves, and left convinced they had seen the real thing. Schoolgirl Joan Simmonds wrote in her diary, 'It is a piece of the battle ground just as it was left. Tin helmets and wire cutters, rifles, rum jars, and the trenches and dug outs and duck-boards. The trenches are in such a bad state that they are not fit to get into and have to be looked at from the top.'[118] 'G.C.', who accompanied the 1935 Public Schools Battlefields Tour, was rather less impressed, deeming it 'now rather too much of a museum, although it appeared instructive to the boys'.[119] By contrast, other veterans found much to admire. 'Here is the real thing,' one wrote, 'The water is as deep as ever, the duckboards have the old trick of getting up and hitting you in the face when you tread on them, and the whittled stumps of the old wood still stand in front. Here, rather than at Vimy Ridge, is the place to get some idea of trench life.'[120] At this point the trenches at Sanctuary Wood were still close to their original condition, albeit regularly maintained by the owner, and not preserved in concrete as at Vimy. For this veteran, the difference seems to have been crucial.

HILL 60: THE PLAYGROUND OF THE BATTLEFIELDS

From the sites at Sanctuary Wood, many then made their way to Hill 60, the greatest attraction on the Ypres battlefields. A mound created by the spoil excavated for the construction of the railway line, it was the site of intense fighting throughout the war. By the end of the conflict it was a mass of trenches and craters, of remnants of tunnels and tunnel entrances, and it was topped by an impressive British pillbox constructed by Australian engineers. Due to the almost continual combat at Hill 60, it gained much media attention, and by the Armistice had become an icon of the British Empire's endurance and dedication to the defence of Ypres. As Beatrix Brice's 1927 *Battle Book of Ypres* commented: 'Through the wild days of furious battle ... the contest for the hill had been an epic of valour, when man met man in desperate fight; and the British soldier established his ascendancy over the Prussian and once more proved his capacity to stay it out to the bitter end, though tried to the uttermost.'[121] With so many units having fought on and for the hill, it was a site that attracted memorial activity and resulted in monuments erected by the Queen Victoria's Rifles and 1st Australian Tunnelling Company.[122]

Additionally, in the ravaged, treeless countryside, Hill 60 stood out starkly, and being so close to the Menin Road and Sanctuary Wood, almost inevitably became a magnet for visitors. It seemed an authentic chunk of the old battlefield, and its lingering dangers added to that sense, with its mess of shell holes concealed by clumps of grass and undergrowth, broken ground, unexploded ammunition, and even human remains, which remained for many years. 'On our path are a few human bones,' wrote a visitor in 1923, 'and the day before two or three skeletons had been discovered'.[123] In 1928 a group of Scouts arrived in time to see a local resident uncover the remains of a German soldier while doing some gardening.[124] During the same year a party from Perth was 'warned to avoid four or five live shells which still stood on the hill'.[125] Hill 60 was a sight worth seeing, a 'must see' location on the old Western Front.

With so much to see and so many wartime obstacles making exploration of the site a challenge, a web of paths and tracks emerged. It was made more complicated by the fact that the hill was not one parcel of land but divided between many owners, each of whom realized they had a money-spinning opportunity and so sought to control access to their own section while enhancing its attractions. Tunnel entrances were excavated and made accessible, dugouts were propped up, at least one small museum and associated cafe were opened, an observation tower was erected, and souvenir stalls flourished.

Hill 60 inspired a variety of reactions, from wonder to disgust, with veterans as likely to take opposing views as any other visitor. When Captain Mee visited in the summer of 1922, he found a real slice of the war he knew in the trenches, tunnels, and other relics liberally scattered around. Its status as a tourist site, and with it a place of money-making, did not seem to concern him at all. Visiting a nearby cafe, he was much more interested in the stories of the owner, who was doing her best to survive while living in little more than a shanty. He judged Hill 60 'a shattered remnant, left untouched save for the monument on its highest point'.[126] The veterans on the 1933 Ypres League Easter pilgrimage happily explored the trenches, tunnels, and 'numerous other war materials, all of which reminded us of other days'.[127] For W.J. Baumgartner and his fellow pilgrims, Hill 60 was the place where they could act like tourists with no qualms, which included a good cup of tea following which 'cameras were much in demand', with the Caterpillar Crater attracting particular attention.[128]

For those with no service history, Hill 60 was living museum and playground rolled into one. In 1927, the Toc H pilgrimage visited the site. One of their number, a former commander of a Machine Gun Corps company, gave them a talk on the Battle of Messines, 'with the aid of old trench maps ... from this vantage ground'. The party then dispersed to explore the 'strangely disfigured' hill. On the north side, it was noted, there was a dusty lane 'lined with Belgian "souvenir" stalls; the cheap-jack with his or her rusty bayonet or tasteless crucifix made of rifle ammunition and shell-case brass, waylays the visitor with a poor bargain; and at the top ... several painted bungalows dispense rival teas and mineral waters'. The siren lure proved too much, as the Toc H chronicler noted: 'A cup of tea was welcomed indeed to all of us, but it was not easy to shake off the Bank Holiday touch and to appreciate the significance of this rising ground or its untold human tragedies of fourteen years ago.'[129]

People could relax and have fun at Hill 60, but the atmosphere did not strip the place of all significance. A party of Scottish pilgrims visiting in the summer of 1923 felt they had seen the real landscape of war. The trenches, pillbox, shell holes, and the Caterpillar crater left them amazed, especially the contrast created by the clumps of cornflowers waving in the breeze amidst the debris. Bits of kit were scattered everywhere and one of the veterans among them came across a boot still containing a foot. (Bones from the leg and foot seem to have been the most common human remains found by visitors.)[130]

A group from Nelson in Lancashire was shown over Hill 60 by an excellent local guide who gave them 'a graphic account of the struggle for the Salient' while they looked over the 'glorious panorama' provided by the high ground.[131] Exploration of the 'excellently preserved' trenches and dugouts meant it was 'not difficult even for the inexperienced to picture the days when they were inhabited by communities of laughing, jesting, but fated humanity', recorded a member of a party from Sunderland visiting in 1935.[132]

The Ypres League August Bank Holiday pilgrimage that same year enjoyed the 'realistic picture' created by the trenches at Sanctuary Wood, as did those at Hill 60, where some also climbed the tower to look through the telescopes at the top.[133] A party from Coventry, also visiting in 1935, took the opportunity to scramble through the main tunnel, where torches had to be carried 'so that we could see the wire hammocks, store-rooms, and holes for the gun barrels. There were planks crossing the underground streams.'[134] What kind of tunnel contained apertures for gun barrels is not explained, and it is likely the party either gained the wrong impression or were misinformed by an ill-prepared guide.

Others were far from impressed by what they saw. Henry Williamson was disgusted and angered by the children waving collecting boxes trying to charge every visitor a fee for simply wandering along the track towards the hill. Many complained at the sale of fake souvenirs and the equally fake trenches.[135] Brice described it as 'now desecrated beyond any place in the Salient by horrible erections of booths and shanties'.[136] Among those booths was the 'Queen Victoria Rifles Rest House' run by Mrs Moon, 'late of the Old Kent Road', as she proclaimed proudly on her advertisements, and with equal pride, 'the only British concern on Hill 60'.[137] Close by was the 'No Man's Land Canteen', which boasted 'British ale' among the many delights it could provide for the visitor.

The tourist hubbub that so disgusted Brice and Williamson was treated by others in a rather more sanguine manner. A Scottish pilgrim, who had first visited Hill 60 in 1923, found it much changed by 1928, but was not that concerned by its transformation into a tourist site, accepting it with resignation: 'It has, in the interval, been adapted for sightseers, a process that, though regrettable, is doubtless inevitable,' he wrote.[138] A member of the 1931 annual Sheffield battlefield pilgrimage was equally laid back, noting without rancour how many children flocked round the party, attempting to sell them posies.[139]

Slightly more controversial was the nature of the souvenir trade, especially as any item deemed unlikely to sell was instead valued as scrap, making the place seem like a manic mine of distasteful activity. Veteran 'Mr X' was bombarded by souvenir-hawkers trying to sell him all sorts of memorabilia as he walked up the footpath to Hill 60, which he did not appreciate at all.[140] Rather more sardonic were the veterans of the Leyland Motor Company; the souvenir hawkers were lampooned in the tour booklet, which included a cartoon of a slightly bedraggled woman standing behind a stall full of tat labelled 'Real Hill 60 Souvenirs'.[141] Another veteran was equally arch, remarking that the thoroughness of the scrap metal seekers on Hill 60 went as far as picking out the lead from the inscriptions on the Australian tunnellers memorial.[142]

The veterans of the 2 London Regiment association watched Belgians at work digging for scrap metal on Hill 60. When questioned, the men said their work could uncover items useful for the identification of bodies. However, this was more than likely a good cover story providing a respectable excuse to search for anything of resale value.[143]

In September 1930, the *Daily Herald* carried a front-page story telling of excavation works at Hill 60 designed to open up the tunnels and dugouts and bolster tourism. However, any such work, the article noted, ran the risk of disturbing bodies.[144] The *Daily Herald* being a newspaper closely linked with the Labour Party and keen to support the idea of learning from the waste of war, ensured the article was infused with a sense of moral superiority and offence, particularly with regard to the sanctity of bodies, but it failed to explore its own logic deeply. If the work parties did discover human remains, and if they were British, then they would be decently and honourably buried in an IWGC cemetery, and possibly identified, too, bringing comfort to their relatives.

A flurry of Fleet Street activity ensued. Within a few days the *Daily Mirror*'s correspondent was on the spot, writing of the distasteful souvenir trade, including the offering of socks, gloves, and scarves as allegedly genuine articles of military clothing. The trench inspected was in perfect order, and the journalist's suspicions were confirmed on being 'informed on good authority that it was not there last year'. Hill 60 was being 'desecrated' by these activities, and they encouraged young British tourists to act in irreverent ways: 'They come in their bands, pick up and don rusty and decayed steel helmets, seize worm-eaten rifles, and strike fanciful postures while one of their number photographs them to the accompaniment of shrieks of mirth.'[145] Having slated British youths, the correspondent returned to complaining about the Belgians, claiming they boycotted the bar run by a British veteran (whom he identified as an ex-artilleryman from Manchester with a Belgian wife). Thus, as long as a concern was British, and in particular run by a veteran, it was perfectly fine to run a business on Hill 60. Some kinds of tourist activities were, therefore, acceptable, and presumably this business, along with Mrs Moon's, was exempted from the otherwise general condemnation: 'Stallholders should be swept off Hill 60; the fraudulent trench should be filled in; and the Hill itself, which, I understand, now partly belongs to Mr. T. T. Calder,

Figure 7.3 Playing at soldiers. Hill 60, August 1935.

the English brewer, wired round and treated with the respect due to it.' A couple of days later, 'D.H.B' wrote in to agree with these sentiments and urged an end to the commercial exploitation of the site.[146]

The Calder referred to did indeed purchase the hill, after which he presented it to the IWGC. Calder's altruistic act required much patience, as he negotiated with the owner of the largest single plot,

Lieutenant-Colonel E.P. Cawston, proprietor of the Battlefields Bureau tourist agency, who clearly felt unable, or unwilling, to do much to control the various commercial activities clustered on and around the hill.[147] The IWGC was then made responsible for the site, and one of its first acts was to consider visitor access, with the intention of taking people away from the trenches and tunnels, presumably for reasons of safety and the doubtful authenticity of some, while still allowing exploration and contemplation.[148]

Curbing the touristic activities at Hill 60 completely was, however, almost impossible, as the IWGC did not own the entirety of the hill. Those sections still under private ownership remained committed to giving the tourist a good time. According to the *Liverpool Echo* in 1937, local police patrols had been stepped up to move on hawkers and stop children pestering visitors, but the cafe (Mrs Moon's business) and war museum were still there 'both in English ownership', which presumably made them immune from criticism.[149] Moon had pitched her credentials wisely: British, and therefore utterly trustworthy, was the message, which also meant any souvenirs purchased from her were the real thing.

THE SOUVENIR TRADE

Obtaining souvenirs (be they macabre, tasteful, fake, or authentic), and whether to buy them from vendors like Mrs Moon or perhaps slightly less respectable characters, was a major preoccupation of the tourist. As the first summer visiting season drew to a close in late August 1919, the *Daily Mail* published an article condemning members of the Chinese Labour Corps for selling souvenirs to visitors. Lambasting these '"Chinks," as Tommy Atkins calls them,' for spending too much of their time waylaying visitors so as to proffer souvenirs for sale, the article archly remarked, 'Presumably these "Chinks" do legitimate work sometimes, though the greater part of their day seems to be devoted to selling shell cases to souvenir-hunting visitors.'[150] The racist and snobbish overtones of the piece meant a deeper degree of analysis was missed entirely: if Chinese workers were diverted from their usual tasks, it could only be because there was a market to be satisfied, and that market was British visitors who were extremely anxious to collect such souvenirs. But price, provenance, and authenticity were not the only things the earnest seeker had to consider.

As all the guidebooks warned, especially those published soon after the war, souvenir-hunting demanded extreme caution. Coop said he understood the desire to search for a souvenir with its additional value of 'having "picked it up myself"', but the dangers far outweighed the thrill. 'Accidents have already happened, and will for years continue to happen, through the reckless handling of things which an experienced soldier would leave severely alone,' he added, making clear the difference between the educated serviceman and the naive visitor. All shells and nose-caps should be left well alone, he warned, before ending with the exhortation, 'Take nothing heavy home!'[151] Lowe, in providing his warning, added a moral layer, referring to a woman who picked up an old army boot only to find it still contained the remains of the foot and leg. Here the souvenir-hunter had not only been appallingly shocked but also inadvertently taken part in a desecration. But, like Coop, Lowe understood the desire to collect a souvenir and recommended old shell cases, which 'can be polished up brightly, and make beautiful vases and flower-pots'.[152]

Displaying his usual curmudgeonly attitude, Williamson took a perverse pleasure in the nature of the souvenir trade. Dud ammunition gave him all the ammunition he required. Teetering on the edge of Evelyn Waugh-like biting bitterness, he recounted with dark glee the explosion of the ammunition dump at Goldfish Chateau caused by locals desperately trying to recover the driving bands from shells, which were 'much in demand by American tourists as souvenirs'. Americans also gave him the chance to luxuriate in caricature. A brattish American boy, complete with horn-rimmed spectacles, was badgering his parents for a pistol from one of the souvenir hawkers at Hill 60. Clearly unwilling to purchase something that looked capable of going off unexpectedly, the father tried to distract the boy with offers of a bayonet or badges:

> 'Oh gee, you said I could have a gun.'
> 'Wal, take a baynit.'
> 'Nix on baynits. You said …'
> 'Will you take a button [badge] or nuthin'?'
> Again the neat cut-away of the hand.
> They passed on to the large, irregular heap of
>   fawn-coloured earth, the boy stifling back
>   his disappointment.

Having described one scene with all the skill of a music-hall sketch writer, Williamson then turned his attention closer to home and lampooned a bishop who could not resist buying a shell case for use as a flower vase. Rather tartly, he added there was little need to purchase it as an act of charity, as the seller seemed 'in sound economic position; but he was certainly pleased with the flower-vase'. For Williamson, the bishop was at best naive, and at worse a daft, irrelevant, unimaginative old man who had no conception of what it was actually like to fight in the trenches and debased himself still further by treating a visit to the former fighting zone as an excuse to purchase a souvenir.[153]

In trying to divide the thoughtless tourist from the veteran, Williamson had oversimplified things. Veterans collected souvenirs with as much glee as any other visitor. Wilfred Hyde returned home with quite an impressive stash of souvenirs: the swords of a British and a German officer, two Prussian Guard helmets, seven shell cases, and two bayonets.[154] One of the easiest, and relatively safe, souvenirs to collect was the ubiquitous shrapnel balls, which could often be found embedded in trees. Both the veteran H. Channing-Renton and the schoolgirl Jean Simmonds had great fun digging them out with penknives.[155]

Some managed to build up impressive and diverse collections. When Lady Florence Garvagh died in 1926, she left the helmets, pistols, rifles, and water bottle she had 'picked out of the trenches in the English lines at Ypres with my own hands' to her grandson.[156] The battlefield came home and became part of the myriad scraps, physical and mental, that make up a family history and heritage. Souvenirs transfigured into relics.

*

Pilloried as unthinking, insensitive, and ghoulish, the battlefield tourist was, in fact, something of a myth. Veterans, the bereaved, and the curious visited much the same sites, and none acted entirely and consistently according to the stereotypes created and discussed in public and private realms. The landscape of the battlefields, and the huge range of material left behind after the tide of battle had receded, made the former Western Front a world of bizarre, surreal wonders inspiring curiosity and awe. After a day in which mind and body were exposed to this alternative world, this mad strip running through France and Belgium, the visitor relaxed by drinking beer or sipping tea in bars and hotels. Whether visitors liked it or not, at some point in their wanderings across the old Western Front, they became tourists, but they were tourists exploring a unique landscape.

8

Postcards from Ypres
(and Its Salient)

What Ypres has lost of its architectural wealth and beauty, what has vanished of its ancient memorials and its people, it has gained in the priceless symbolism conferred upon it by the Great War. No fire or steel or poison-fumes can rob it of its new glory in the annals of the British Empire ... What Jerusalem is to the Jewish race, what Mecca is to the Mohamedan, Ypres must always be to the millions who have in that long conflict lost a husband, son or brother, slain in its defence and who sleep their eternal sleep within sight of its silent belfry. Ypres and the expanse of earth spread out eastwards is in truth the 'Holy Ground of British Arms'.

Henry Beckles Willson (1919)

This is a ghastly city.
   Scribbled on a postcard depicting the ruins of the Cloth Hall (1919)

When they reached the Menin Gate the mothers were completely bewildered. Elderly women, wrinkled and bent with a life of housework, they had travelled from far parts of England, Scotland and Wales on a night journey to this legendary town of Ypres – this city of death. That in itself was an enterprise which dazed them.
James Dunn, *Daily Mail*, on the unveiling of the Menin Gate (24 July 1927)

Ypres was the alpha and omega of the Western Front battlefields to the British, and indeed to the entire British Empire. It was the place of martyrdom during the war and became the Jerusalem after it. In the years after the war, particularly the first three or four, it was also, according to some, the desecrated temple of the holy city, a place of money-changers and hawkers deserving of expulsion and scourge. Every pilgrim, every veteran, and every tourist made their way to Ypres. Every pilgrim, every veteran, and every tourist left Ypres knowing they had seen the most important relic of the battlefields. Ypres *was* the Western Front. All 'felt drawn to this place, Ypres, by some invisible power', as one veteran wrote after returning from a pilgrimage.[1] Ypres was the place 'more than anywhere else [that] the desolation of war, and the splendid courage and endurance of the men who held on to the last, can be seen and realised', declared Sybilla Kirkland Vesey of St Barnabas Hostels.[2] Ypres was the essence of the battlefields.

Ypres was a legend long before the Armistice ended hostilities on the Western Front. Around it five great battles were fought, drawing in troops, combat, and ancillary from every corner of the Empire. It was deemed the British Empire's equivalent of Verdun, with the same aura of sacrifice and martyrdom for a sacred place in a sacred cause. Renowned as the gateway to the Channel, Ypres was declared the outpost of Britain's security, and therefore of the entire Britannic world.[3] And it was precisely this strategic status that placed the city at the heart of British battlefield visiting in the twenties and thirties. Close to the coast and the French frontier and sitting just in front of a crease of high ground stretching from Cassel in France, Ypres was also the converging point of important road and railway links. In other words, Ypres was easily accessible from Britain, and although the ground to its east was shattered and pulped beyond belief, reaching it from the coast was relatively easy, especially once the Franco-Belgian Channel ports, from Boulogne in the south-west to Ostend in the north-east, had fully reopened for passenger traffic. Ypres was transformed from the gateway to the Channel into the gateway to the battlefields of Flanders and beyond.

But it was a city in a pitiful condition. Smashed from end to end by constant bombardment, Ypres was a pile of ruins made haunting by the tiny glimpses of its former grandeur. Every sign of the immense wealth of medieval Flanders was displayed in lavish exuberance before the horrors of modern war arrived: a huge market square surrounded by fine gabled houses and shops, and a soaring cathedral, while the Cloth Hall, Europe's largest Gothic secular building and an unparalleled wonder, dominated the city. All these were damaged in the regular frenzies of artillery fire, but the cathedral and Cloth Hall, being so massive, could not be entirely erased. Complete disintegration would perhaps have left less of an impact on the mind. Far, far more hauntingly, bits of them survived to stick up as forlorn ruins. Here was just enough for the imagination to work, for the mind's eye to recreate the lost glory, inspiring pity and wonder equally. James Kerr-Lawson, one of Canada's official war artists, caught this precise quality in his huge canvas 'The Cloth Hall, Ypres', painted during 1918 and placed in the Senate Chamber of the Canadian parliament. As the art critic Paul Konody realized, the effect of Kerr-Lawson's work resided in its ability to conjure up what was missing, 'The ruins, as depicted by him, still retain a good deal of the erstwhile magnificence and much of the exquisite Gothic details for which the building was cherished by generation after generation'.[4] Fragments of the past possessed the present. All that was required to turn the Gothic into the Gothick was some trailing ivy. Kerr-Lawson made his depiction of an exquisite mess more haunting still by the inclusion of cavalry trotting past. Cavalry, always glorious, always dominating through the combined height and bulk of man and horse, was reduced to toy soldier scale when set against the huge backdrop of the Cloth Hall's ruins.

Ypres haunted through the very obvious spectres of its history. It meant the ruins of Ypres exerted an awful grip on people, particularly in the early post-war period when these were so extensive. One visitor was unashamedly mesmerized, noting: 'It so fascinated me that I could not tear myself away.'[5] It also meant Ypres was 'the peep-show of the war', as Sir Philip Gibbs noted on watching hordes of Japanese, French, and American tourists wandering through the market square.[6]

The tourists viewing those ruins added to the increasing hustle and bustle of the city as it started to come back to life, no matter how precariously. Starting with a trickle, but quickly gathering pace, the original inhabitants returned to rebuild their homes and livelihoods. They were joined by plenty of others with no previous connection but a very keen eye for the main chance. On the minds of all was the issue of reconstruction, but the opinions expressed on it differed hugely. For some, this was the perfect opportunity to do something radical and innovative. Ypres and its surrounding villages, including the farms, could be built according to modern design principles, hugely improving the lives and conditions of all. For others, including the vast majority of local people, the only solution was to rebuild their beloved city and countryside exactly as it had been, accepting the odd good idea and improvement suggested by the modernizers along the way. Providing an overview of the damage and destruction inflicted on the major historic buildings of French and Belgian cities along the battlefront in December 1919, the *Architectural Review* posed the main question of the moment: 'Shall they be reverently and completely restored, or shall they be allowed to remain in their present condition as so many memorials of the barbarity of the enemy?'[7] As the debates – and many arguments – started, another question was being considered. Was it possible to do anything on a site so severely damaged?

In the spring of 1920 the artist and critic Julius M. Price made a tour of Belgium to assess conditions for himself. After carefully surveying the vandalism, conscious despoilation, and results of the wholesale theft committed by the Germans in the occupied zones, he turned to the battlefields. At this point came the true measure of destruction. 'All this devastation, however, pales into insignificance when compared with that of West Flanders within the zone comprising the Ypres salient,' he wrote, 'for there the whole area has practically to be procreated, and it will take generations to achieve this, as the very soil has been annihilated.'[8]

The process of annihilation had elevated Ypres to a higher status. As an altar of imperial sacrifice, Ypres was no longer a place of quotidian human activity but a holy place. Its ruins were stark testimony to the endeavours of the British Empire in defence of Belgium and civilized values. This concept underpinned the school of thought on reconstruction and was the one that appealed to many in the British Empire: Ypres should be left in ruins. Rebuilding would be sacrilegious.[9] Winston Churchill, the secretary of state for war, was convinced the city should remain in ruins as a testimony to the experiences of the British Empire. 'I should like us to acquire the whole of the ruins of Ypres,' he told an Imperial War Graves Commission (IWGC) meeting in January 1919, continuing, 'A more beautiful monument than Ypres in the afternoon light can hardly be conceived.' He then concluded on a note of typical Churchillian rhetoric: 'A more sacred place for the British race does not exist in the world.'[10] Henry Beckles Willson, who before founding the Ypres League was a journalist and writer before serving on the Canadian general staff during the war, was equally convinced. By 1919, he was the senior British military authority of the town.[11] Concerned at the pace of reconstruction, even if it was in the form of temporary buildings, he sought to persuade the Belgian government of the importance of Ypres and the dignity of its ruins.[12] Taking his argument to the wider public, he published a guide to the city in 1920 under the resonant title *Ypres: Holy Ground of British Arms*. By this time the Belgian and British authorities had taken steps towards preservation of the ruins. The cathedral and Cloth Hall sites were fenced off and notices posted announcing their sacred nature and warning against the removal of the piled rubble as souvenirs. Such steps were designed to influence public behaviour in the centre of Ypres, ensuring respectful contemplation in what was known as the 'Zone of Silence'.

As all realized, maintaining this position indefinitely was unlikely. Every instinct was for rebuilding, and the main square was already surrounded by a collection of huts and cabins, dedicated mostly to servicing visitor needs. Ypres was going to be rebuilt, and it was going to be rebuilt largely as it was before it was caught in war's inferno, a notion that the city architect, Jules Coomans, supported entirely.

With the preservation of the ruins no longer a certainty, an alternative plan for British imperial commemoration was devised in consultation with the Belgian authorities, local and national: a memorial on the site of the old Menin Gate. Planning for a major memorial in Ypres had commenced in 1919, when Sir Reginald Blomfield, the eminent architect, had been asked to visit the city to consider potential sites and forms. After a close study of the ruined city, Blomfield drew up his conclusions and initial designs. As he did so, the permanent and temporary buildings of the new Ypres were mushrooming everywhere.

For Beckles Willson the force of this tide was to be resisted fiercely. As well as his book, he poured out articles for British newspapers, venting his fury at the desecration of the sacred site. Two fears haunted him. First, that the local officials were set on tearing down the ruins and rebuilding the city as quickly as possible, and second, that the current population was unrepresentative of true former residents, being outsiders who had flooded in to make money from tourism. He labelled such people 'commercial Vandals' who had created an 'ignoble environment' in this 'most hallowed ground on earth to Britons'. Continuing his passionate onslaught against those he believed were cheapening and coarsening a revered site, he threw down a challenge: 'If the Belgian *cafetiers* and souvenir merchants can be bought off, let us, in God's name, and the name of our dead, buy them off.' This had to be

Figure 8.1   The Cloth Hall, Ypres, 1923.

done, for Ypres was, indeed, 'the Holy Ground of British Arms'.[13] Similarly, in an article for the *Daily News*, he complained about the road to the Menin Gate being lined with 'cafés and dramshops'.[14] Another thunderous article was rumbled out in the *Times,* in which he vented his anger on a particular obsession, a cafe on the main square called the 'British Tavern'. For Beckles Willson, such a name was almost blasphemous, and its tawdry commercialization epitomized what the city had become.[15]

Stephen Graham was equally appalled by Ypres, and yet he also perceived its special, unique spirit: 'This Ypres is a terrible place still [because] there is no life when night comes but tavern life ... Death and the ruins completely outweigh the living ... There is a pull from the other world, a drag on the heart and spirit. One is ashamed to be alive.'[16] It had become a city of tourists, and in doing so had lost its soul, making it feel desolate. The strangeness of Ypres by night that Graham had noted was felt by many: after the rude vitality of the day, it became a place of deep darkness. Sounds emerging from the wooden huts and cafes echoed around the ruins, adding to the sense of eeriness. Walking the streets, the visitor had to be careful not to fall into one of the many holes and ditches while fighting off the sensation that ghostly soldiers still inhabited their old billets.[17]

Figure 8.2 The British Tavern, Grand'Place, Ypres, c. 1920.

THE BUSYNESS AND BUSINESS OF YPRES

During the daylight hours the market square was full of visitors staring at the ruins, sitting at cafes, and starting or returning from tours of the surrounding battlefields. It was also the time when the souvenir and postcard hawkers were busy. Their activities annoyed and insulted some, causing them to sympathize with Beckles Willson's sentiments. Returning to the city on a visit to the battlefields in 1920, Major Charles Fair felt it was cheapened by its very obvious appeal to the tourist, especially in the number of children clamouring to sell him postcards and chocolates.[18]

One of the most famous of the souvenir sellers was Henri Duprez. Made armless in a pre-war industrial accident, he sold souvenirs from a tray slung round his neck. Doubtless many visitors thought he was a disabled war veteran, and doubtless Duprez did little to disabuse them. He was, of course, a perfect butt for Henry Williamson's arrows. The latter archly referred to this well-known Ypres character as a man in 'robust health and ... sound economic position', clearly implying that he was not an impoverished veteran worthy of support from well-meaning, but gullible, tourists.

The resentment veterans expressed against such people was accentuated by their sense of ownership. Having been the almost sole inhabitants of Ypres for four years of war, the locals seemed like the interlopers to them. This led to an almost obsessive observation of all local habits and practices. Williamson was convinced Ypres bar owners charged British visitors far more than they did locals.

Seemingly desperate to record another calumny against the locals, he claimed a veteran and his wife entered a bar in Wytschaete, a village a few miles south of Ypres, where they encountered a group of young Belgian men who revelled in trying to embarrass the couple by reeling off all the filthy English words they had learned as young boys during the war, before facetiously asking what the terms meant. The veteran had eventually got so angry that he assaulted all three men and managed to scare them off. After recounting this nasty little tale, Williamson did at least soften it by then transforming it into a story about local pride and identity, for when the man and his wife reached Messines they were greeted warmly and told it was typical of the common people who lived in Wytschaete.[19]

The Canadian veteran Will Bird also felt slighted by the attitude of locals. 'Most of the men one meets in the Salient seem averse to conversation, [they] are not friendly', he wrote. One man confronted him and angrily blamed the British for the death of his son when a dud shell finally exploded in 1921. Bird asked how the man knew it was British and not German. The man replied it was common knowledge that the British made poor-quality ammunition. But, like Williamson, Bird also glimpsed the reverse side of the coin, and once again it was the people of Messines who rectified things, for the locals drinking at the bar made him very welcome. He was treated to drinks and 'all insisted on shaking hands with me after each drink they had' as they joined in a sing-song of the wartime favourites.[20]

Ferdinand Tuohy was less interested in balancing out stories and experiences, being far more committed to prosecuting his case against the people of Ypres and its surrounding villages. In a waspish piece for the *Sphere*, published on the eve of the unveiling of the Menin Gate in 1927, he condemned the locals for rebuilding Ypres and in the process wiping out a sacred site. Further, he believed that in doing so the inhabitants had swept away a unique relic of the war and with it the chance to make it a tourist attraction without parallel. If the ruins had been maintained, the tourist buses would have stopped, he argued. The visitors would have been 'invited to descend and *walk* where the old front line of the Salient ran'. Instead, they now zoomed past, intent on completing their hectic schedules before rushing back to Ostend in time for tea. It left the sacred sites the possession of hordes of 'flappers giggling over kodaks and "snapping" each other up against the ruins of the Cloth Hall'. Having failed to preserve the single most important aspect of the city, the ruins, Ypres had ruined its own fortune, he claimed.[21] Declaring the city as stagnating economically, but without truly justifying the claim, Tuohy's article descended into a flood of recriminations, inuendoes, and bitterness. In his disgust at the new Ypres, he turned the local people into a mixture of the vainglorious, the exploitative, and the foolish. At the same time, he failed to understand the ironies and contradictions in his own position, for he admitted that everyone in Ypres was welcoming and always very friendly to the British.

Further, like every other British veteran or pilgrim who found Ypres in some way uncouth or reprehensible, there was very little acknowledgement that the city had taken its current form largely because of the British. The British created the tourist market; the demand for beer, tea, cake, chips, somewhere to sleep, postcards for friends and family, and a nice souvenir to bring home created commercial opportunity. Short of camping and using nothing but provisions brought from home, British visitors could not help but require local services and amenities. If there was a problem in Ypres, then the British were very firmly part of it.

Fortunately, not all were so lacking in perception nor so quick to blame the locals. Isabella Marie Imandt, a pioneering female journalist based in Dundee, wrote a glowing piece on Belgium's commitment to reconstructing itself. Focusing on Ypres and its tourists, she rejected the usual condemnation of commercialization, and instead praised the vitality, independence, and initiative shown by locals. Probably referring to the Excelsior hotel, she declared the business 'a lesson in the art of overcoming difficulties' typical of the spirit shown in the city. For Imandt, the people of Ypres were to be admired for their enterprise and certainly not condemned as exploiters of the misery of others.[22] And many of those people of enterprise were working with pride, intent on giving their British visitors a warm welcome while displaying respect for British sacrifice. The Belgian antiquary G.H. Stinglhamber was a leading example of those committed to a high-quality tourist industry. He held a series of meetings in Ypres aimed at improving facilities for visitors. He urged a common set of standards among tour operators to ensure safe and reliable motor vehicles and properly informed guides.[23] Typical of the many businesses working to these standards was the Tea Room on the market square. Such was the owners' determination to make the British at home, they committed themselves to that most homely of luxuries, for their advertising proclaimed, 'If you want a good cup of strong English tea have it at The Tea Room, 9, Grand'Place, Ypres.'

Sitting alongside these many Belgian businesses catering for British visitors was a substantial expatriate community whose livelihood also depended on visitors, whether they be pilgrims or tourists. British status was then used as a badge of honour and quality. Anyone who may have felt that Belgian-run businesses were, possibly, not quite pukka could opt instead for a British alternative. Thus, in Ypres could be found Wipers Auto Services, founded and run by the Parminter brothers, and deemed 'valuable allies' by the Ypres League.[24] (Presumably Beckles Willson found it possible to forgive Percy Parminter for being the son-in-law of Mr Vercruysse, proprietor of the despised British Tavern.) On the Rue de Stuers could be found the Empire Tea Room, which proudly proclaimed its British proprietorship.[25] British Motor Tours, based at the London Garage, Menin Gate, offered daily tours around the salient driven by 'British ex-service' employees.[26]

Among the most vibrant of the British expatriates was Leo Murphy. An Old Contemptible who had served with the Queen's Regiment before transferring to the Intelligence Corps, Murphy was in Béthune during the German spring offensive in 1918. Assisting with the evacuation of the town as the Germans advanced, he helped rescue a family trapped in a cellar, which sparked a friendship, and in 1919 he married one of the daughters. Demobilized in the same year, he and his wife settled in Ypres, where he displayed a distinctly entrepreneurial spirit, starting a range of businesses.[27]

Among his portfolio of activities was the British Information Bureau on the market square. This was a somewhat grandiloquent term for a small bookshop and store supplying everything a visitor might need for a thorough exploration of the battlefields. Murphy was also the proprietor of the Ypres Salient War Museum in the basement of the Meat Hall, which faced the Cloth Hall. He was quite a showman, and this jack of all trades proudly showed off the testimonials provided by his highly satisfied, 'spellbound' visitors. No less a person than Graham Seton Hutchison declared, 'A conducted tour by Mr L.N. Murphy, the curator, round "The Ypres Salient War Museum" at Ypres is a cross between a Prime Minister's Speech and a First-Class Music-Hall turn.'[28] Tuohy was also a great admirer, finding in Murphy a tourist operator of the highest standards. Clearly Tuohy felt tourism was fine when overseen by a good British veteran. And not just a veteran, but an Old Contemptible too.[29]

Far less celebrated, indeed quietly forgotten after his flying start, was Lieutenant-Colonel E.P. Cawston. Seemingly a fine officer during the war, there were, nonetheless, some queries about his behaviour, and those queries increased when he transferred to the Directorate of Graves Registration and Enquiries Unit at the end of the war.[30] After a rapid departure from the DGRE, Cawston then cropped up as the proprietor of the Battlefields Bureau Ltd, with a head office in London and a local one at Brielen, a few miles to the north-west of Ypres. At the same time, he acquired at least a portion of Hill 60. Over the next few years, the precise nature of his stake in the famous visitor attraction was the question of much discussion and confusion, especially as the site was subject to increasingly critical comment. Eventually, he disappeared from Ypres, only to reappear in South London, where he was implicated in an arson case. A man of mystery, and perhaps mischief too, Cawston was a colourful figure, but one the guardians of British probity seemed happy to let slide into obscurity.[31]

STAMPING BRITISHNESS ONTO YPRES

Ypres, and the district around it, was home not only to those who directly facilitated battlefield visiting but also to a large number of IWGC staff. The explanation for this fact lay in the brutal accountancy of the military struggle around the city, which led to the creation of 169 IWGC cemeteries and burial plots in fifty-four square miles: the Ypres battlefields represented an immense portion of the IWGC's global commitments.

As the staff settled into their routines, the married men brought their families from Britain, while the unmarried often settled with local women. By the mid-twenties this meant the area was home to a considerable number of school-aged children, and questions arose regarding their education. IWGC staff expressed the desire to have their children educated in a British manner, which, of course, often ignored the fact that many of them were Anglo-Belgian or Anglo-French. Granting their staff a sympathetic hearing, the IWGC persuaded the British government to despatch an inspector to investigate the situation. Revealing a good deal of British parochialism, he sympathized with the demands of the men, being particularly concerned by the number of children unable to speak English, having been brought up with French or West Flemish Dutch as their first language. He thought it particularly important to ward off the 'insidious influence of the foreign mother-in-law' and concluded a school for the children of Commission employees was vital.[32] Moreover, he identified Ypres as the only suitable location, being the hub of the IWGC's activities on the old Western Front battlefields.

Fortunately for the Commission, the Ypres League, with its influential members Field Marshal Lord Plumer and Lieutenant-General Sir William Pulteney, was equally interested in the subject. Using their status as Old Etonians, Plumer and Pulteney brought the question before their alma mater. Within a short space of time Eton College committed itself to building a school in Ypres as a memorial to the 342 Etonians killed in the Salient. The commemorative idea was simple but powerful. Eton College had produced men capable of faithful service regardless of personal cost, inspired by a beneficial education in a nurturing environment, so in founding a new school it would honour their memory and produce new generations of fine Britons in a location sacred to the Empire. Opened in 1931, the school soon proved popular, and by April 1933 112 children were on the roll. Regardless of precise background and culture, all were exposed to a strict British diet of the English language, Anglican morals, and veneration of Empire Day and Armistice Day.[33]

This act of physical and cultural colonization of a corner of the battlefields deemed forever England was only one element in a bigger project involving physical construction on spiritual and emotional foundations. Among the most important of the physical and spiritual elements was St George's memorial church. Acknowledging the spiritual needs of the expatriate British community, as well as those of visiting pilgrims, the idea of an Anglican church in Ypres had emerged in 1919 and, due once again to the energy of the Ypres League and its patron, Sir John French, who backed the idea with great enthusiasm, fund-raising commenced that year. However, it was not until the mid-twenties that the project matured, and on

24 July 1927 Lord Plumer of Messines laid the foundation stone after presiding at the unveiling of the Menin Gate earlier in the day.

Sir Reginald Blomfield, aware of Roman Catholicism's dominance in the region (an awareness heightened by the chosen site lying almost directly opposite the west gate of St Martin's cathedral), designed a gable elevation for the East End facade, allowing the church to blend in with the surrounding buildings. Construction and furnishing were complete by late 1928, and the dedication service was held by the Bishop of Fulham, under whose administration the church came, on 24 March 1929. Among those present was a group of poor pilgrims, funded by the Ypres League, who had lost loved ones in the battles for Ypres. Their presence made the ceremony even more moving for many; one person present deemed it 'the most beautiful piece of work the League has ever done'.[34]

By the time of its dedication St George's was not just a memorial in itself but also host to a huge number of memorials, including 79 brass plaques, 12 memorial windows, and 176 memorial chairs. Virtually every fixture and fitting had a memorial function, from the altar linen worked by a 'Miss E. Booth and friends' to the tower paid for by Sir James Knott in memory of his two sons killed in the conflict.

The church, under its first incumbent, the Reverend G.R. Milner, now became the hub of two communities: the British residents, mainly IWGC employees and their families, and the huge number of pilgrims passing through Ypres. The usual services were held for the resident community, and special memorial and commemorative services for pilgrimage groups. Playing host to a wide range of pilgrims and organizations allowed St George's to live up to its professed desire to act as an ecumenical place of worship and was by no means the sole preserve of Anglicans.[35] For most visitors, attending a service at St George's was an extremely important and moving experience. One pilgrim wrote: 'Up betimes on Sunday morning and to the church of St George, for Holy Communion. Our hearts are filled with pride as we entered the warrior's church, its windows filled with the insignia of gallant regiments who fought and suffered to keep this old town inviolate.'[36] For the British residents of Ypres, St George's served not only as a place of worship but also as the hub of many community networks. The church held a library of English-language books, and the vicar and parish officers undertook a large range of social activities, including support of the local sporting clubs founded by the IWGC employees and members of the local British Legion branch. At the same time local people adopted it into their community. In 1929 *Het Ypersch nieuws* began publishing a 'Notices and Forthcoming Events' column in English, listing services and other activities of interest to British residents.

Another component in this British domestication of Ypres, firmly interlinked with the life of the church and the school, was the British Club. Home to the Ypres branch of the British Legion, an IWGC employees club, and the administrative offices of the numerous sporting and leisure activities run under its auspices, the club was vital to the well-being and joy of many. In the winter of 1924, A.W.K. Hawking, the club's treasurer, appealed for contributions, in particular books for the library and money to re-cover the billiards table. Hawking stressed that the vast majority of the members were gardeners and IWGC employees, as well as veterans, and thus underlined their special status as worthy of support and attention.[37]

The impressive collection of school, church, club, and other buildings and possessions gave the British community a significant profile in the city. Indeed, this was recognized by the Belgian government in the early thirties, when it granted special legal status to what was labelled the 'British Settlement in Ypres'. A small imperial enclave in West Flanders had emerged.

THE MENIN GATE

At the heart of this enclave was the greatest memorial of the Western Front, the Menin Gate. Although nowhere near as complex architecturally as Lutyens's Thiepval masterpiece, and indeed precisely because it was nowhere near as complex as that undertaking, Sir Reginald Blomfield's Menin Gate was the memorial most deeply revered by the British people. And while the dominions had their own special sites across the Western Front, the entire Britannic community also recognized and acknowledged its importance. Unlike the other combined memorials to the missing and battle exploits, the Menin Gate was the only one to be truly pan-imperial. On its 1,200 individual stone tablets, arranged into sixty panels, were inscribed the names of the nearly 55,000 missing of Australia, Canada, India, South Africa, the West Indies, and Great Britain, while New Zealand's contribution was acknowledged in a special commemorative panel (New Zealand opted for the original IWGC solution to the missing by commemorating them in cemeteries).

As Blomfield's greatest work of public architecture, the Menin Gate was the concentrated essence of his inspirations, ideals, and style.[38] As with his design for St George's church, Blomfield drew upon the history of Ypres and the precise history of the site. The result was a Wren–Vauban fusion, a bastion gateway for a ramparted, fortress city with the delicacy of St Bride's and the solidity of St Margaret's Lothbury or St Vedast's Foster Lane. Whereas the Thiepval memorial says, 'Stop. Wait in awe-filled (awful?) wonder,' the Menin Gate embraces and sucks in the visitor. This was partly achieved by Blomfield having the advantage of working on a roadway site, allowing him to create a tunnel in the form of an elongated arch 115 feet long by 66 feet wide, which he described with great poignancy as the 'hall of memory'. A coffered ceiling punctuated by three circular apertures to allow light and ventilation also emphasized the size and span of the arch, drawing the eye up and around. Doric columns topped by entablatures give way to a pedestal carrying sculptures by William Reid Dick. At the western (city) end the sculptural work consists of a sarcophagus draped in a Union Jack flag. Reid Dick gave the flag the appearance of a winding sheet and thus implied that the dead had risen and departed, creating a literal cenotaph.

At the eastern (Menin road) end a British imperial lion couchant was sculpted. Impassive and at ease after its labours, the lion personified the sense of stoicism and endurance displayed by the British imperial forces, carefully avoiding any sense of triumphalism or hubris. The impression is reinforced by the simple, bold inscription: 'To the armies of the British Empire who stood here from 1914 to 1918 and to those of their dead who have no known grave.' Along the northern and southern sides run elevated loggias, supported by Doric columns. Here Blomfield created graceful promenades, demanding a slow, deliberate tread while bringing the pilgrim into close contact with the name panels. Names stare out from floor to ceiling. Then, by using brick and Euville stone, Blomfield knitted the great gate into the landscape. The brick created harmony with the ramparts; the creamy white French Euville, with its allusions to Portland, linked Britain, France, Flanders, and the war cemeteries together. However, lest anyone forget the precise order, hierarchies, and allegiances of the dead and missing, tablets bearing the legends 'Pro Patria' and 'Pro Rege' made it absolutely clear that the British Empire did, indeed, stand here between 1914 and 1918 and was here to stay. Blomfield had made a remarkable statement: majestic and magisterial, yet somehow friendly and welcoming, sorrowful and yet uplifting, great and yet intimate, the Menin Gate fully reflects the spirit of the Commission's wider architectural project.

The unveiling of the gate on 24 July 1927 was a remarkable event. Pilgrims poured into Ypres. Standing supreme among them, according to so many British press reports, were bereaved women. St Barnabas Hostels, assisted by the Ypres League, had collected over £2,000, enabling them to take nearly 600 people, filling every available space on their specially chartered steamer. General Pulteney was amazed at 'the fortitude displayed by the pilgrims, especially the mothers' on seeing them arrive in Ypres. For him the secret of the nation's strength and glory lay in these people: 'With such parents, one ceases to wonder at the heroism and endurance of their sons who laid down their lives for King and Country.'[39] Covering the ceremony for the *Daily Mail*, James Dunn told of a woman leaning against a wall, exhausted by her long journey and the hot sunny day, but now happy: 'Hers was the radiant happiness of the hundreds of pilgrim mothers who have made the long journey from various parts of Great Britain to this little Belgian town, and here have appeased the ache of years by seeing the memorial on which is engraved the name of a soldier son long dead but most vividly and tenderly remembered.' Using purple prose, he described the reactions of others standing at the Gate who saw the name of their loved one, lacking a known grave, but finally commemorated with dignity and beauty: 'As their faces shone with this satisfied mother-love, and an ache which had hurt them for years was in one glorious moment swept away, leaving an absolute calm, these tired women seemed to become young and proudly erect.' One woman told him she was going to kiss her son's name when she it saw on the panel; ever since his death in 1914 she had 'wanted to tread where he trod', she added, before concluding, 'I am happy.'

During the service itself emotions ran high and 'woman after woman standing near me broke down and sobbed unrestrainedly'. Poignantly, pathetically, many had brought wreaths of flowers, carefully picked at home and wrapped in tissue paper, which were then placed on the steps of the memorial.[40] For the *Daily Express*'s correspondent, the famed journalist and author H.V. Morton, this was the most moving part of the ceremony. Portraying the women as naive innocents who were 'a little dazed by the foreign turmoil' of arrival in Ypres, when they sat down in their special enclosure at the

Gate 'they did not know that they were sitting on the road to Hell Fire Corner'. Then came the moment to lay their wreaths: 'They came to it bravely, so bravely and calmly, holding their little posies of English flowers. All one could do was to put one's arm in theirs, as if they had been one's own mother, and pointing high up on Menin Gate, spell out a name to them.'[41] For so many people denied a place of commemoration for so long, the time and place for true remembrance had come at last. The pilgrim with nowhere to go had finally found a journey's end. They were made equal with those who had graves to visit. As Lord Plumer famously stated during his unveiling speech, 'He is not missing. He is here.'

THE MENIN GATE AS THE KEY TO UNDERSTANDING YPRES AND ITS BATTLEFIELDS

Few doubted the significance of the site nor the wonder of the memorial. Among the few doubters was Siegfried Sassoon, who famously referred to it as a 'sepulchre of crime' in his condemnatory poem 'On Passing the New Menin Gate'. The Reverend Neville Talbot had none of Sassoon's anger, just a nagging regret. Already utterly disoriented by his return to the battlefields, to him the Menin Gate lacked soul and spirituality. 'I don't think it is beautiful ... I think Pro Regi [*sic*] and Pro Patria just ludicrously inadequate inscriptions,' he wrote in his diary.[42] But such criticisms, made publicly or privately, were in the tiny minority. Rather, the general reaction was that of T.A. Hannam, correspondent for the *World's Pictorial News*, who declared the Menin Gate a perfect memorial for commemorating all. Although each corps and regiment were listed separately and according to rank, there was 'no weakly vain selection of any rank or arm or contingent for prominence. The massive memorial gate and hall seem eloquently to typify death the leveller equally with a proud symbolism of indomitable strength, sternness, and soberness in victory.' For this reason, it would be 'viewed by the world with respect, admiration, and awe – the last because of the seemingly interminable lists of the names of men who were and are not, and as at once the world's finest cenotaph and grimmest warning'.[43]

The unveiling of the Gate was also a time to emphasize, once again, the overwhelming importance of Ypres to the British Empire. Journalist, author, and former artillery office Boyd Cable summed up the feelings in an article for the *Graphic* titled 'Memories of the

Figure 8.3 The Menin Gate unveiling ceremony, 24 July 1927.

Menin Road'. Ypres was far, far more than a battlefield; it was an idea and an ideal:

> For Ypres came to mean from the first days of the war more than a mere town to be held or lost, a square of the great chess game to be yielded or taken. For the Empire and the whole world, 'Wipers' became a sign and a symbol. Dearly enough in the first year we had paid for its holding; none in the latter years would have believed any feeble tale of 'strategic retirement' or 'according to plan' if we had lost it or withdrawn from it. In 1914 the enemy had strained every nerve to take and we to hold it. A prize so highly paid for must, in the minds of the multitude, be a prize worth the price. Through the years the belief strengthened that Ypres must be held at all costs, and if it had fallen, if once that Menin Gate had been forced, the world would have trembled to the crash of the fall. If only for that reason, the Gate had to be held; and for that, or any other reason you like, it was held.[44]

Writing the introduction for the St Barnabas Hostels' souvenir book commemorating the 1927 pilgrimage, Ian Hay, writer and veteran, felt similarly (even if he also denied the strategic significance given to the city during the war):

> Ypres is a particularly appropriate spot for such a memorial, for Ypres was to the British Army what Verdun was to the French, and more. More, because Verdun was strategically necessary to France: it barred the way to Paris. If Verdun fell, France might fall. Ypres barred the way to nothing save a few barren sand-dunes and the North Sea. But we had said to the Belgians: 'So far as the British Army and the British Navy can compass it, this last remaining corner of your country, within which your gallant King still reigns and fights amid the remnants of his Army, shall not be torn from you.' And to the sincerity and completeness of that splendid gesture, fifty-five thousand names of British Missing alone, carved upon the Menin Gate, bear witness to-day.

Little wonder then that the Gate and the Menin Road were described by John Buchan as 'the *via Dolorosa* of our troops, and also the way of triumph. It is the entrance to the Salient, which is the Holy Land of British arms.'[45]

INFORMING THE WORLD ABOUT THE GATE

Those unable to attend the unveiling were given the chance to be present by the BBC's first, large-scale live broadcast from the continent. The newsreel cameras were also present, but the most potent way in which the Gate was made known across the Empire was through the Australian landscape painter Will Longstaff.

A camouflage officer during the war, Longstaff had direct knowledge of conditions on the Western Front and so stood at the Gate as a veteran as well as an artist. Unable to sleep after the emotional odyssey of unveiling day, around midnight he rose and went for a walk. At the Menin Gate he had a vision in which he saw the dead soldiers of the salient rise and march out towards the battlefields. Unable to escape the power of the scene he decided to commit it to canvas and rapidly completed his most famous painting, *Menin Gate at Midnight*. With equal rapidity it achieved a legendary status encapsulating the mystical-spiritual value invested in Ypres by people across the Empire.

Riding high on a wave of publicity, the painting was bought by Lord Woolavington for 2,000 guineas, the highest sum paid for an Australian painting at the time, and was donated to the Australian government. It was initially displayed at Australia House in London, where a group of VIPs, including Blomfield, took the chance to study it. The king subsequently requested the privilege of viewing it at Buckingham Palace. Duly lent for private scrutiny, it was reported that the king and queen returned to the painting daily. On leaving the palace the painting commenced a tour starting in Manchester in April 1928 before moving on to Glasgow, where 2,000 people saw it daily during the two weeks of its exhibition. When it finally reached Australia, public curiosity was even greater. Over 35,000 people viewed the painting at Melbourne Town Hall during a three-week period in February 1929. It then commenced a tour of major Australian cities and in Sydney was displayed alongside a scale model of the memorial. Such was the interest that a poster reproduction was soon printed, which proved hugely popular, helping to fund the Australian War Memorial in Canberra. In this form, the Menin Gate and Ypres reached hearts across the world.

THE ADVENT OF THE LAST POST CEREMONY

Within a year of its unveiling, the aura of the Menin Gate was enhanced still further by the inauguration of what subsequently became a daily ceremony. On 2 July 1928 the 'Last Post' was sounded under the arch of the Menin Gate by local men. An entirely unofficial event, it emerged, apparently spontaneously, from within the soul of the city. The fact that the buglers dispensed with pomp and circumstance, turning up in their ordinary work clothes, gave the ceremony a special poignancy. Some seventy people were present at the original ceremony, but within the space of a year attendance had risen to between six and seven hundred.

In the first two years the ritual was performed between July and September, clearly to coincide with the tourist season. From 1930 the ritual took place throughout the whole year, although turnout dropped noticeably during the winter cold. A new, deeply moving, remembrance ritual had been inaugurated, and Henry Benson ensured his readers knew exactly who to thank for it. The organizers had 'emphatically laid down that no British visitor shall be asked to contribute towards the fund to pay the musicians, and the other incidental expenses, as the Yprois wish it to be their personal tribute to our glorious dead'.[46] Nonetheless, the stigma of Ypres as the centre of venal exploitation and disrespectful practices remained. A *Sunday Express* article published under the barbed headline 'Belgium's Tribute to the Ypres Dead Is Bought with British Money' led the charge. Seemingly based on information supplied by Captain H.H. Chanter, a man with a grudge against the IWGC, the sensationalist diatribe threw a range of accusations at the Belgians and French. The British had to pay the locals to sound the last post at the Menin Gate through subscriptions raised by veterans' groups, it was claimed, and even with this subsidy the service was often carried out only by one or two men wearing their scruffy work clothes rather than the full ceremonial uniform.[47] A complete travesty and almost wilful misinterpretation of the facts, the story nonetheless revealed how easy it was to spread calumnies.

Fortunately, the many pilgrims to Ypres saw the reality of Belgian reverence for themselves. Standing at the Menin Gate for the Last Post ceremony one evening in September 1935, the veteran Rex Sargeant felt a strong surge of emotion: 'As the clear, piercing notes ring out one is conscious of searching emotions and a swelling heart that seems too big for one's chest, bringing contact across a gap of twenty years.'[48] Wilfred Hyde was convinced that the men whose names were inscribed on the Menin Gate could hear the Last Post as it was sounded every evening in their honour.[49] The ghostly guardians of Ypres were saluted and acknowledged by the community they had so valiantly defended. 'Many, many times have I heard the Last Post,' wrote the Canadian veteran Will Bird, 'but never as you hear it at the Last Post of the Menin Gate. It is perfect. It is lovely, exquisite, yet indescribably tragic – the death of another day, a music that seems to finger the very chords of your heart, stifling all else. All veterans go to the Gate in the evening, and one meets men from all corners.'[50]

EXPLORING THE YPRES SALIENT AND BATTLEFIELD

The veterans who passed through the gate stepped out into the old battlefields as if the memorial was not just a physical portal but also a temporal one. On his first night in Ypres, immediately following the Last Post ceremony, 'A.C.K.' and his chums set off for a walk along the Menin Road. While tramping along they 'tried to imagine the scene as we used to know it and then to bed, tired but satisfied after our first day'.[51] Once out of the city and wandering along the roads, the salient the veterans knew was literally beneath their feet.

In the early autumn of 1931, the *Ypres Times* reported a catalogue of collapsed dugouts around Ypres, starting in 1925 when paths along the ramparts had suffered subsidence due to the rotting of timbers; the canal bank near Essex Farm Cemetery had caved in for the same reason. Things were then relatively calm until the winter of 1930, when there was a major collapse on the Ypres–Lille road as a German dugout system gave way. Soon after it was the turn of the Menin Road at Hell Fire Corner, when a tunnel under the road buckled. This was immediately succeeded by a collapse at the junction with the old Cambridge Road trench, the site of an aid post, causing a chasm some 50 square feet in size to open up. Continuous heavy rains then made repairs almost impossible, requiring the diversion of both the road and tramway. The rains were probably also responsible for the collapse of German tunnelling galleries at Clapham Junction, creating a huge trench along the road and removing a vast quantity of rotten timber.[52] Restoration truly had been no more than skin-deep in many places.

The intensity of action around Ypres combined with the lingering marks of war meant, for veterans, it was *the* battlefield, the place set apart from all others. Standing in the market square in the summer of 1922, Captain Mee was gripped by his memory: all the old feelings of gut-wrenching fear came back to him.[53] In the autumn of 1935, the Reverend S. Gibson, former chaplain of 63 Division, contributed five articles to the *Sunderland Echo* about his recent battlefield tour. The series culminated in an Armistice Day piece on Ypres, and Ypres was *the* destination throughout: the only suitable subject for an Armistice Day contribution. Gibson admitted as much in his opening paragraph: 'Ypres and the immortal Salient! For a week or more the route had lain through old, familiar places, and every inch of ground covered had awakened memories and stirred the emotions. Yet Ypres had been present in our thoughts from the outset, and when at last it was entered an atmosphere peculiar to itself immediately became apparent.'[54] In fact, the presence of Ypres had haunted all five articles, for each carried at least one photograph of cemeteries, memorials, or locations in the salient.

Graham Seton Hutchison believed Ypres was a crucible in which the very essence of the imperial armies had been made molten, and in the process found a quality missing from those who never served on its battlefields. Ypres was not only the most horrific battlefield of the war, but also a sealed world incommunicable to those who had not experienced it. The searing experience forged a nobility due to the immensity of the suffering patiently endured. Such intensity – mental, spiritual, and physical – of action impressed itself into the fabric of the landscape. 'Every stone, every tree ... became for us haunted with memory,' Seton Hutchison argued, which meant that 'the Ypres battlefield, both in its spiritual and topographical features and formation, is ... wholly different from the other scenes in Flanders and near-by Picardy'. As a result, 'no man of British blood can but glow' at the defence of Ypres.[55] Will Bird came to a similar conclusion. To him the men who endured the conditions of the salient were set apart and through their actions bested the Germans, who, despite holding every tactical advantage, could never overcome the sheer fortitude of the forces of the Empire.[56] Such was the mystical aura of Ypres, perceptible to veterans and others prepared to tread its battlefields with care.

As visitors, whether tourist, veteran, or bereaved pilgrim, made their way across those battlefields, along the main roads radiating out from Ypres, they came to the other sacred sites of the salient, although in the period immediately after the war identifying them could be a challenge. Atherton Fleming warned of Passchendaele itself that 'there will be some difficulty in finding [it], although there ought to be a bit of the church left'.[57] Once on top of the Passchendaele ridge the view was spectacular, but also sobering, containing, as it did, the immense Tyne Cot Cemetery. For a veteran like Henry Williamson the view from Passchendaele ridge induced silence as he mentally wandered back to his former self.[58] H.A. Taylor, dropping his perspective as an experienced veteran, told people to go to Passchendaele 'as a pilgrim and not as a strategist, for to stand on the ridge and look with a critical eye over this area in which so many thousands of our best lost their lives, is only to ask questions that will never be answered'.[59] Fleming made the misery caused by the battle into a moral burden on all who visited Passchendaele. 'I think the very limit of human endurance was reached during the Passchendaele "stunt," for if ever man had a foretaste of hell it was surely there,' he wrote with stark frankness, before continuing, 'There are many who have reason to remember this place. When you go there try and realise what it must have meant – and cost – to storm the ridge with the weather conditions at their very worst, with no cover worth mentioning, and the enemy fighting all he knew.'[60] The visitor could never be flippant in Passchendaele.

If Passchendaele provoked silence, thinly disguised trauma, and even hints of disillusion, then St Julien was the place of composure and balm. Here was the memorial to the Canadian Corps' defence of Ypres in 1915. Frederick Chapman Clemesha's beautiful sculpture, in the form of a tall obelisk from which emerges the head and shoulders of a helmeted soldier, rifle reversed, chin tucked down in the traditional military pose for respectful remembrance, moved all who saw it.

Particularly affected were veterans who found in the sober, sombre, reflective 'brooding soldier' (as he was labelled) the perfect memorial. Its simplicity and strength, its sense of being irrevocably rooted in the landscape, like an Excalibur driven into the stone by a great warrior, while also being free from bombast, had a hypnotic effect on veterans.

Williamson was both transfixed and transfigured by the memorial, calling it the 'genius of the salient'. He saw in it the essence of classical statuary's ability to reveal universal human truths and emotions: 'It mourns; but it mourns for all mankind. We are silent before it, as we are before the stone figures of the Greeks. The thoughtless one-sided babble about national righteousness or wrongness, the clichés of jingo patriotism, the abstract virtues parasitic on the human spirit, fade before the colossal figure of the common soldier by the wayside. The genius of Man rises out of the stone, and our tears fall before it.'[61] It was a conception of the memorial Williamson shared with the *Evening Standard* correspondent reporting on the unveiling in August 1923. Describing it as 'a monument which affected me beyond the power of stone' reflecting the very 'soul of those who fell', the journalist went on to analyse the architectural merits, which were matched by the inscription, which he felt had almost the power of the Greek epitaph on the Spartans' defence of Athens ('Go tell the Spartans, stranger passing by, that here, obedient to her laws, we lie,' states Simonides' epitaph on the Battle of Thermopylae). 'There is a mysterious power in this brooding figure,' the correspondent noted, deeply moved by the magnetic pull of the memorial, seeing in it the power of all great architecture to move the spirit deeply while being a perfect fit in its own landscape. Indeed, its perfection was such that it worked with the landscape and elements, blending time, space, and place: 'It is conceivable that a grey day might add to the spiritual significance of this memorial; in the blazing August sun its shock is overwhelming.'[62]

H.A. Taylor was another who believed in the unique qualities of the memorial. 'The man who can look upon this beautiful memorial without being moved must be singularly unemotional,' he told his readers. 'Before the completion of the Menin Gate, no memorial in the salient came close to the impression it made on the imagination,' he added.[63]

Unsurprisingly, for the Canadian Will Bird the site was 'the most impressive place in the Salient': 'No words will ever describe the brooding figure that keeps watch there above those 2,000 who died to make Canada a name as respected as any in history. It is wonderful, gripping, forceful. One stands a long time there, no matter what his race or creed, and they told me that when there is a land mist many persons will come in the early morning to look at the memorial. The mist hides all the foundation and, rising from the fog, is the soldier, leaning on his rifle, pondering, waiting, brooding. The effect fills the watcher with awe.'[64] The 'Brooding Soldier' was the genius of the salient, as Williamson claimed.

EXPLORING THE CITY

Back in Ypres, a restored part of the city's historic fabric, the ramparts, exerted a pull of similar emotional strength. Williamson was certainly beguiled by languid walks along the top of the Vauban fortifications. Lured twice in three days, his first excursion came on a fine evening. Listening to the jolly sounds of summer, children playing, cars trundling towards the Menin Gate, suddenly he felt as if he had encountered the ghost of his wartime self. Unlike the sense of dislocation Williamson experienced when visiting other places that his old persona knew intimately, on the ramparts he found it possible to balance the past and present. In this realm of peace, calm, and *rus in urbe*, as he watched an old man contentedly fishing in the moat, Williamson could visualize old scenes while also orienting himself in the present.

A day later he returned, this time accompanied by a fellow veteran. Lounging on the grass and taking it easy, the two looked out towards the ridges they had fought over that had since become the stuff of history books. Now, red-tiled houses dominated a landscape that had once been smashed beyond belief, and everywhere the natural world inspired. Among the flowers, animals, and birds of the ramparts Williamson experienced a rare moment of rapture, redemption, and mending of sorts: 'How sweet a thing it is to be alive and free in the sunlight among the fair grasses of summer, watching the swallows' wings gleaming blue above the water.'[65] The ramparts inspired this dream of summer, akin to a tone poem by Delius, in many other veterans. 'At midnight under a full moon ... I strolled up to the Ramparts for a moments' quiet with the "boys." It was a wonderful experience and the thought that, could this be the scene of so much slaughter but a few short years before, and now all so quiet and peaceful on this summer night,' recorded one.[66]

It is highly likely that this veteran had visited the Ramparts Cemetery (Lille Gate). The cemetery's architect, G.H. Goldsmith, working in partnership with Blomfield, had indeed created a rural idyll, a small jewel box of a cemetery overlooking the moat. Irish yew trees and a yew hedge created something between a rural churchyard and a scale model of a garden designed for a fine country home while overlooking a Flemish landscape. Visiting it in 1933, R.H. Henderson-Bland, a veteran of the Gloucestershire Regiment, was moved to poetry. For him, like his fellow veteran, the cemetery was a place best appreciated on a placid night:

> Calm and lovely is the night
> And the graves are lovely too;
> The moon rides high as if it rode
> With deep intent to strew
> Its beams upon the water,
> When peace is born anew.
> It is well with you, my brothers, it is well
> Sleeping in the shadows of this immortal place
> That saw your comrades pass, and pass again,
> And was the silent witness of their grace,
> And all their holy pain.[67]

Not great poetry, but the sense of inspiration welling up from communion with a place perceived to be special is strong. For Henderson-Bland, standing in the Ramparts Cemetery, the monstrous anger of the guns floated away on the breezes and drifted out on the ripples of the moat, and the war was detectable only through this ennoblement of death.

Away from the cemetery it was difficult to realize war had ever touched the gentle city. As one veteran wrote: 'To walk on and around the ancient ramparts with its surrounding moats and many backwaters the picturesque pre-war appearances of Ypres could now be readily visualised, indeed, it seems hard to imagine that the foul breath of war ever blotted out the pretty scenery one finds at the southern edge of town, the aspect is quite like that of an English park with the summer-clad boating parties wending their way past shady banks and overhanging trees.'[68] Through fire, Ypres, the slaughtered and devastated innocent, had transformed into a beautiful phoenix.

*

Ypres was special. Ypres was an idea as well as a place. In Ypres everything was experienced and expressed more intently, whether complaints about unscrupulous commercialism or the looming presence of the dead. The ghostly home of myriads, Ypres became the commemoration place of all, regardless of their battlefield. It was the Western Front, the concentrated essence of all its battlefields. Lord Plumer was right in more ways than one when he intoned, 'He is not missing. He is here.'

9

Postcards from Arras

Shattered by four years of bombardment at close range, Arras holds a place second only to that of Rheims among the 'martyred towns' of France.
Muirhead's *North-Eastern France* (1922)

The early sixteenth century Hôtel de Ville, with its Gothic façade and belfry, were reduced to a heap of stones. The picturesque Flemish houses, nearly three centuries old, and built with arcaded façades and gables, were all grievously damaged by 1917.
B.S. Townroe, *A Pilgrim in Picardy* (1927)

Arras was rich in architectural wonder, the *Grande Place*, with its market centre, being one of the most picturesque in the world. But Arras was bombarded day and night during the Great War.
Graham Seton Hutchison, *Pilgrimage* (1935)

For the British, Arras, with its quaint colonnaded squares, gabled shops and houses, fine cathedral, and town hall, stood for a certain vision of France. It lacked the sophistication of Paris, in particular the capital's rather more louche side. Rather, Arras was less challenging to Anglo-Saxon sensibilities while still being intensely French. It was the city of historical romance, of lace jabots and cuffs, great kegs of wine and brandy in deep, musty cellars, and the chink of spurs on cobbles. In short, it was a city seen through the eyes of Alexandre Dumas, rather than those of Émile Zola or Marcel Proust. This conception of a tourist picture postcard place thus greatly influenced the way it was perceived by those who arrived after it had been reduced to ruins.

The martyrdom of Arras was a protracted affair. The war arrived early, with German forces entering the city on 6 September 1914.[1]

It was a brief stay, as they were forced out three days later, but only after large quantities of goods and stores had been requisitioned. The fighting around the city continued, and as the line settled in the autumn, Arras was left at the apex of a bulge in the front that bowed back to its north and south. The city's exposure to shellfire was made worse by the fact that it sat in a depression; the Germans, occupying the surrounding high ground, were able to direct artillery fire on to it with ease.[2] This associated Arras with that essential technical military term of the conflict – the salient – but even more important was its intertwining with the essential geological term of the Great War – the ridge.

Of course, Arras lost the struggle to be thought of as *the* salient city to Ypres, its Flemish cousin to the north. When it came to the ridge, though, it was more than able to hold its own. Thanks to the titanic struggles fought for their control, the ridges of the Western Front became totemic for the British. They loomed down on British and Empire troops throughout much of the war, and have loomed over its history and memory ever since. To the distant north of Arras there were those of Aubers (ones only for those visitors with a committed interest), while the Ypres salient was bound by Messines and the only other true contender for the title of 'ridge of the war', Passchendaele. To the south, on the Somme, there were those of Thiepval and Le Transloy, but these were more often perceived in terms of the villages sitting on them, rather than the spine-like crests rippling through the landscape.

None of these could quite compete with that just beyond the Arras city limits: Vimy Ridge. Sitting on the north side of the city and running in a diagonal from north-west to south-east, Vimy Ridge not only dominated the sightlines to Arras, but also ensured mastery over the heart of French industry in Lens and the coal mining districts surrounding it. And, of course, from the autumn of 1914 the Germans held it, as the British found when they took over the sector in the spring of 1916.

Overlooked, and with the enemy no more than two-and-a-half miles away in places, there was no way Arras could escape severe punishment. The first phase of heavy bombardment came in October 1914. Deserted by its civilian population, Arras was left to the soldiers, some in charge of the materials piled ready for movement up the line, and others transiting to and from duty in the front line trenches. Although shelling was a daily occurrence, only occasionally did it lurch into intensive, concentrated fire, which created weird anomalies in the city's condition.

On the eve of the great British offensive of April 1917, of 4,521 houses (excluding the suburbs) only 292 had not been damaged in some way. Some 926 buildings had been totally destroyed, 1,595 damaged beyond repair, and 1,735 hit but capable of restoration.[3] And yet, within all that mess, the railway station went almost completely unscathed. A photograph taken in that same month shows the fine wrought-iron trapezium covering the main platforms in good condition; only the glass shattered from the lattice-fronts at either end and the grass growing between the tracks give it away as a place where life is not quite as usual. Sir John Singer Sargent captured the same element of incongruity in his painting 'A Street in Arras'. The result of his visit to the city as an official war artist in the summer of 1918, everything about 'A Street in Arras' is beguiling, but odd, starting with the Scottish troops sitting on the ground, slouched back against a very high wall. The soldiers are part of a modern war, as revealed by their steel helmets and rifles, and yet there is the archaism of their kilts. Overhanging the wall are the fronds of trees, and the architectural features of the wall, including the fine portico and pediment surrounding a shut gate, reveal it to be late seventeenth or early eighteenth century in origin. But the dominating feature is the rent in the wall caused by shellfire. This has revealed a lean-to coach house immediately behind the wall; containing a fine carriage covered in dust caused by the collapse of masonry and timber, it is damaged but not beyond repair. An elegant and civilized remnant of Arras's history has suddenly been dragged into industrial warfare. It has been disfigured by war, but not quite desecrated entirely. Instead, the bombardments of great regularity but varying intensity have turned the city into a strange open-air museum of fine buildings rapidly deteriorating due to the extraordinary conditions. Wraiths from the past rub shoulders with a transient military population, conjuring up a strange world dominated by its own distinctive sense of time.

WARTIME ARRAS

For the British, 9 April 1917, the first day of their great spring offensive, was completely different to 1 July 1916 on the Somme. This time a remarkable advance rewarded the British efforts, and the

Figure 9.1 Arras station, 1919.

early phases of the Battle of Arras saw the Germans pushed back considerable distances (by Great War standards). It also resulted in an enduring legend for Canada in the storming of Vimy Ridge.[4] The seemingly invincible fortress dominating the region was finally captured, and the victors standing at the top of the ridge were given a stunning view stretching way into Belgium, and even as far as the coast on the clearest days. Within the immediate eyeline was the wreck of Lens and the myriad slagheaps marking the coal mines of the district.

With the Germans pushed back, the way was opened for citizens to return, and Arras suddenly flourished again, with shops, cafes, and restaurants recommencing a trade brisk due to the huge numbers of soldiers milling around. For returning veterans, this memory of Arras as a fragment of decent living in still recognizably beautiful surroundings was strong. H.A. Taylor recalled it in terms similar to Singer Sargent's painting. He remembered the wartime city as a place of walled gardens and courtyards containing chestnut trees in which birds sang merrily. Other ranks could treat themselves to beers in the estaminets, and for officers there was a club complete with a bathing pool.

More luxuriously, there was the Hotel du Commerce, 'where, on all but a few occasions, it was possible to get a dinner which seemed like a banquet, and where a flaxen-haired ma'mselle at the cashdesk might bestow a brave smile on you, if you found favour in her eyes'.[5] Atherton Fleming had equally happy memories of the hotel, recalling it as a place where a good meal at a good price was always possible, no matter what the Germans were flinging at the city.[6] Its praises were also sung by B.S. Townroe, who expressed his admiration for the restaurateur who kept his establishment going throughout, sitting his guests at tables surrounded by wall mirrors splintered by shellfire and gaping holes in the walls.[7] The Reverend J.O. Coop heaped similar praises on a bar frequented by British officers near Ypres, 'the "Skindles" of Arras', bringing out another trick for any discussion of Arras – constant comparison with the Ypres region and battlefields.[8]

Tragically, for those who dared to come back and start afresh, the war wasn't quite over for Arras, as the 1918 German spring offensive brought a second evacuation. At least this time it was a relatively brief exposure to danger, as the counter-offensives in August pushed the Germans back once and for all. However, the disruption caused by these sudden retreats and advances left Arras eerily quiet at the end of the war.

Driving through on his way to Mons in the freezing February of 1919, YMCA worker John Hastings Eastwood experienced this disconcerting landscape: 'Then we came to the wrecked homes, the destroyed machinery and guns littering the wayside, the little cemeteries and often single crosses, until finally we rode into Arras, which is almost like a city of the dead: skeletons of houses and buildings everywhere, and even in those which still stood, not a piece of glass, but everything boarded up.'[9] Another YMCA official noted the squares, with their 'curious relics of the period of Spanish domination in the seventeenth century' and the examples of 'Hispano-Flemish' architecture dominating the city centre. All were now in ruins, he added, and were likely to remain so 'as a tribute to German kultur and a memorial of the war'.[10]

It was the appalling gap between what Arras once was and what it had become that inspired shocked and pitiful comment. An early arrival was the Canadian journalist, John W. Dafoe, who saw the city in the spring of 1919. For him it was a 'pitiful spectacle of a huge collection of uninhabitable houses – domestic shrines from which the fire has gone cold and can never be revived'.[11] Two years later, and despite the commencement of reconstruction work in the city, the state of Arras retained its ability to shock British visitors. Charles and Amelia Jones, who were following in the footsteps of their son, who was killed in 1916, declared Arras and the villages in the vicinity to have 'absolutely disappeared except for heaps of rubble and debris'. Only a very few people had returned, living in former army huts.[12] In 1922, the journalist Henry Benson found the city a web of ruins still, and told his readers that it was one of the worst hit cities in France.[13]

It is little wonder that the West sisters, touring the battlefields in May 1919 after wartime service in France, believed the city was going to be left in ruins by the French government.[14] That was hardly likely to happen given the importance of Arras to the local economy, but no matter how much effort was put into reconstruction, the sheer scale of destruction across a sizeable city meant the effect of the war was bound to linger on. Arriving in the early autumn of 1926, the veteran, W.A. Michell, discovered a place far from fully restored. 'Arras is by no means reconstructed,' he wrote, 'On every hand builders are hard at work still, removing the debris of ruined buildings and constructing new ones in their places. Many, too many, of those which remain standing show only too clearly the marks of shelling.'[15] As late as 1930, a pilgrimage group found houses in ruins, the roads in a poor condition, and little to commend the city 'to the sightseer', although its charms must have grown on the chronicler, W.A. Francis, secretary of the 2 London Regiment association, for he concluded his account by thoroughly recommending a visit.[16]

ARRAS: THE YPRES OF FRANCE

When considering the destruction and reconstruction of Arras, the yardstick and comparator all reached for was Ypres. Thiepval could have been the British Empire's Ypres in France: it had all the right elements of utter devastation, but in the end, it was its sheer inaccessibility, its utter lack of anything resembling anything, that ruled it out. Instead, that status was conferred on Arras, a city whose road and rail links, and beautiful, historic architecture, had been destroyed. However, unlike the obliteration of Ypres, just enough of the fine architecture of Arras was left standing to provide a reminder of its former glory. 'The wonders of Ypres were powdered: those of Arras almost suffered the same fate,' declared Graham Seton Hutchison.[17]

The existence of such an extensive skeleton provided the framework for the imagination, allowing visitors to envisage what it once was. In turn, this could make Arras a more evocative site than the monotony of the flattened and pulped battlefields.

Dafoe wandered along the wide colonnades of the Grande Place and found it 'a dreadful, heart-rending ruin'.[18] For R.H. Mottram there was more tragedy in the destruction of Arras and Ypres than there was in the devastation of the countryside. In his view, the sheer power of natural forces engendered its own form of recovery, whereas 'with the works of man it is different. A blade of grass can be replaced by a blade of grass, and a poplar by a poplar, but a beautiful and historic town like Arras (or Ypres for that matter) can only be reproduced in facsimile'. Mottram regretted the fact that the reconstruction of Arras had, by necessity, included many entirely new buildings, but he did concede that where the old had been replicated it had at least been done well, with 'the most valuable of the old buildings, those that serve as specimens of the architecture of an age ... well reproduced'.[19]

Taylor felt similarly. Commenting on a new apartment block on one of the main boulevards, which was furnished with all the latest amenities and fittings, he couldn't hide his shock: 'Flats in Arras! One almost wished for another war to blow them away. It seems incredible to think that along this very boulevard, in 1915, barbed-wire separated the French and German trenches.' However, Arras also inspired something akin to affection in him because it was a 'patched' city. Spared the utter destruction of Ypres, it required not wholesale rebuilding and reconstruction from end to end, but a mixed process requiring gradual repair and restoration. Arras was a city heavily marked by war, but not erased by it. The remaining original fabric then suffered when the legacies of the war exposed themselves. And so, every time an old mine gallery under the city began to crumble, 'subsistence occurs. Drain pipes are broken, and other underground services are affected. That does not happen in Ypres. Everything is completely new and recently installed.'[20] Given that such a modernization process was due to the obliteration of the city, Taylor doubted the citizens of Arras were jealous of those in Ypres. Showing slightly more subtlety about the opinions of local people, Townroe noted some were 'glad that the old street lines ... with their quaint turnings and their unexpected vistas' had been retained in the rebuilding, but others maintained that 'by not planning the new town on more modern lines' a great opportunity had been missed.[21]

Arras was thus remembered as the city of decorative architecture, which had held on to the vestiges of civilization by a thread as delicate as the Flemish lacework for which the city was famous. Unlike other places on the Western Front where the war created its own history, akin to a previous geological era, leaving the visitor nothing to be remembered of any pre-watershed existence, Arras never lost the sense of its glorious past. Nearly every writer-veteran making a pilgrimage to Arras found themselves unable to resist defining the city by reference to that past. For Graham Seton Hutchison, Arras was a place of 'old houses, arcades and sculptural glories'. Townroe agreed, describing it as a fine example of the seventeenth-century world of Spanish Flanders.[22] Given his obsession with the history of the Flanders region, R.H. Mottram took the opportunity to stress the importance of Arras, and how it sat at the intersection of French and Flemish cultures with the curious admixture of the Habsburg Spanish.[23] It was H.A. Taylor, however, who provided the richest paean to that past:

> History can suggest few contrasts more dramatic than that provided by these drab phlegmatic warriors of the twentieth century, with their gas-masks slung over their shoulders, and those young Gascons of Cyrano de Bergerac's company who, in days when war had some romance to redeem it, swaggered through the cobbled streets of Arras arrayed in gay doublets and plumed hats. With delicately fashioned rapiers, these bold cadets counted themselves adequately armed for the worst that might befall them in tomorrow's battle on the hills of Artois.[24]

In carefully comparing and contrasting the war he knew with these cavaliers of the past, Taylor revealed an unwitting similarity with Singer Sargent's interpretation of Arras as a crossroads where past and present met. Clearly, the ghostly footsteps of Arras exerted a powerful effect on the imagination.

If this was the world of Dumas's cavaliers, then Arras was also home of one of the great roundheads, which again few writer-veterans could resist mentioning. Robespierre 'was a native of the city', wrote Seton Hutchison, a fact also pointed out by Townroe, Taylor, and many others.[25] Veterans making their way back to Arras were therefore as likely to point out important details of local pre-war history as any reputable tourist guide.

EXPLORING THE ARRAS BATTLEFIELD

Arras attracted veterans and visitors not just because it was an important battlefield site in itself, but also due to its strategic position as a focal point for exploration. The writer-veteran Stephen Graham defined Arras as sitting at the hinge point of the British line: Flanders looked down to it; the Somme looked up to it.[26]

In a series of articles published in the spring of 1920, the *Dundee Courier* recommended an itinerary. The 'ideal trip' would start at Ostend, allowing for a tour of the Ypres salient while also providing the opportunity of visiting Brussels. Then Arras would become the base for exploration towards Cambrai and 'the scenes of victorious fighting in the late summer of 1918'. The city also allowed the Loos, Vimy, and Lens battlefields to be covered, while, to the south, Bapaume, Albert, and Péronne were accessible, and the Somme battlefields provided 'fodder enough upon which to sate a tourist's appetite'.[27]

Visitors using Arras as a base navigated a world far more devastated than the city itself. In the months immediately after the Armistice this experience could be sobering and haunting in equal measure. Having set off to look at the battlefields around Cambrai, Marjory West and her sister hitched a ride on a French lorry. They found the ride an eerie experience in a silent landscape: 'Mile after mile with only trenches and devastation all around and nothing living but ourselves.' So isolating and disorienting did they find the landscape that they longed to be back in Arras, where there was at least some semblance of life. Once at their destination, the remains of Bourlon Wood, they were disturbed only by a series of explosions a long way off, evidence of salvage companies at work. 'Otherwise there was a deep silence, a stillness that could be felt, save when, blessed sound, a lark suddenly rose singing up to heaven, nature's innocent messenger to the horror-stricken waste.'

On their way back to Arras they traversed the Hindenburg Line zone in darkness and the sights turned into 'one of the most uncanny experiences of our whole time ... [We] expected to see rising the ghosts of the German dead which were said to be lying still unburied at some parts of the line.'[28] Ghostly visions affected many. Seeing traces of the military activity around Arras, particularly in the vaults and cellars, where there were still plenty of old tins and candles, a YMCA official had visions of soldiers 'sitting on their beds smoking their Woodbines at their game of cards, or writing to Mummie or Polly, perhaps their last letter before being hurried off with their company ... to their last fight on the fields of Artois, in whose cemeteries all that remains of many of them are now lying'.[29] Standing in the Grand Place, Sir Philip Gibbs felt the presence of a 'crowd of ghosts, whom I see everywhere in this country of the war, and who, in khaki and kilts, swarmed in all the broken houses' of the city.[30]

The eerie battlefield experience even influenced the visitors' perceptions of the natural world. Henry Benson wrote of the 'clusters of gas-poisoned, leafless trees' surrounding Arras 'standing out on the horizon like grim ghosts'.[31]

As with so many visiting the wasteland of the fighting zones, Dafoe had the problem of trying to communicate the nature and effect of the sight in front of him. Viewing the Arras battlefields from Vimy Ridge on a cold March 1919 day, he found it 'bleak and cheerless beyond the power of words to express'.[32] The French railway guide stated of the same vista, 'The pen refuses to describe what is offered to our eyes.'[33]

Relating the level of devastation was the next issue to tax the visitors. As Olive Edis moved from one shocking site to another in her early 1919 visit, she was left trying to make sense of them, as well as the interrelationships between them. Driving along the crest of Vimy Ridge, she saw 'an unparalleled scene of devastation' in the direction of Lens, and continued: 'As far as the eye could see [was] nothing but chaos, hopeless ruin.' Perhaps realizing the pointlessness of the endless moving between sites of devastation, Edis added, 'And this is but a drop in the ocean of France's desolation – a very small fraction of the bill which no indemnity can cover.'[34] The extent of the damage allowed for no hasty restoration capable of wiping away all remains of the war. In 1928, Taylor took the road from Arras to Souchez, where he found a world thick with evidence in which 'traces of the War mingle with the fruits of reconstruction'. Wire, stakes, shell holes, telephone cable, nose-caps, rusting mess tins, and rifle stocks were strewn everywhere.[35] A photograph of the road between Arras and Tilloy taken in 1936 for the part-work *Twenty Years After*, which looked back on the battlefields 'then and now', described it as 'a typical view of the melancholy highway nowadays'. The silent world was the war's lasting legacy.[36]

THE LOCAL COMPARATOR: LENS

In the reckoning of wrecking Arras was compared not only with Ypres, but with its nearest neighbour, Lens. The recapture of Lens, a hub of industry, was much desired by the Allies, but attempts to do so, in another of the war's ironies, ensured the utter destruction of the productive capacity it possessed. It also meant that Lens was the total antithesis of Arras: if Arras was the world of Paul Delaroche history paintings, Lens was that of L.S. Lowry, with its smoking chimneys and factories in a region often deemed the French 'Black Country' by British soldiers and visitors.[37] Every early visitor to the ruins was amazed by it, and each sought to say the same thing: they had never seen anything quite like it.

Arriving in Lens in 1919, Dafoe had absolutely no doubt where it stood in the ledger book: Lens was 'the acme of destruction so complete as to almost blur the sense of human association.'[38] Sommerville Story was equally awed by the mess. 'There is no scene of ruin and desolation on the whole battle front to be compared with this big town', he wrote in his 1920 guidebook, 'One passes through street after street where scarcely a brick is left on another, all being ground to dust and white powder'.[39] 'A rambling place of 38,000 inhabitants,' J.H. Sowden informed the audience at a public lecture in his native Shipley on his return from an officially organized visit, adding that 'there was not a wall 8ft. high left' in attempt to conjure up a sight imaginable by his listeners.[40] Stephen Graham thanked God that Arras was not damaged to the same extent as Lens, with its 'sinister stare' over the region.[41] Deploying his usual blunt frankness, J.O. Coop declared Lens 'more completely devastated than any large town on the whole British front'. Trying to make the extent of devastation comprehensible, he wrote that, compared with Lens, 'Ypres is a smiling village'.[42] Marjory West also drew a comparison with Ypres. Again, much of her wonder was at the sheer scale of destruction, for, as she noted, Lens was not a small, compact medieval city like Ypres, but a sprawling industrial centre, and yet it had still been smashed from end to end.⁴³ Ypres was the obvious yardstick for devastation across the battlefields.

When the earliest visitors contemplated the wilderness around Lens, they often considered the possibility of human life and activity recommencing in the region. For some, the presence of people deepened their sense of despair, exposing the seemingly impossible task of ever recreating normality in such an environment. Dafoe saw a 'few of the original inhabitants' who had 'crept back and can be seen standing in little disconsolate groups around the dust heaps that were their homes'.⁴⁴

Arthur Yapp, a senior YMCA figure, saw the terrible mess of Lens in the late summer of 1919, but on returning in early 1920 he took consolation in seeing something of the hustle and bustle reviving.⁴⁵

The strengthening pulse of human activity could also be disconcerting. In the spring of 1923 Henry Benson found it slightly incongruous to see a fair in full swing among the ruins of Arras. Especially difficult for him to fathom was the sight of French women, still in mourning black, participating with great enthusiasm. Unlike many British visitors, who could be rather po-faced about any sign of local festivity in what they perceived to be not only hallowed ground, but *their* hallowed ground, Benson declined to criticize them, and indeed admitted, 'I must plead guilty to having spent last evening at the fair', but he then underlined his status as the British outsider who saw beyond the immediate and felt the press of recent history, 'although I was amongst the merrymakers it could not be truthfully said that I was one of them'.⁴⁶

When R.H. Mottram returned in the mid-thirties, he was very pleased to see the romance of Arras had returned, even if he could find precious few people willing to talk about the war.⁴⁷

As ever, Henry Williamson was the one veteran determined to find not one redeemable characteristic in the local inhabitants. According to him, they were a miserable, petty-minded bunch, contemptuous of British visitors, utterly unable to drop their violent anti-German sentiments, and masters of the dark arts of the swindle, fuelled by their easy access to reparations funds. After bringing this welter of charges and accusations, he did present some mitigating evidence, noting the lingering fear of Germans engendered by the recent history of France, and conceded that his own very brief visit to the city may not have provided a fully rounded view. However, having got that small concession out of the way, he returned to the onslaught with vim, as his readers were given the treat of his raw diary notes. Taking no prisoners, he judged Arras 'a dirty, pimply, spattered, pocked place … Arras is a cess-pit'.[48]

Some ten years later the situation had changed, allowing another veteran to pointedly refute Williamson's claims. While Williamson was not referred to by name, each of his accusations was stated and rebutted: this veteran was met with a friendly reception, had a good night's sleep in a decent hotel, and greatly enjoyed seeing the city so neatly restored. Indeed, the spirit of Arras gave him 'the sort of feeling which makes people at peace with a place – and a people'.[49]

COMMEMORATING AND MEMORIALIZING THE BRITISH EMPIRE

Having adopted Arras under the British League of Help scheme, the citizens of Newcastle-upon-Tyne felt a special affinity with the city. One of the most successful of the pairings, Newcastle raised £20,000 for various restoration projects.[50] As an industrial giant in its own right, Newcastle was far more like Lens than Arras, but perhaps the latter's air of civilized manners and pleasures appealed. Newcastle's mayor was certainly at pains to point out the similarities, while also admitting the differences: 'Arras, like Newcastle, was the shopping town for mining villages round about and the centre for a large area of agricultural land held and worked by small but real owners. Though small compared with Newcastle, it was a town a man might feel proud to acknowledge as his native place.' Sharing the platform with the mayor in launching the appeal, the Sheriff seconded these comments, adding 'by "adopting" Arras, they would be adopting a town that would become the dignity of Newcastle'.[51] It could all have been read the other way, though. Arras, through its history and charm, helped to dignify the smoke, hammer, and crash of Newcastle.

Arras attracted the visitor's attention due to the intensity of the battle fought around it in 1917. Although the shortest of the British Expeditionary Force's major offensives, it was also the bloodiest of the war when measured by casualties per day in action. This short but bloody reputation made the appeal of the antique town of Arras to the visitor even more sphinx-like: it never had quite the emotional grip of the Somme, and was usually understood only in contrast to Ypres, and yet it could not be ignored; the visitor's attention was drawn back to it again and again.

Despite the Battle of Arras playing a minor supporting role compared with the star tragic roles of the Somme and Passchendaele, Taylor understood how its roots had entangled themselves into the imaginations of communities across the Empire. He pointed out that the cemeteries around Vimy and Arras contained the graves of men from the entire Britannic world: Canadians, more Scots 'than ever fought under Robert the Bruce', South Africans who 'did wonderful work near Bailleul and about Fampoux', Welsh, Irish, Australians. In fact, the only grouping under-represented was, according to Taylor, the English: 'England's numerical contribution to the battle was naturally, greater than that of any other component. But apart from the cemeteries, only one memorial, at Monchy, speaks of an exclusively English achievement.' Actually, Taylor was not quite accurate, for a few English memorials were clustered around Arras, but he had stumbled onto something: despite a relatively compact battlefield, the entire imperial host could be met in its cemeteries and memorials.[52]

Among the English who felt strongly about the Arras battlefields, the community of Kingston upon Hull stood out prominently. As with almost all associations between British communities and particular sites in France and Belgium, the reason was sanguinary. When the men of the 92nd (Infantry) Brigade, 'the Hull Pals', attacked at Oppy Wood, just north of Arras, on 3 May 1917, they suffered heavy casualties, including the Hull rugby star Jack Harrison, who won a posthumous Victoria Cross. Soon after the cessation of hostilities, Edward Peak, an old school friend of Harrison, and a lieutenant in the Tank Corps, wrote to the *Hull Daily Mail* to say he had explored Oppy Wood and seen the large collection of East Yorkshire regiment graves, but had found no trace of Harrison's. Lieutenant Harrison's body was never recovered, and his name was inscribed on the Arras memorial to the missing, but the sense of Hull's connection with Oppy grew in strength. The first sign of this maturing relationship was the establishment of an anniversary memorial service, which was instigated in May 1923. In May 1926, on the ninth anniversary of the battle, it was announced that Hull was to build a memorial in Oppy itself. A year later, on the eve of the tenth anniversary, the memorial design, by the principal of the Hull School of Art, J.J. Brownsword, was reproduced in the *Hull Daily Mail.*

Suffering was at the heart of the memorial, for its form consisted of a central crucifix bearing an image of Christ, with the inscription, 'He died to make men holy, these died to make men free', enclosed by two flanking walls. The memorial's iconography made Hull's sons fellow travellers to a 'hill without a city wall'. At the same time, the imagery referenced the many wayside calvaries and shrines of France, blending Hull's contribution to the landscape with local traditions, emphasizing a symbiotic relationship between the two communities. The aesthetic qualities and scale of the memorial allowed Hull to express its civic pride and significance. It was the sort of place which not only saw fit to erect a memorial at home, but could also boast one on its own little patch of France. Perhaps, unlike the gentrified burgers of old Arras, the mayor of Hull saw no reason to avoid false modesty. On the unveiling day in Oppy, he ensured all knew the significance of the event, as Hull gained 'the credit and honour of being the first Yorkshire city to commemorate its fallen sons with a memorial on French soil'.[53] It is highly probable that the comments were a not so subtle attempt to cock a snook at Sheffield, with its well-publicized interest in Serre on the Somme.

THE WORK OF THE IWGC

The National Battle Memorials Committee (NBMC) and Imperial War Graves Commission (IWGC) also recognized the importance of Arras, identifying it as an appropriate location for a major memorial. A number of memorial schemes were then considered. Given its architectural heritage, restoration of the city's impressive town hall or cathedral was suggested, as was the addition of a new building of value to local people, such as a covered market. Apart from the costs involved, such plans ran into the psychological and emotional barriers of not seeming quite fitting as a monument intended to commemorate the great acts of sacrifice by British and imperial forces. Of much more interest were the French plans to free up some of the land around the old fortress for future development, which included the widening of a major boulevard, the Faubourg D'Amiens, through the site. In addition, the French had decided to remove the graves of its soldiers from a Franco-British cemetery on the boulevard for reinterment at the Notre Dame de Lorette French national cemetery. With the British graves left in situ, there was the chance to adopt the entire site. Discussions with the local authorities produced a highly favourable response to the idea of a British memorial, which appealed far more than any contribution to the general reconstruction of the city. However, lest the British indulge in a scheme of grandiose scale, the Prefect added that a great arch or gateway (he clearly had the Menin Gate in mind) could face 'criticism and comparison with the arches of the Napoleonic era; it was quite possible that people would say, "Ah yes, a monument to the British conquest of France during the great war!"' By the same token, he was clearly much excited by the prospect of a major British memorial in Arras, and very much enjoyed imagining the kind of unveiling ceremony it would demand.[54]

As with the plans to erect a memorial at St Quentin (see chapter 10), the NBMC and IWGC found local representatives keen for the British to honour their city. And, as with those plans, the Commission found the Paris government far more sensitive to the idea of a memorial arch than the Arrasiens. The architect was also the same, Edwin Lutyens. In September 1919 the Battle Exploits Memorial Committee (the predecessor to the NBMC) had suggested Lutyens be sent to France with Arras on the list of sites for consideration. Although the committee had kept Vimy Ridge in mind as the possible location for a memorial, the knowledge of Canadian interest meant alternative options required exploration.[55]

Fabian Ware, vice-chairman of the IWGC, felt Lutyens was the perfect man for this task. With Blomfield agreed for the Ypres memorial, Ware doubtless felt the pressure to ensure a suitably prestigious project for Lutyens. Arras provided Ware with an answer to the problem, and also allowed him to justify it on supposed suitability of task and person. Ignoring Blomfield's expertise in classicism (he was a great authority on Wren), Ware said his style made him more sympathetic to working in the Gothic environment of Ypres. By contrast, Lutyens's well-known genius for developing classicism in new ways was felt to be more in tune with the fabric of Arras.[56]

Lutyens certainly let his untameable imagination loose when he considered the opportunity with which he was presented. His first design was a long, high wall forming an L-shape with a shorter colonnade, and from the nature of its shape, it seems likely that Lutyens was thinking of it in relation to the Faubourg D'Amiens cemetery. Flanked by pylon towers at the terminals of the colonnade and the main facade, at the centre of the main elevation was a soaring, stepped pyramid bell tower, sitting atop an ascending arch. It was clearly related to his designs for the St Quentin (later Thiepval) memorial, making use of red brick with Portland stone string courses, porticos, pediments, and coping. Like the St Quentin plan, the French central authorities were scared stiff by it. Responding to French concerns, Lutyens altered his design and dropped the height of the central tower by some 30 feet to 124 feet, but even this was still not enough to reassure them. Another issue causing the French to blink was the probable location. If sited near the cemetery, it would be on a major new road sweeping into the city on a north–south axis and linking on to another crucial point of arrival, the railway station. Whether Lutyens would be in the mood for a further set of alterations, having already altered the height of the central tower, concerned H.R. Robinson, the IWGC's Director of Works. With Lutyens's designs for a memorial at St Quentin under equal scrutiny from the French, Robinson was probably worried by the prospect of a double rejection being delivered to the great architect.

With impasse looming, Arthur Ingpen at the IWGC returned to an earlier idea he had suggested for a cloister wall to surround the cemetery on the Faubourg D'Amiens site, which was discussed in the Anglo-French mixed committee. Ware snatched at it. Robinson and Ingpen set off to investigate the site and found it suitable for a memorial capable of carrying at least 70,000 names of the missing 'without undue overcrowding'. Using his friendship with Lutyens, Ware was then able to persuade the architect to revise his designs to suit the scale of the site and the tastes of the French.[57] Within the confines of the requirements, Lutyens managed to retain some elements of his original plans. The clearest survival was the 380 ft wall running along the main road, consisting of red brick encased in stone. Other retained features were the three classical arches supported by columns, acting as entrances to the memorial, and the Cross of Sacrifice, which sat in an apse facing the road, making obvious the cemetery space enclosed within the walls. Lutyens slid his memorial – tasteful, delicate, classical, and yet bastion-like in its solidity – into the real and imagined space of Arras.[58] It was as if Lutyens had restored the broken wall that had so captivated Singer Sargent.

By contrast, the 9 (Scottish) Division memorial at Point-du-Jour, a few miles outside of Arras, brought Scotland to Artois. Standing at a bend in a road on the crest of a hill, the memorial took the form of a traditional highland cairn, with heather, broom, and whin-bush (gorse) planted round its base, creating a tiny Scottish moor. The reason it was sitting outside Arras was to mark victory first and foremost: the dead were honoured by celebrating military achievement. General Furse, who had commanded the division for much of the war, made that clear in his speech on the unveiling day, 9 April 1922: 'The reason why we finally settled on this spot was that five years ago to-day the Ninth Division covered itself with conspicuous glory over the battle-ground we see to the west of us. This was indeed its "Point du Jour".'[59]

Similarly, 12 and 37 Divisions marked their great advances on that day with memorials around Arras. Twelve Division put up a replica of the York Minster cross at Feuchy on the main Arras–Cambrai road, giving it a commanding position close to Windmill British Cemetery.[60] A short distance away sat the village of Monchy-le-Preux. Perched at the top of a distinctive hill on the plateau-like space to the east of Arras, Monchy attracted many memorial schemes. Here, 37 Division erected its monument, consisting of three soldiers finely sculpted by Lady Feodora Gleichen, sister of the division's wartime commander.[61]

'The other memorial in Monchy is peculiar', wrote H.A. Taylor, 'in that its plinth is the ruin of an old house, including a fine old wooden fireplace standing four-square among the tumbled bricks. In this pedestal is all that remains of the old village'. Sitting on top of this battlefield relic was a sculpture of a Newfoundland caribou, one of five such Newfoundland memorials that mark major its engagements on the Western Front. Unlike the other divisional memorials around Arras, this one was associated less with victory and far more with tragedy, being the place where Newfoundland suffered its second highest losses after those sustained attacking Beaumont-Hamel on the first day of the Battle of the Somme. The combination of the ruins of human civilization topped by a magnificent specimen of the wilds of Newfoundland made a striking memorial. W.A. Francis, a 2nd London Regiment veteran, certainly felt its power, declaring it 'the most effective memorial we had yet seen' in his account of the regiment's 1930 battlefield pilgrimage.[62]

ALL ROADS LEAD TO VIMY

Once outside the confines of Arras, there was one destination which drew most visitors: Vimy Ridge. Replete with knowledge from wartime coverage, every visitor to Vimy knew it was a place of importance and that it was especially important to Canada. Such was the weight of association, it seemed as if Canada's footprint had been set on this particular patch of France centuries earlier. Seeing it in early 1919, Olive Edis described Vimy Ridge as a place 'of such inspiring tradition', as if it marked a battle of antiquity rather than one that had not yet reached its second anniversary. Commenting on the understanding that France had ceded the ground to Canada, she added it was a fitting site for a great memorial 'for every foot of this ground had been won at the price of Canada's best blood'.[63]

Standing at the top of the ridge gave the visitor an amazing panorama, which never failed to impress and brought home the wartime significance of the ridge even to those most oblivious to the laws of military geography. 'The immense strategic importance of the ridge will be appreciated by the tourist as soon as he reaches it', Coop stated in his 1920 guidebook, in his trademark headmaster style.[64] Sommerville Story spelt out the significance in clear terms, just in case his (mostly American) readers could not grasp it: 'The strategical importance of a place like Vimy is apparent, as it is an unsurpassed observation post over the country for fifteen miles around, while in the sunken road between the two sides of the Ridge battalions of men could be massed almost unperceived.'[65] Sixteen years later, the veteran and enthusiastic cyclist Bernard Newman encouraged readers of his cycling guide to Northern France to attempt the climb up Vimy Ridge, adding 'it needs no military genius to realize its tremendous importance.'[66]

A contrast to these straight-forward descriptions was provided by the normally phlegmatic Lieutenant-Colonel Lowe, who indulged in a dramatic shift of gear when describing the prospect from the summit. 'A walk along the crest of the Vimy Ridge will reveal to the tourist many things, principally, perhaps, the extent of the observation which was gained by its capture', states his guide, in its clipped, business-like manner, before delivering the pay-off with all the vim of an excited boy-scout, 'The view from the crest is a truly astonishing one'.[67] The writer Wilfrid Ewart, a veteran officer of the Scots Guards visiting in the spring of 1922, identified something else, a unique quality in Vimy due to the astonishing contrasts in the landscape. Suddenly, from a countryside 'as flat as our English fens', a huge fold rose up, providing an excellent vantage point from which the pace of reconstruction was measurable. He could tell Lens was still in a terrible mess, but 'Hazebrouck hives' and Armentières still 'had the basis of a prosperous town'.[68]

Like the shattered remnants of Thiepval sitting on its ridge, Vimy gripped the imagination because it took so long to recover from its wartime state. In 1919 it was impossible to miss the trenches and fortifications. Sharing the same tone as Coop, Atherton Fleming told his readers in no-nonsense terms: 'a thorough examination of the surrounding district will well repay the trouble entailed'.[69] 'Grey and terrible is Vimy Ridge with its line of block-houses and the masts of Farbus Wood', wrote Stephen Graham, unconstrained by the strictures of the guidebook author.[70] These slowly crumbling concrete pillboxes provided Henry Williamson with a platform from which he could study the scene. Even in 1927 it was easy for him to see the traces of trenches 'revealed as blurred lines of whitish chalk-lumps levelled with the brown top-soil of the fields'.[71]

Four years later, the Canadian veteran Will Bird described Vimy as 'deserted, a wilderness [where] everything is untouched'.[72] Though just a short walk away from Arras, and a few miles from Lens, Vimy felt remote. Doomed because of its geography, tunnels and trenches covered Vimy as thickly as underground railway lines in the centre of London. Restoration of a land pounded by incalculable weights of artillery shells and mined consistently, was a fearsome challenge; the additional necessity of hauling everything required for the job up to a naked strip of chalk made the task look near impossible. It's no wonder people stayed away. By March 1921 only 49 per cent of the population of the entire Vimy *canton* (a French administrative district made up of a number of communes) had returned. Unsure as to how much could be done around the restored Vimy village, the problem was passed to the French office for water and forests for the simplest solution available: afforestation. Fortunately, its plans coincided with those of Canada, and 107 hectares were ceded for tree planting as part of the memorial scheme for the ridge. Landscaping the ridge in this way both obliterated and preserved the impact of the war.[73] Viewed from afar, the crest would be topped by a fine wood, adding to the sense of height through the tall reach of the trees combined with the soaring twin pylons of the memorial. Viewed up close, the ground revealed every shell hole and every trace of trench. Before the trees matured, the only thing to be seen was the desolation of the battlefield.

The inclusion of this maze of trenches in the memorial scheme was due to the actions of Brigadier-General H.T. Hughes, the chief engineer to the Canadian Battlefield Memorials Committee, in consultation with his colleague, Lieutenant-Colonel M.N. Ross, the landscape architect to the Committee. As Western Front veterans, both had a strong sense of the importance of battlefield space and its rightful role within a grander memorial scheme. In securing the trenches, their preservation was guaranteed in the short term, but as every ex-serviceman knew, trenches required constant labour to ward off the effects of nature, especially during winter. It was in early 1926, when construction on the memorial was held up by supply problems, that Captain D.C. Unwin-Simson, assistant chief engineer on the project, found the time to consider the condition of the trenches. Seeing their deterioration, he decided to preserve a short section using concrete-filled sandbags and duckboards. With the tunnels also becoming increasingly treacherous, he began work on the Grange system of subterranean passages and chambers, making them safe and introducing electric light.[74]

The veteran H. Channing-Renton visited the site just as the preservation work was reaching its conclusion. His reaction was everything Unwin-Simson could have wished, for Channing-Renton saw in the trenches genuine remnants of the battlefield, creating 'a memorial far more vivid than any group of monuments or statues'.[75] This interpretation made the actual ground, with its deeply incised scars in the forms of trenches and shell holes, a literalized battlefield memorial, doing away with the need for any form of metaphysics, allegory, and iconography. In 1927 the excavation of the Grange Tunnel at Vimy was completed, opening up the entire system for visiting.

With the construction of the main Vimy memorial progressing at a very slow rate, these trenches and tunnels provided the only accessible part of the overall commemorative scheme for a long time, ensuring their status as the focal point of the site. The famous journalist and writer H.V. Morton was one of the earliest to explore the preserved trenches and tunnels. On being taken into the tunnels by Unwin-Simson he felt himself in touch with the past. 'It was as though we had been switched back to April 1917 ... Nothing had changed', he told readers of the *Daily Express*.[76] Amazed to find the candle-niches carved in the walls still black with soot, quantities of grenades, helmets, and ration tins, and the myriad number of names and messages scribbled on the chalk, Morton felt the presence of those 'Canadian boys' who were 'joking, laughing, waiting, quite unconscious that they were carving not only their own names, but also history'. It was the realm of ghosts: 'no matter whether they are alive or dead, their personalities live beneath the soil of France

7. - VIMY-RILGE. — Tranchées Allemandes de 1re Ligne.   German First Line Trenches.

Figure 9.2   Preserved trenches at Vimy c. 1926.

so vividly one expects to meet them round the next corner'. Unable to shake off this impression, Morton went on: 'Here in this dark tunnel, and here only do we seem to meet the men who fought and died. Here only do we seem to see again in the long chalk passages those well-known faces.'[77]

The trenches and tunnels made a similar impression on many others. A veteran who conducted a 'proud mother from Sheffield' round the trenches said they had a sobering effect on her, particularly when she was shown the spot where a shell had landed, damaging the parapet. The sight left both wondering 'whether the trench was occupied at the time and whether any human being went west'. Having seen this 'amazing relic of Armageddon', the woman took 'another scrap of war information ... back with her to Sheffield'.[78] Mrs E.A. Smith, a West Country teacher, joined her husband on the 1928 British Legion pilgrimage and was deeply moved by the entire Vimy site. Her experience started with the approach to the memorial park. At Vimy station the pilgrims were offered the opportunity to walk up the slope or be taken in buses. Opting to walk, Smith detected a particular emotional atmosphere. 'There seemed an air of quiet happiness everywhere and mothers and wives seemed to me to forget their sorrow in their pride at being there, just walking along where many hundreds of lads had walked from 1914 –18'. But this was just the foretaste of what was to come. On reaching the top of the ridge she saw the signboard pointing to the Grange Tunnel and followed it to the restored trenches. With great fascination she described how they led 'back to a series of shafts leading down to a marvellous system of underground tunnels which extend for miles'. Entering the tunnels caused her to descend even further into a different world, in which communion with the former inhabitants was very strong:

> While going through one is constantly reminded of the stoicism of the men, who spent days there and when on duty scratched their names in the walls here and there. There are many remains there still, boots, shrapnel helmets, rifles, bully tins etc. and thanks are due to the Canadian Govt. in allowing guides to take pilgrims through … Here it was very noticeable how groups formed after being guided through. They just gazed, never saying a word. Then, women old and young, seemed to picture our lads within speaking distance of the enemy, and such an enemy, the prospect left them dumb. Pilgrims left this vicinity with bowed heads. It seemed as if no one thought of doing anything else.[79]

Pilgrims were struck dumb, awed into silence: the weight of memory was felt intensely in the confinement of those tunnels.

The effect created by the trenches and underground passages was not restricted to those with no direct experience of life at the front, but also captivated veterans. H.A. Taylor was lured into a world of shadows. Accepting the need for concrete in the preservation process, he saw beyond the artificial solidity, knowing weathering would soften the crisp outlines of sandbags, bringing back something of their original state. For him, the use of such methods was but a small price to pay for retaining the overall authenticity of the site. Passing through the trenches 'fascinating though [they] are, there is something even more fascinating yet to be seen', reveals Taylor, who, like Morton, was enchanted by the tunnel system, the highlight of his tour of the site. He too felt the same thrill in seeing collections of the strewn quotidian objects, such as bully beef tins, wire beds, and rum jars, as well as the graffiti inscribed into the walls. Indeed, he wondered whether this subterranean world might remain the most dominant impression regardless of the broader memorial scheme:

> So full is the tunnel of sights, sounds, smells and memories, so fascinating is the whole spectacle, that one wonders whether the massive stone memorial higher up the ridge, the memorial which is to proclaim the sacrifice of the Canadian Forces here, will be regarded as the important feature of the park.

Taylor answered his own question immediately by explaining the awesome combination of precise site and memorial design. Its vantage point, the impressiveness of the excavations, and emerging plinth all provided more than enough hints at the grandeur of what was to come when building was completed.[80] Others were also impressed by the results of the initial construction phase and could envisage the final effect.[81] 'This wonderful piece of work, only begun towards the end of 1926, will ever stand as a credit to Canada, and a tribute to the undying feats of her sons', commented the secretary of the 1/7 Northumberland Fusiliers association after visiting the site in the summer of 1928.[82] Mrs E.A. Smith's imagination was equally fired by the prospect: 'I should think [it] will be one of the largest memorials in France. It is being built by the Canadian Govt. as a great national memorial, and they have even made a beautiful new road to the spot for the purpose of conveying the materials for building. It was only in its early stages when I saw it but the position could not have been better.'[83] In complete contrast, a pilgrim exploring Vimy in 1932 did not say a single word about the extensive construction site and memorial gradually arising from the works, but instead remained firmly focused on the trenches and tunnels, stating that a visit to Vimy Ridge 'is not complete without seeing the Grange tunnels beneath the Ridge'.[84]

Hampered by delays, the construction of the Vimy Ridge memorial was a long drawn-out affair, and it took many years for the wonder of Walter Allward's design to become apparent. By the early thirties visitors could begin to imagine the full, stunning, effect of Allward's vision. Framed by a purposely created amphitheatre lined with eighteen larger-than-life allegorical figures, and dominated by two central pylons soaring nearly 100 feet high resting on a plinth 170 feet long, the Vimy memorial was (and is) shock and awe in stone. It dwarfs the visitor and makes them master of all they survey at one and the same time; there is a deafening silence, broken only by the wind, which batters the site even on the calmest days, whipping and bucking the flags around the flagstaffs, and rendering speech impossible.

Allward's Vimy memorial, as complex as Lutyens's masterpiece at Thiepval, shared with it a complicated evolution, including the collusion between design and place not envisaged at conception. Where Vimy trumped Thiepval was in its easy accessibility from the major communication hub of Arras. This allowed Vimy to become a tourist-pilgrim shorthand for the entire Arras battlefield, as perceived by a British veteran visiting in the later thirties: 'In Arras, then, the word "battlefield" connotes Vimy, and it is easy to find the reason for this belief. It is in the (now) old story that only in the neighbourhood of that hamlet are any tangible traces of the War of 1914–1918 still easily to be found. For the average visitor this is true.'[85]

Vimy offered another kind of accessibility, an emotional accessibility, through the sculptures. These powerful works provided a human face to the memorial, enabling an empathic relationship in a way Lutyens's riddle inside an enigma at Thiepval did not. Central to this humanizing effect was the figure 'Canada Bereft', which stands on top of the plinth wall containing the names of Canada's missing. Thirteen feet in height, and carved from a single solid thirty tonne block, 'Canada Bereft' is a statue of a woman in a hooded cloak, head tilted forward in grief suggestive of a 'pietà', clutching a laurel branch in her hand. Allward told Edna Moynihan, the model for 'Canada Bereft', that 'he wanted a mother figure with shoulders wide enough to carry the sorrows of her dead sons'.[86] She certainly has the effect of drawing all eyes, as well as all emotions, to her, soaking in tears of sorrow, gratitude, and pride equally.

Taken as a commemorative whole, the Vimy memorial is magnificent and triumphant without any hint of bombast or pretention; it is immensely great and yet immensely humble at the same time; it is the height of inspiration but suggests the utter depth of despair. However, its late arrival on the memorial scene – it was not unveiled until 1936 – also meant that, like Thiepval, it became a part of a schedule rather than the major destination of a pilgrimage for many British visitors. Admiration was often expressed for its beauty and scale, but it was also equally likely to be praised for providing such a fantastic platform for panoramic views.[87] All visitors ticked it off their list, and commented on its wonder and commanding views; all were glad to have seen it, but not all penetrated its depths of meaning or engaged with it in the way they had with the tunnels and trenches.

There were very rare apostates about the memorial park. R.H. Mottram felt the trenches were mere shadows of the real things; they had become replicas of reality. 'There is something wrong', he wrote, 'It is not our trenches, nor even those of the enemy of all those years. The sandbags are too neatly piled, the duckboards too unbroken, the little ledges and cupboards in the walls too neat and empty of grenades and tins of bully. There is lacking that smell of explosives and sewerage, of hasty burials and brazier smoke.' Most importantly of all, it lacked the genuine atmosphere of the trenches, of men going about their routines, of their spirit and humour. Unable to detect these ghosts, he could not 'tune in' to the site. 'Whatever it was that makes us not unwilling to remember those days and revisit those places, it is not present at Vimy to-day', he went on. However, he concluded by noting that it was the only issue he had against the otherwise 'admirable work' of the Canadian authorities at Vimy.[88] As he seems to have realized, Mottram was disappointed by the omission of something no one could provide. This is slightly odd, given his status as both a veteran and a writer: he of all people should have been capable of making the imaginative leap, or mentally recreating the world he experienced, as so many other veterans did, when visiting Vimy.

Much harsher in judgement was the anonymous veteran who contributed the chapter 'Arras and Vimy Revisited' for the part-work *Twenty Years After*, published between 1936 and 1938. He saw the Vimy memorial as an unnecessary hyperbolic statement; its size and scale were desperate attempts to match the significance of the original military achievement. It was the antithesis of the sober simplicity and dignity of the Cenotaph. Less, for this veteran, was most definitely more. He underlined this by noting the appeal of the memorial was at its strongest when seen from afar, when the sun softened the lines and accentuated the warm creaminess of the stone. In other words, Vimy made its deepest impression when nature produced its own *sfumato* effects, blending the memorial into the landscape and making it one with the old battlefield. Although this reaction was utterly atypical of responses to the memorial, he was far more conventional when it came to the other famed aspect of the park: 'No visit to Vimy would be complete without a descent into the maze of reconstructed, or preserved, deep trenches and dug-outs which honeycomb the slopes of the Ridge just as they did in 1917.'[89] The Vimy trenches were the essential mnemonic of the war experience and its essential memorial.

BEYOND VIMY

British visitors did not progress much further than Vimy, which meant relatively few took the time and trouble to appreciate another vast commemorative project, the French national cemetery and memorial at Notre Dame de Lorette, even though it was clearly visible from Vimy Ridge and contained stained glass gifted by the British.[90] Consisting of 20,000 graves (expanded to 40,000 since the Second World War), eight ossuaries, a chapel, and a tall lantern tower spread over some thirteen hectares, Notre Dame de Lorette overwhelmed in its sombre commemoration of French death and suffering. Equally difficult to fathom for the British was Louis-Marie Cardonnier's architectural fusion of Roman Byzantium and the Third Republic. This combination of the basilica and the plots of white crosses and headstones (there are separate plots for Muslim and Jewish soldiers) created a commemorative landscape utterly unlike anything produced by the IWGC.

British visitors tended to focus on the lantern tower, with its beacon (which revolved five times per minute) lit at nightfall, shining a light across the landscape to a range of forty-five miles. It made a focal point for brief stops in crowded itineraries, provoking postcard comments: 'Souchez and Notre Dame de Lorette were next visited. The wonderful French Memorial, a lighthouse and a chapel, and the view obtainable from the ground evoked exclamations of surprise and approval,' wrote the chronicler of one pilgrimage.[91] Taylor summed up it in an equally succinct manner as 'the wonderful lighthouse-memorial ... the French have built on the heights of Notre Dame de Lorette', before moving on to consider another monument.[92]

Occasionally (very occasionally in fact), someone went into greater detail and attempted a deeper engagement with the site. A member of the British Legion 1928 pilgrimage explained the significance of the memorial: 'To the French the name of Notre Dame de

Lorette conveys something like the same meaning as Vimy does to the Englishman and Canadian, and recalls associations of the same character,' before letting his countrymen off the hook by adding that it was 'naturally less known to English visitors to the battlefields than the British cemeteries and memorials in the adjoining district of Vimy Ridge ... The two can easily be seen in the same day.'[93] It was, at least, an attempt to annotate and interpret a site so very foreign. It was the ultimate irony of British battlefield pilgrimages: French memorial architecture and sentiments were deemed strange despite being rooted in the very French soil on which the pilgrims trudged. Instead, the IWGC's prim, neat Portland stone and red brick, or rough-cut stone walls enclosing cheery plantings forging with their flowers an English heaven, were thought to be the commemorative markers most truly in tune with this corner of France.

*

Arras offered the British a place of great practicality and great fantasy. Easily reachable by train, full of good restaurants, cafes, and hotels, it was the ideal place to stay. Rich in history made palpable in its sumptuous architecture, it offered a realm of romantic dreams. Beyond its city limits lay another land of more recent historical adventures, allowing veterans to recapture something of their former selves and others to try to imagine that past reality. As the traces of the battlefields disappeared through reconstruction and restoration, sites such as Vimy Ridge gained an even greater status. A day walking those Vimy trenches could then give way to an evening strolling the arcades and boulevards of Arras. It was quite a combination, providing contrasts and continuities. Will Bird's meditative judgement on Arras perfectly encapsulated the effect it left on battlefield visitors, whether veteran, pilgrim, or tourist: 'It is a quaint city. Its old-fashioned narrow streets, twisted, and with house drains and dirty gutters, are offensive at first glance; then one feels the lure of the whole surroundings, the age, the mystery, the unknown history of those dark ways. An impressionable person will never forget Arras.'[94]

10

Postcards from Thiepval

It must once have been a lovely and romantic glen, strangely beautiful throughout. Even now its lower reach between a steep bank of scrub and Thiepval Wood is as lovely as a place can be after the passing of a cyclone.

John Masefield after visiting Thiepval Wood (1917)

The road to Thiepval, which branches off to the right, is only passable for about 1.5km. From this point the tourist should go on foot to Thiepval.

Illustrated Michelin Guides to the Battle-Fields (1914–1918):
The Somme Battle-Fields, vol. I (1919)

So completely devastated is Thiepval that no permanent rebuilding has been attempted in its ruins. No bright new home farmhouses or tilled fields are to be seen here, nothing but uneven earth, rank grass and weeds. Here and there, seeking carefully, you may find a brick or two denoting the original building line of the main street of the village, but these are the only bricks to be seen. Such structures as exist are wooden shanties, and old army huts. While Beaumont-Hamel rises from its ruins, and Pozières prospers abundantly, Thiepval remains a series of grass-covered mounds ... Thiepval is likely to remain as it is.

Captain H.A. Taylor (1928)

In 1914 Thiepval was a large and prosperous village and commune. At its heart was a handsome chateau with fine gardens, which sat at the highest point on the Thiepval ridge. By the war's end absolutely nothing was left. Endless bombardments had smashed and pulped all evidence of centuries of human activity. All that remained was the natural geographical feature of the ridge, bare and windswept, visible from miles around. Although only three miles from the main Albert–Bapaume road, Thiepval was an island in a sea of mud and devastation. The once fine chateau belonging to the de Bréda family was reduced to nothing more than a peculiar stain in the muck, and no one was sure whether the family would ever return and attempt to rebuild their old home. Without them, no one was sure whether economic, social, and cultural recovery was possible.

It was, of course, Thiepval's geography that did for it. Sitting on the heights of the ridge since the late summer of 1914, the Germans could see right across the Ancre river as it made its sluggardly way along the marshy valley bottom. They could see down to the British rear positions in Albert, and across to the southern side of the Albert–Bapaume road, which ran along a ridge dissecting the battlefield. The Allies were determined to capture Thiepval Ridge. Of course, the Germans knew that sooner or later the Allies would come for them, so they fortified it. In fact, they fortified it with a dedication rare even for an army passionately devoted to fieldworks. At the heart of the defensive system was a complex circuit of trenches known to the British as the Schwaben Redoubt. And yet, on 1 July 1916, the first day of the battle of the Somme, this awesomely powerful position was stormed by men of 36 (Ulster) Division, who, with the speed of their advance, completely overwhelmed the German defences. Agonizingly, as no British units were able to advance on their flanks, the Ulstermen found themselves surrounded and were forced back to their start lines before the day was out.[1] From this Greek tragedy of endeavour – so nearly successful, so dreadfully undermined – a small Picard village was turned into a legend for Ulster. And after the war, it was the people of Ulster who staked a claim on Thiepval, indeed doing so more rapidly than any local inhabitant bold enough to venture into the quagmire. But that was the future. First came the fighting.

As efforts to capture the village ground on, bombardment after bombardment turned the ground into a soupy mess. Finally, in September 1916, the British managed to wrest the remnants of Thiepval from the Germans, following a barrage of approximately 100,000 shells.[2] Fighting returned to the area in 1918, but with nothing like the intensity of 1916, as the British fell back before the German advance in the spring, and then just as quickly pushed them back out in the summer. What the resumption of hostilities did was further ensure the separation of Thiepval from the rest of the landscape.

In the wake of the German retreat in the early spring of 1917, some civilians had returned to the Somme district, and the French government, supported by the British forces, had attempted to restore farms and fields, recommencing agricultural activity.³ But not at Thiepval. Nothing could be done at Thiepval. John Masefield recognized this peculiarly, irretrievably dead atmosphere in Thiepval when he was taken on his official visit in 1917. He imagined a place 'which may once have been lovely' but that was now a mess of ruins framed by 'blasted, dead, pitted stumps of trees' and, more than any other corner of the Somme battlefield, was a place of mud.⁴ Given the conditions on the Somme front, this was some accolade, even if a rather appalling one.

THE LINGERING FOOTPRINT OF WAR

The Armistice opened the way for locals to return and for visitors to explore, but Thiepval remained a special case. In 1919, the *Illustrated Michelin Guides to the Battle-Fields (1914–1918): The Somme Battle-Fields* reported the zone from Thiepval to Albert 'useless for agriculture for many years to come, and a scheme to plant this area with pine trees is now being considered'.⁵ In the same year, the Touring Club of France in conjunction with the Railways of the Nord advised walking to the Thiepval heights after leaving the train at Albert. Walking was the only way to get there. No other form of transport could get through.⁶

It is little wonder the pine forest option was being considered. The de Bréda family certainly showed no interest in rebuilding their own home – 'about as much interest as a fish in an apple' according to one embittered former resident – and even if they had the desire, the logistical difficulties probably scared them off. With many of the villagers employees of the family, or in some way connected with it, that decision largely sealed the fate of Thiepval.⁷ The people did not return.

Instead, those who came to Thiepval were visitors, and they required a good degree of commitment and perseverance to manage it. The first thing they noted was the increased sense of desolation, even in that wrecked region. However, trying to account for the condition of Thiepval led some to a crisis in communication, totally at a loss as how to describe it. Canon J.D. Pierce, chaplain to the forces, whose son had been killed at Thiepval, visited its ruins soon after the Armistice: 'Thiepval as it was early in 1919 was indescribably the most complete scene of desolation anywhere on the Western Front,' he wrote, 'It burned itself into my soul. I spent some days there trying to find the grave of my dear lad, but with no success.' (The body of Second Lieutenant R.H.M. Pierce was never found. He is commemorated on the Thiepval memorial to the missing of the Somme.) Councillor Percy Lawson of Tonbridge in Kent believed his town should adopt Thiepval under the British League of Help scheme. Having visited the area in April 1919, he was questioned about his experience by a local newspaper reporter. Lawson's responses provided an eloquent testimony, their nonsensical quality relaying his bewilderment: 'Thiepval to-day does not exist. It simply is not there. In fact, if you can imagine it, it has a minus quality.'[8] A veteran of 49 Division followed the paths of his old unit on behalf of the *Yorkshire Post* in the summer of 1921. On crossing the Ancre valley and beginning the ascent to Thiepval Wood, he was astonished how little things had changed since the battle. Once he had reached the highest point, the site of the old village, the effect was intense:

> It is impossible to stand on that ridge of Thiepval, and to look around, without feelings of deep emotion. For it was across these quiet fields, now so still and peaceful in the soft sunshine, that the great 'tide of living valour' rolled on the foe in 1916. Over these hills, day after day, the dread clamour of the wrestle of mighty nations rose and fell. And now, for miles, there is not, save ourselves, a living soul in sight, and the only sound is the song of the grasshopper. But beneath and around lie the legions of the departed brave.[9]

The following summer, another veteran made his way to the Thiepval ridge, and although his account lacks the emotive power of his fellow ex-officer, Captain Mee made the same point. The landscape bore '[all] too plainly the marks of a terrible conflict'.[10]

'Few other stretches of the battle-zone can show such a scene of the ravages of war as this valley of the Ancre and the heights of Thiepval', wrote Nina Stephenson-Browne of her visit in September 1923. Extensive remains of the trenches were still much in evidence, combined with life forms she did not expect: on clambering down into a dugout, she disturbed a host of frogs and lizards.[11] The smashing of the Ancre's banks by artillery fire during the war caused the intense mushrooming of the natural marshes and ponds that marked its course across the battlefield, and with it came frogs. For Henry Williamson, the frogs did not so much inspire him with thoughts of the triumph of nature over destruction as disturb him with their sheer profusion and otherness. Staying in an estaminet near Aveluy, at the bottom of the valley, Williamson found sleep impossible due to the profusion of thoughts and memories. On flinging open his bedroom window to seek escape and distraction, he was instead plunged deeper into his own psychological marsh: 'Those frogs! Millions of muttering percussions as the cold-blooded creatures sent out their ghastly cries ... Rust and mildew and long tangled grass and frogs.'[12]

As the years went by, and the Somme district slowly recovered, Thiepval was left even further behind as a strange revenant of an almost mythical past. It existed, but only in its ghostly, former self. When W.A. Michell returned to his old battlefield in the late summer of 1926, Thiepval was almost exactly as he remembered it. Tramping around 'up here, on the ridge, one is suddenly taken back to 1916 and 1918'. The site was so completely familiar that the years seemed to 'fall away as if they had never been'. He fell into a reverie in which 'the old life lived in the immediate present without a future, and all its old standards and values came back with such force that one felt almost compelled to get off the parapet and then take cover'. The former battlefield left a deep impression on Mitchell, and it took him several days to recover from its grip.[13] A year later, the writer-veteran Ferdinand Tuohy saw a landscape barely altered by the years. 'Viewed from the Ulster Tower, this area of the Somme is desolate, maimed and horrible still, [being by] a long way the saddest natural survival' of the battlefields.[14]

Unsurprisingly, 1928, the tenth anniversary of the war's end, saw a surge in battlefield visiting. In particular, the numbers were swollen by the British Legion, which led its most ambitious pilgrimage since its foundation. Yet, ten years had done little to alter the condition of Thiepval or provide any sign of life returning. The Ulster and Irish Free State contingents accompanying the pilgrimage declared Thiepval to be in the same state as it was at the war's end. An enquiry made to a man in Albert as to the fate of Thiepval caused him to shrug his shoulders, turn out his hands, and cry 'Pouf!' It turned out that he was the mayor of what had once been Thiepval.[15] The 1/7 Northumberland Fusiliers association noted that not a single house had been reconstructed, attributing this to the impossibility of digging firm foundations due to subsidence caused by incessant bombardments.[16]

Writing under his pseudonym of Raymond Bridgeway, Captain H.A. Taylor filed a report for the *Derby Daily Telegraph* describing Thiepval as nothing but a few shacks, whose inhabitants appeared to do little other than reclaim metal from the battlefields to sell as scrap, adding, 'The scrap-metal merchant who bought the Thiepval area struck a lucky patch.'[17] As a veteran, Taylor knew the reasons behind the depressing scene: 'There were many places on the battlefield which suffered elimination as complete as did Thiepval, but there was none that had longer or more tempestuous bombardments ... No ground on the Somme was cheaply gained in 1916, but Thiepval must rank among the points for which we paid the heaviest price.'[18]

The Imperial War Graves Commission (IWGC) clearly believed the death of Thiepval to be a historical and administrative fact. When the Mill Road Cemetery was completed in 1927, the annotations in the cemetery register stated, 'The village of Thiepval was destroyed and the former territory of the commune has now been distributed among the neighbouring communes.'[19] Georges Carpentier's survey of economic recovery in the Somme district a year later stated that only 526 hectares (from 367,000 in total) remained in their wartime state of devastation: almost all were in the old commune of Thiepval.[20] It was 'the unbuildable town ... the orphan of the Somme' battlefields.[21] When the celebrated and successful writer-veteran R.H. Mottram toured the area in 1935, so sparse were the houses and signs of human life in the Thiepval district that he felt the British Empire memorials and cemeteries outnumbered every other indication of reconstruction.[22] But within this land of the dead there was a tiny vestige of human life, and it was provided, and sustained, by Ulster.

NORTHERN IRELAND AND THIEPVAL

At the war's end few doubted Thiepval was not only *a* battlefield site connected with Ulster, but *the* battlefield site, and that it was going to be the location of a major war memorial for the province. Deeply aware of the need to stress Northern Ireland's credentials as a new and distinctive component of the United Kingdom and yet united with the rest of the Kingdom in its wanting to pay homage to the dead, the province's political elite erected the Ulster Tower at Thiepval with incredible rapidity, completing the project in November 1921.[23]

H.A. Taylor noted, 'In all its hard fighting, the 36th (Ulster) Division wrote no more glorious chapter of history than it did here, at Thiepval,' adding, 'And it may well be that, a hundred years hence, Britons will still come here to meditate upon deeds of valour and of matchless self-sacrifice.'[24] Whereas Ypres might exert a particular grip on the British, stated the *Larne Times*, for the Ulsterman 'the call of the Somme is insistent to anyone familiar with the story of the war'.[25] The *Belfast News-Letter* told its readers that 'when it was decided to erect a monument in France as a memorial to the men from Ulster … there was no doubt as to where the site should be. The hallowed ground of Thiepval was the one spot fitted for such a memorial. There was absolute unanimity on that point, and this small plot of ground will in the future be one of Ulster's proudest possessions.'[26] It was for this reason that Sir James Craig, first prime minister of Northern Ireland, rejected a suggestion from the Battle Exploits Memorial Committee that the site might be shared with other units wishing to erect a memorial in the area. It was impossible to imagine Ulster making a concession to any other claimant.[27]

What was described by Taylor as 'a magnificent tower of stone' and by another veteran as 'a beautiful and worthy memorial' was a replica of Helen's Tower on the Clandeboye estate in County Down, built by Baron Dufferin as a memorial to his mother.[28] In the early days of the war, the units of 36 (Ulster) Division had trained on the grounds of the estate, and this inspired the decision to replicate its graceful form. In its Scots Baronial Gothic, the architecture of the tower was also a clear statement of Protestant identity, and thus made a firm statement about the culture and nature of the new Northern Ireland. The memorial was unveiled in a grand ceremony on 19 November 1921 and was portrayed as a precedent by a *Northern Whig* correspondent: 'To-day's was only the first of many pilgrimages of which Thiepval Tower will be the shrine in years to come'.[29]

Pilgrims from Ulster certainly did come. When Nina Stephenson-Browne of Portstewart set out for Picardy in September 1923, her desire to see Thiepval overrode any other interest in the battlefields. She said for her, and all other Ulster people, it would 'always be first in our thoughts, that part known as the "Red Zone," the battlefields of the Somme'. Use of the official French term, 'zone rouge', for the most devastated sections of the former battlefields was a rarity for British battlefields visitors, and reveals her knowledge that the degree of destruction around Thiepval was even more intense than the general devastation of the battlefields. Stephenson-Browne's excitement and emotional engagement with Thiepval are palpable in her accounts of the pilgrimage written for her local newspaper. During the drive from Amiens, she watched as the levels of destruction increased. Passing through Albert, she saw the collection of wooden huts that was Hamel village, 'and then I saw, on the top of a long lofty ridge, a white tower standing clear and distinct against the sky-line. No need to ask the driver its name: I knew Thiepval too well'. The first stop was Connaught Cemetery. Wandering around among the graves acting as an entrance way to the tower, Stephenson-Browne saw 'how in this corner of Picardy there will always be one spot forever Ulster, where our "boys" sleep quietly till the great Reveille'. Then it was on to the tower itself. For Stephenson-Browne, the real heart of Thiepval had been reached: 'The car stopped at the entrance gates, and with mingled feelings I alighted. This to me was my goal; this was the place in France I had longed above all others to visit, and now I stood on the place made for ever famous by the deeds of the immortal Ulster Division.'

The feeling of communion with her home was very strong. It was a sensation enhanced by the platform at the tower's top, from which she gained a 'magnificent view over thirty miles of country'.[30]

She concluded her series of articles with an audit of memories as she went through her mental list of postcards from France and the battlefields. Despite having visited many other sites, including Paris:

> Of all the pictures memory calls up four will always remain clear and vivid. This is what I see –
>
> A tall white tower crowning a lofty hill, standing out clear and distinct against a deep blue sky, near by a carefully tended cemetery with its rows of flower-decked graves, and a white dusty road winding down into the valley below.
>
> A little hill jagged and torn with shell-fire, behind it half buried in the rough grass the remains of a British tank, and across the road a little group of wooden crosses.
>
> A white road dipping down into the deep valley, a few hundred yards away a group of shattered leafless trees standing stark and grim against the sunset sky; on the

sky-line the tall white memorial to the Anzac troops [the Australian 1st Division memorial at Pozières].

A grave amid many others decked with lavender and flowers, overshadowed by the Cross of Remembrance, lying on the slope of a quiet cemetery above a busy port [Le Havre]. This is what my memory calls up.

I am glad I have seen these scenes. One understands as never before the sacrifice of our men, and to me as to all, Thiepval and Bapaume, Flers and Albert, Villers and Senlis, are names which stand for all that is noble and valiant and glorious in the history of our country and our Province.

To us in Ulster this corner of Picardy will be for ever sacred, for beneath her poppies lies the flower of our Province. That white tower on the heights above the valley of the Ancre is the outward memorial of the undying gratitude enshrined in all our hearts to the men of Ulster, who have for all time made the name of Thiepval synonymous of a glorious and imperishable memory.

O valiant hearts who to your glory came
Through dust of conflict and through battle-flame
Tranquil you lie, your knightly virtue proved,
Your memory hallowed in the land you loved.[31]

Stephenson-Browne's love-letter to Ulster was dominated by Thiepval. Three of the four scenes frozen in her memory were of Thiepval or its immediate vicinity. Ulster was in Thiepval and Thiepval was in Ulster. She wrapped up this paean to chivalrous devotion to sacrifice and selflessness by quoting from the hymn 'O Valiant Hearts', which became the anthem of battlefield pilgrims. Truly, she had replicated Tennyson's lines carved on both the original tower in Clandeboye and its Thiepval replica: 'Ulster's love in letter'd gold'.

THE ULSTER TOWER AS VISITOR ATTRACTION

As well as its handsome, accessible architectural form, which attracted and impressed visitors, the Ulster Tower had the huge advantage of a viewing platform on its roof. This made it a multifaceted building. Not only was it was a memorial in the sense of enshrining the memory of Ulstermen who were killed during the war, but it was also a more literal memorial for veterans, in that the precise geographical location of the tower along with its viewing platform transformed it into an antenna or triangulation point for memories: it located the veteran in the landscape, allowing him to reimagine the scene with great clarity, while, for the non-veteran general visitor, the rooftop was an ideal tourist attraction, which provided a view of, and insight into, the scarred landscape. As an ex-officer, H.A. Taylor drew upon his military memory, identifying the rooftop lookout as one of the tower's great qualities ('from the top ... can be gained a view of the battlefields such as will hold the interest of any soldier for an hour or more'[32]), while Stephenson-Browne, standing by the tower and looking out over the countryside, 'required no great stretch of imagination to picture how this scene looked on that first of July day when the Ulster Division stormed up to where I stood with their triumphant shout, "No surrender!"'

Given this combination of literal and symbolic elements, it is little wonder that the Ulster Tower attracted people to an otherwise inaccessible spot. When Henry Williamson described the tower, he alluded to it as the Red Hand of Ulster in architectural form, but at the same time it acted as the key, unlocking his memory:

> On the high ground above Thiepval Wood, where
> thousands of our men perished on July 1st from the
> machine-guns of the Schwaben Redoubt, stands the Ulster
> Memorial Tower, like a giant hand severed at the wrist
> and upheld as a warning. The trenches – where for a few
> hours on that hot summer afternoon the men of the Ulster
> division rested and watched eastwards until the enfilade
> fire drove them back to their old trenches – are like
> mole-runs half-hidden by the long wild grasses of the years
> – Wretched Way, Lucky Way, Tea Trench, Coffee Trench,
> Rum Trench. These half-hidden seams in the hillside
> filled me with indescribable emotion – the haunting of
> ancient sunlight.[33]

In contrast to Williamson's considered prose, the tower visitor books recorded the raw responses of those willing to scribble their thoughts. Veterans certainly seemed to have made contact with their former selves and lives through the tower. In the summer of 1925, a man who served with the Royal Irish Rifles wrote, 'Brings back memories of old faces.' G.O. Mallows's entry, 'hereabouts 1/7/16', was laconic but wonderfully tantalizing in its suggestion of intense experience. In August 1931, J.B. Cormac noted, 'A wonderful memorial.'

Ulster identity and pride were sometimes expressed more forcefully. 'No surrender, 1688,' wrote a Londonderry resident. Others expressed sentiments of grateful remembrance. Mary Halloran of Sydney, Australia, called it 'a worthy memorial to many worthy Ulstermen'; Mr and Mrs McLean, also of Sydney, and possibly also of Irish background, wrote, 'Well done Ulster in memory of our brave men who fought and died for our freedom and liberty. Beautiful.' Another married couple, from Guildford in Surrey and bearing no obvious Ulster connection, were also inspired by the memorial. The comment was obviously written by the husband, for he revealed his military background as a former artilleryman, recording: 'Well done Ulster, Gunner.'[34]

In providing a lookout over the entire country, and as a startling example of something new and pristine in the landscape, the tower emphasized the sea of destruction all around it. The contrast between Thiepval and the rest of the region only increased as the years passed. Visitors, whether they ascended the tower or surveyed the horizon from its grounds, saw the huge difference between the immediate locality and the district it dominated; they became aware of Thiepval's status as a unique battlefield remnant. A veteran visiting in the summer of 1921 managed to trace the course of the trenches and the positions of the Schwaben and Leipzig redoubts with ease.[35] The journalists filing a series of articles for the *Belfast Telegraph* and *Larne Times* soon after the unveiling of the tower in November that same year found the trenches and dugouts of the Schwaben Redoubt 'in a remarkably good state of preservation', some of the military bridges over the Ancre still intact, sandbags, cartridge cases, and old rifles everywhere, and even trench name noticeboards still in evidence.[36] Two years later, Nina Stephenson-Browne saw quantities of spent cartridge cases, rifles, shell fragments, and other bits of equipment scattered in the fields around the tower.[37] Visiting the Ulster Tower in the summer of 1928, the men of the 1/7 Northumberland Fusiliers association were impressed by the splendid view gained from the top, as well as the neat planting and maturing of the trees around the memorial site. The party then realized that the horticultural scheme around the tower was the only glimpse of cultivation 'amid the rank grass' still surrounding it.[38]

THIEPVAL AS THE GAUGE OF DESTRUCTION

The condition of Thiepval created a yardstick by which to measure other sites. When Western Front veteran W.A. Michell undertook a comprehensive tour of the battlefields in 1926, he was deeply impressed by Vimy Ridge. He saw the construction work on the Canadian memorial in progress and realized how it would dominate the district once completed. Then came the comparison with Thiepval, in which Vimy was granted a slightly lower status: 'In a somewhat lesser degree Vimy ridge resembles Thiepval ridge, except that parts of it have been cleared of the wreckage of war, whereas Thiepval is almost exactly as we left it.'[39]

In fact, by the mid-twenties the nature and sheer extent of the field fortifications in Thiepval contributed to the process whereby it consumed itself. Being riddled by trenches, tunnels, and dugouts, the ground was much prone to collapse and subsidence, particularly after winter weather assailed it. This meant trenches began to disappear as their sides caved in, but, as soon as one fell in on itself, another chunk of land shifted, revealing more fieldworks. The Ulster Tower itself was built on top of the German front line and was surrounded by the remains of the German defensive system, which were clearly visible for much of the year, particularly in the early autumn when farmers ploughed the chalky fields, exposing the traces. In 1925 a major dugout collapse occurred, prompting Captain Stuart Oswald, an expatriate battlefield guide based in Amiens, to contact the Northern Ireland government regarding preservation of the historical sites around the tower. Noting the extreme interest his customers took in these trenches, he was convinced many would contribute to any fund raised with the intention of securing the remaining examples and offered to collect any donations. The idea came to nothing as the memorial committee, which administered the tower and its fund, did not have the resources to undertake the project.[40]

This was not quite the end of the story, for Oswald's shrewd observations on visitor interest meant the issue would not fade away. The Northern Irish grandee Colonel W.A. Lenox-Conyngham wrote to

Sir Wilfred Spender, permanent secretary to the province's Ministry of Finance, in May 1928. He included a letter from his sister, who had visited the tower. She stated:

> On Monday last (May 14) we were at Thiepval and visited the Ulster Memorial and were shown round by the caretaker (Mr McMaster). He showed us the deep German trench and dugout at the back of the Tower a short distance from it. The ground all around is being reclaimed and cultivated by the owners and very soon unless steps are taken to prevent it the trench and dug out will disappear. This would seem a great pity as it is such an interesting memorial of what Ulster men did on 1st July 1916. The caretaker Mr McMaster has no power to interfere. He said he had written to Belfast but had received no reply. We promised to try and bring the matter before the Committee. To buy the piece of land, less than an acre should not cost much.

Spender was enthusiastic about the proposal. His response revealed his firm belief in the landscape as a commemorative actor in its own right; it was the perfect complement to the tower: 'I certainly think it would be a very great pity if this very interesting Memorial – in some ways more important than the Tower itself – were to be scrapped and that posterity would blame us for not making an effort to retain it.' The Northern Ireland prime minister, Sir James Craig, supported the idea and wanted to know whether the land could be purchased for 'somewhere about £50'. The IWGC was approached for its assistance, and its director of works in France, Colonel H.F. Robinson, duly inspected the site. He found it to be in a state of dilapidation and believed the costs to restore it, especially as the dugout was at least thirty feet deep, would be considerable. Concrete injection as at Vimy Ridge was not thought a suitable option, for he warned it was by no means the maintenance-free option the committee desired. More significantly, Robinson also doubted the historical and commemorative value of the procedure, 'as the true character and originality of the earthworks would thereby be obliterated'. Here Robinson asked the essential question: at what point did a battlefield relic transubstantiate in the wrong direction, as the genuine object became fake entirely through the laudable, but perhaps misguided, intention of preserving it? Preservation unwittingly became artificial replication, and in the process the veracity and integrity of the site were lost. Whether this element was more influential than the cost is not obvious, but the Thiepval Memorial Committee decided to let the matter rest, and no further action was taken.[41]

It was still not quite the end of the matter, for it was brought fully to public attention via a letter to the *Belfast Telegraph* two months after Lenox-Conyngham's original enquiry. Ulsterman James Gracey, who fought with the Canadians during the war, joined a party from Northern Ireland visiting the battlefields in 1928. At Thiepval, the group saw the trenches and tunnels still visible close to the Ulster Tower; acting on his own initiative, the custodian had wired off these remnants of the battlefield, even though the land was beyond the tower site. Gracey then took up the matter, equally unsuccessfully, with the Northern Ireland government. Comparing the site to the conserved trenches at Vimy Ridge, he believed a restoration fund should be launched to ensure the ground 'sacred to the peoples of Ulster' could be retained as a memorial. He urged speed 'as water and rain are soaking through the wood of the dug-outs and tunnels, causing a rotting of everything. The parapets are sagging and falling in, and the stairway of the tunnels are choked with wet soil so that they cannot be entered.'[42] As Gracey's description revealed, regardless of its condition, the trench complex was extensive and still relatively easy to access, apart from in a few places.

It is possible someone attempted a form of preservation for these ubiquitous remains, for R.H. Mottram wrote:

> As late as 1929, the summit of the Thiepval ridge had been kept in its last state, if one may call it. Here, as at Vimy, some hundreds of yards of weed-grown and partially collapsed trenches had been bolstered up by concrete revetments cunningly imitating sandbags and pit-props, dug-outs, coils of now very rusty wire, rifle-barrels with the wooden grips and butts, helmets perforated, pieces of equipment, tools and grotesque trench weapons strewed the ground. Amid these appear the striking memorials now built above this partially cleared-up wilderness.[43]

No other evidence of artificial preservation can be found, and so Mottram may have confused genuine concrete defensive works with signs of post-war intervention. Evidence of German use of concrete was certainly noticed by the IWGC. In 1923, when work commenced on Mill Road Cemetery, adjacent to the Ulster Tower, the architect J.S. Hutton made it clear on the construction plans that the reinforced concrete German dugout had to be removed first.[44] Regardless of the origin of the concrete reinforcements, Mottram's comment revealed a landscape still heavily marked by the fighting over a decade after it had ceased.

Getting to Thiepval took Mottram along a genuine wartime route, via 'the ugly and still scarred track from Ovillers' and the Aveluy Wood–Mill Road approaches, the places immortalized by the poetry and prose of Blunden and Masefield.[45] By walking these tracks and this ground, other veterans also communicated with the ghosts of their past. Despite initial disorientation due to the scale of reconstruction, 'J.B.M.' (he signed his account with his initials only) descended into his old world once he had followed the path into Thiepval Wood. Here the memories came thick and fast, as 'Elgin Avenue, Gordon Castle – all the old trenches and dugouts and duckboards reappeared. I was no longer alone – I was marching step by step in the ranks of the Deathless Legion' of old comrades.[46]

The Ulster Tower therefore brought life of a kind to Thiepval; people came to see it, climbed up to its roof, and marvelled at the view, and they wandered around the ruins, but they did not stay long. Access from the Albert–Bapaume road could be tricky, because for many, many years the roads simply petered out into tracks. In 1921, a veteran reported Thiepval as 'remote from towns, and not on the beaten track'.[47] This meant pilgrimage itineraries made Thiepval a brief interlude on somewhat hectic tours of the Somme battlefields; it was somewhere many wanted to see but few wanted to linger over.

Even the Ypres League, an organization very much opposed to the idea of the Cook's tour of the battlefields, wedged Thiepval into a crowded schedule. A typical example was its pilgrimage of 1927, during which the party detoured off the Albert–Bapaume road to visit the tower and stand on the ridge, where it was 'possible to appreciate the dominating positions held everywhere by the enemy' before setting off again, this time for Beaumont-Hamel and the Newfoundland memorial park.[48]

THE LONELINESS OF THIEPVAL: WILLIAM MCMASTER AND LIFE AT THE ULSTER TOWER

Such rapid comings and goings must have made life a very strange kaleidoscope for William McMaster, the caretaker. A former NCO in the Royal Irish Rifles, McMaster not only looked after the tower, but also lived in it along with his wife and daughter. However, it seems that, at least in the earliest days, his family returned to Northern Ireland for the winter months, leaving him there alone. Marooned at the highest spot of the Somme battlefields, exposed to the wind, snow, and rain in that treeless environment during the winter season, his must have been an extremely difficult, lonely existence. His loneliness was a condition that certainly concerned the few people with whom he did come into contact outside of the spring and summer visiting period. Captain Malcolm Cockerell, formerly of the Royal Army Service Corps, and by 1920 the owner-operator of a battlefields touring company based in Station Yard, Albert, was one of those few. In September 1924, Cockerell wrote to the Thiepval Memorial Committee in Belfast about McMaster, expressing anxiety about the effect of the forthcoming winter. The 'lifeless heights of Thiepval' were not a good place to be 'during the winter [when] visitors are few and far between. The days are short and the evenings exceedingly long and must perforce be very lonely,' he explained. Cockerell had a suggestion: knowing that the BBC's new Belfast station had commenced broadcasting, he suggested a wireless set, which would allow McMaster a link with his home. As an expert in wireless himself, Cockerell added that he was happy to assemble and install the set in the tower.

Back in Belfast, no one was quite sure whether the memorial fund could be used to purchase 'luxuries of this kind'. In describing the wireless as a luxury, the committee revealed little insight into McMaster's lifestyle, and perhaps they expected a monk-like devotion to his sacred duty to carry him through. However, no one objected to a new fund devoted to this purpose. It was believed great sympathy would be shown 'from the people who appreciate <u>our outpost</u>'. The underlining in the original revealed much about Ulster's sense of ownership of the empty spaces of Thiepval, a province of a province. Sir James Craig then made a plea for funds, which was broadcast using the services of the new BBC Northern Ireland studios. Cockerell was delighted and wrote back, 'I have heard with very great pleasure

Figure 10.1 William McMaster, custodian of the Ulster Tower, c. 1925.

"Belfast Calling" at the official opening [of the Belfast station]' and was going to tell 'McMaster of the reference to Thiepval made in his (Sir James') speech'.[49] Visiting in 1928, H.A. Taylor was able to see the success of the appeal: 'Driving away, instinctively one looks back over this hallowed ground. As the top of the Ulster Tower disappears from view, with its wireless mast alert to convey the news from 2LO [Station] to the exiled McMaster.'[50]

Taylor, like many others, was much impressed both by McMaster himself and the conditions in which he lived and worked. He was 'a tall, wiry man, dour of expression but kindly of heart', and when Taylor met him had not been back to his Northern Ireland home for two years. McMaster had conducted thousands of people around the tower and the remaining trenches, and Taylor noted how, when he did so, he did 'not talk the conventional "patter" of the professional guide. When he speaks of the attack, his eyes light up, and he talks with a fervour that suggests that he is living over again those hours of hell in the old redoubt.'[51] What Taylor appeared to have uncovered was a quasi-voluntary captive of the war, imprisoned in the Ulster Tower. McMaster could have given up the job and gone home, he could have sought to suppress or forget his war experiences by literally and metaphorically walking away from it all, but he did the complete opposite: he cocooned himself in his war experiences. In such cases it is difficult to know whether a veteran such as this was exorcising himself, enjoying himself, or caught in some form of enslavement to the war.

Indeed, McMaster and the tower seemed to merge into each other, as both faced problems caused by a landscape frozen in its wartime condition. For the tower, standing alone on high ground in a treeless environment, the wear and tear on the fabric was a difficulty. One of the first problems to be noticed was the speed with which the Union flags, flown from a mast at the top of the tower, deteriorated. The Ulster Women's Unionist Council agreed to purchase replacement flags and set up a fund, collecting £24.[52] It was always necessary to have two flags to hand. The larger was there for fair, calm weather and the smaller for the harsher weather 'when the heavy winds characteristic of the Thiepval district are blowing', as the *County Down Spectator* stated.[53] Such was the strength of the wind, replacements were required three to four times a year. By 1928 the flagpole was the part of the tower causing most concern: because it was so big and bulky, the vibration caused by the wind over the previous six winters was placing a strain on the building. McMaster was concerned 'that the fabric may be gradually weakened by the constant vibration', and it was suggested that the current flagpole be cut in two to reduce the degree of stress on the tower.[54] So exposed was it on the roof that when Ferdinand Tuohy was taken to the top by McMaster they had to yell at each other in order to be heard in the stiff wind.[55]

By the early thirties there were further signs that both man and memorial were feeling the strain imposed by a highly distinctive location, even by the standards of the former battlefields. In June 1934, E.J. Osborough, secretary to the Thiepval Memorial Committee, asked a fellow committee member, Major J.C. Boyle, to assess the state of the tower, the horticulture and, through the means of asking for a copy of the descriptive booklet, the demeanour of McMaster. Boyle undertook the role, visited on 16 June, and reported back most thoroughly. In the process, he revealed himself unable to comprehend the sheer grinding boredom of living in such an environment. Instead, he produced a catalogue of complaints and faults, including a condemnation of McMaster's character. 'I was not struck by his manner in any way at all,' Boyle reported. He deemed McMaster to be uninterested in his work, bored, and vexed by visitors, rather than showing pride in the memorial, and probably keen to clear out as quickly as he could. Following up on Osborough's request that he ask McMaster for a copy of the booklet for purchase, 'a certain amount of difficulty' was experienced in obtaining one, but eventually a copy was produced in exchange for the one-franc fee.

Turning to the condition of the tower, Boyle kept up the offensive. The bronze inscription plate was dirty, the memorial room was in need of sprucing up, the battered divisional history needed replacing, and the visitors' book could be better placed and presented. As for the woodwork, he declared it in need of extensive maintenance, and he was none too keen on the use of green paint for the gate, saying no good Ulsterman 'would select green as a colour for the entrance gate of their Memorial'. Given that apple green was the colour used by the IWGC on its cemetery signs and noticeboards, it is highly likely that McMaster had sourced his paint from a friend working in one of the nearby cemeteries. Last on the list was the standard of the horticulture. Boyle felt the lonely atmosphere of the tower was emphasized by the lack of flowers or a genuine planting scheme, including well-selected trees, which he contrasted with the colour and vitality provided by the nearby IWGC cemeteries. Indeed, the work of the IWGC became the general yardstick for Boyle's judgements: 'As compared with the Cemeteries close-by our memorial (i.e., the keeping of the grounds) does not compare and I think it should. I know comparisons are odious, but having first visited some of the Cemeteries adjoining and the British and French Armies Memorial [the Thiepval Memorial], it was impossible not to notice the difference.' This unfavourable comparison was also laid at the door of McMaster, being the unhappy result of his lack of horticultural knowledge.

Implying that it was perhaps time for a change, Boyle referred to another ex-serviceman of the Ulster Division, Charles Smith, a Londoner by birth, who had opened a cafe in Thiepval and was building his own home. Smith was identified as a man to be cultivated and one who would prove useful thanks to his vitality and energy. After such an onslaught, Boyle did reveal some sensitivity towards the precise conditions in which McMaster worked, mentioning an old British prejudice, the standard of French plumbing and sanitary arrangements, and he also raised the possibility that McMaster may have been under the weather when he visited. Interpreting the situation according to the immediate, surface impression left Boyle unable to understand the deep physical and psychological strains imposed by living in the peculiar conditions of Thiepval for over a decade.

Boyle's lack of sympathy and inability to contextualize were revealed by Sir William Spender, who visited soon after him and was pleasantly surprised by the state of the site: 'Frankly, my impression of the Memorial was considerably better than what I expected to be the case after reading Major Boyle's report.' Acknowledging that McMaster was not a trained gardener, had no help, and had to operate within the relatively modest sums allowed for horticulture, Spender was generally happy with the condition of the tower and grounds. By the same token, he found McMaster willing to give up the custodianship if he could be guaranteed a post elsewhere but knew it would be difficult given his absence from Northern Ireland for such a long period of time. Overall, Spender was inclined to agree with the IWGC's judgement that McMaster had done his best in the circumstances, but now was the time to consider a new approach. Faced with the issue of finding the time, money, and expertise to maintain the site, the way forward according to Spender, the Thiepval Memorial Committee, and the Northern Ireland government was to include the IWGC in the discussion, with a view to placing the tower in its care. Following a series of talks and meetings, the IWGC agreed to take on the maintenance, with the Northern Ireland government transferring investments and funds to cover the costs. McMaster then became an IWGC employee until his retirement in March 1938.[56]

MCMASTER'S FEW NEIGHBOURS

Charles Smith, the man Major Boyle spoke of so enthusiastically, was one of the very few people who had settled in Thiepval by the 1930s. During the early twenties, aside from the periods when McMaster's wife and daughter were in residence, there was only one other person nearby. This was Marie Louise Dutart, an old French woman, who had somehow managed to knock up a wooden shanty in what had been the centre of the old village and clung on to what must have been a harsh and precarious existence, eventually selling refreshments from her doorstep.[57] She became known as the 'heroine of the ruins', and was later described as 'without a doubt ... the stuff of which "heroines" are made. What a story, too, she could tell of it, for in all Picardy, Thiepval was Devastated Village No. 1.'[58]

By 1926, three families had returned, all living in shanties. The town of Tonbridge, the would-be surrogate parent of Thiepval under the British League of Help scheme, was eager to help, but was in a strange situation. Having collected sums to cover the costs of a new water supply network as part of a general reconstruction programme for Thiepval, it found there was no sign of such a network ever being implemented. The formal adoption therefore collapsed on the technicality that there was nothing to be adopted, 'as Tonbridge received authoritative information from the Prefect of the Somme' that Thiepval could 'never be reconstructed'.[59] There was therefore little the adoption committee could do other than dish out doles, and so they agreed to send FF100 for each family. It was about the only gesture they could make.[60]

Arrayed against the reconstruction of Thiepval was a host of interlocking problems. To start with, it was not solely Great War military architecture that needed to be contended with: many other underground chambers were discovered dating from the Spanish wars of the seventeenth century.[61] Next was the sheer depth and complexity of the German trench system. If anyone had the time and money to devote to clearing a plot of land and reached the stage of sinking footings and foundations, they immediately ran into the problem of the subterranean labyrinth. Thirteen years after the Armistice, Major Macfarlane, the IWGC officer in charge of works on the Thiepval Memorial site, discovered a whole network of tunnels and dugouts, formerly part of the German second line. In these were found quantities of perfectly preserved ammunition. Once these

Figure 10.2 The Heroine of the Ruins, c. 1920.

were carefully removed, further digging was required to ensure a stable base for the enormous monument: an additional nineteen feet was excavated before the foundations were solid enough for the placing of a concrete raft nine feet seven inches in thickness.[62] Given these problems, it was little wonder that in the days immediately before the unveiling of the memorial in August 1932, the *Edinburgh Evening News* described Thiepval as 'likely to remain as it is. A memory. "Here stood Thiepval."'[63] Four years earlier a veteran on the British Legion pilgrimage had felt something similar, writing, 'Thiepval appears to be one of those places which will only remain a name. The village is still buried, and its original existence signified by a sign only.'[64]

THE THIEPVAL MEMORIAL: LUTYENS'S MASTERPIECE

The utter deadness of Thiepval then seemed to be underlined by the British decision to erect an immense memorial commemorating its 73,000 missing of the Somme on the site of the long-vanished chateau. The origins of the memorial lay in the work of the National Battle Memorials Committee to mark the great engagements fought by British forces. All agreed the Somme was important and was worthy of at least one major memorial. As the committee noted, 'To some extent the Somme stands for France, much as Ypres stands for Belgium in the eyes of the British soldier'.[65] In these discussions, Thiepval was just one potential site among many, with Fricourt, Delville Wood, Pozières, Butte de Warlencourt, Beaumont-Hamel, Guillemont, High Wood, Flers, and Courcelette all mentioned. When the senior generals were consulted, Henry Horne identified Albert and the Butte de Warlencourt, Henry Rawlinson listed Pozières, and only the former commander-in-chief, Haig, mentioned Thiepval. Given one of the issues the committee discussed was the likely impact of the memorial, which would make the precise site of great importance, Thiepval did not perhaps leap out as the most obvious choice. Here, the view of the historian, especially the military historian, imbued with the desire to mark places of significance, clashed with the severely practical:

> From a military point of view, probably the best monument to a battle is at the culminating point of the battle itself; but the most famous battlefields (e.g. The Hindenburg Line and the Somme) are remote from the ordinary routes of travellers, and a memorial which can only be reached by a long motor car journey will rarely be visited by the ordinary man, and if it is only seen by very few French peasants, it will fail to commemorate the sentiment of alliance of which such memorials should perhaps be a reminder.[66]

Such comments also revealed some of the confusions regarding the purpose of the committee's work. Were memorials suitable to mark military action on the sites on the list they were compiling, and thus to commemorate the Anglo-French alliance that had fought and won the war, or were they to be shrines to the dead, or even a combination of both? Eventually, it was decided that the memorials should combine the functions of commemorating the military actions and commemorate the missing. At the same time, the idea of an entirely separate memorial to Anglo-French solidarity was pursued, with Amiens identified as the most fitting home for such a monument.[67]

Among the possible sites for a Somme memorial was St Quentin, the historic city on the edge of the battlefield zone, which had been dragged into danger when the fighting suddenly burst beyond the trench lines in the spring and summer of 1918. Visiting the city in July 1923, the architect Edwin Lutyens was welcomed warmly by the mayor, and pretty much told to select whatever site he wanted. The mayor expressed 'the most grateful thanks of the [city]' for the Lutyens' design proposal, which he felt would 'considerably add to the historic and artistic interest of St. Quentin', possibly code for tourist trade potential.[68] Lutyens was impressed by the city and believed a site straddling the main road into St Quentin was suitable for the truly remarkable memorial he had in mind.

The concept soon ran into difficulties, but those difficulties came from Paris, not at the local level. Although the implication was the difficulty of maintaining a memorial across a major thoroughfare, it seems there was an element of French pride at play. A huge British monument under and through which thousands of local people would pass all day, every day seemed a bit too much like Britain saying it had won the war alone, and the people of the region had the British to thank for their liberation. Major A.L. Ingpen, the secretary to the IWGC's Anglo-French Mixed Committee had very sharp antennae and detected an element of wounded Gallic pride. He reported to Ware

> the opinion that the designs submitted are somewhat exaggerated and too grandiose. Further, and in view of the fact that, owing to the present financial conditions in France, the French government can do nothing to commemorate their own missing, such grandiose monuments will not be understood, or appreciated, by the general public, and may give rise to hostile comment, not only of an international character, but also against the Commission des Monuments Historique itself for having approved such grandiose schemes put forward by a foreign government, for execution in French territory.

Lutyens responded by suggesting scaling down the memorial slightly and placing it on a plinth, allowing it to be built alongside the road.[69]

Ware reported back on the discussions that had been held in the Anglo-French Mixed Committee on the topic of battle exploit memorials and memorials to the missing. Although it was not mentioned specifically, Lutyens's design for the St Quentin memorial was obviously the inspiration behind French references 'to the comparative sizes of the Arc de Triomphe and of some of the memorials designed for the Imperial War Graves Commission'. Sensitive to French opinion, and probably also aware of budgetary implications, Ware raised that suggestion 'that one Memorial on the Somme should take the place of those at St. Quentin, Pozieres and Amiens'. He then brought Thiepval into play as a possible site. He thought that 'for many reasons this [memorial] should be in the Zone Rouge', and they 'had considered a site at Thiepval Ridge', adding, somewhat cryptically, 'This had considerable advantages in many ways.' However, he did not state what those advantages were; instead he outlined some of the challenges regarding land acquisition and designing a memorial large enough to include the vast list of names. Here the argument became circular, as he noted a detailed design could not be drawn up until the precise nature of the location was agreed and known. The chair of the committee, Sir Laming Worthington-Evans, the Secretary of State for War, backed Ware's comments and underlined the need to be aware of French concerns, stating that 'in deference to French opinion the Memorial should not be of too grandiose a nature and he [Ware] suggested that this should be upon the record'.[70]

In June 1926 a party set out for the Somme battlefields to assess sites for a memorial. Desiring the memorial to sit on a commanding site, height was a guiding requirement of the search. Serre impressed the group, as it provided excellent sight lines and the view included a profusion of cemeteries. The believed cheapness of the land was a further, not inconsiderable, advantage. The party then set off across the Ancre valley and climbed up to the Thiepval ridge. 'It would be difficult to imagine a more appropriate site if it is desired to commemorate names of the missing in the Somme fighting in the locality where they fell', reported H.F. Robinson, the director of works. For Robinson, the height of the ridge combined with the sheer emptiness of the commune of Thiepval made it perfect. But, before they could reach a decision, there were three more sites to be inspected. Nearby Pozières did not seem so suitable, as the restoration of farming land around the village had increased land prices. Pushing further up the Albert–Bapaume road, they examined the hugely symbolic site of the Butte de Warlencourt, an ancient, man-made mound, which had marked the limit of the 1916 advance and was topped by a series of memorials erected during the war. Although its height made it attractive, the site was nowhere near big enough a base on which to construct a major memorial, and it, too, was rejected. The party then motored south-west from Bapaume in the direction of the old British front line, but nowhere impressed them as having the right qualities. Thiepval was emerging more strongly as a contender.[71]

Robinson's follow-up report revealed further refinement of his thoughts. As Pozières had been in the running for some time, he made sure to point out the potential difficulty of land acquisition. It is also possible that Robinson was considering the likely French reaction to the suggestion. Pozières sat astride the main Albert–Bapaume road, and if Lutyens was appointed architect, he would be presented with the chance of using his original St Quentin design with the road running through the central axis. Rumours to that effect must have been floating about that summer, for W.A. Michell, a veteran on a battlefield pilgrimage, was told that a memorial arch was to be built across the Albert–Bapaume road.[72] If this was thought a genuine prospect, IWGC officials must have suspected a similar response from the French authorities: a colossal British memorial straddling a main road would not have appealed.

Whatever Robinson's precise thought processes, he proffered Thiepval as a place suitable for a memorial capable of commemorating up to 80,000 names. Further, he believed Lutyens's St Quentin design was ideal for the site. Robinson confidently predicted Lutyens would be sure to understand French concerns regarding the precise scale of the current design and would doubtless prove amenable to making some slight adjustments. Moving swiftly, Robinson made arrangements for Ware to visit Thiepval. Ware must have been impressed by what he saw, for he asked Robinson to invite Lutyens for an inspection. At exactly the same time, Ware was mulling over the issue of the Anglo-French memorial and must have had this in mind when he undertook the Thiepval reconnaissance mission. Indeed, Ware was working with Lord Crewe, the British ambassador in Paris, to ascertain French opinion on the proposal that the major British memorial on French soil should contain an inscription commemorating the deeds of both armies.

Therefore, when the Anglo-French Mixed Committee met in St Omer on 29 July, much preparatory work paving the way towards a final decision had already been undertaken. As a result, agreement was reached on the erection of a memorial making overt reference to the Anglo-French military effort, and the location was finally settled: 'they agreed in principle to a proposal that in a great clearing on the great Thiepval Ridge should be a Monument overlooking the Somme Battlefields and Albert'. But at this point the precise nature of the memorial scheme was not finalized. They discussed the possibility of two separate, but linked, memorials: one to commemorate the joint military effort, and the other to contain the names of the British missing of the Somme as well as those Frenchmen who originated in the *département*. In addition, the second component, which was to take the form of a Campo Santo or cloister, would have a space for the burial of an equal number of British and French unknown soldiers.

Having agreed this plan, the matter was passed back to Lutyens to make architectural sense of the demands. By this time, Robinson had taken Lutyens around Thiepval along with Monsieur Pontremoli, the head of the French equivalent of the Office of Works. Robinson was probably aware of Lutyens's desire to see his St Quentin memorial realized, and so told Ware he thought it important that all were clear about the concerns expressed by Monsieur Verrier of the Anglo-French Mixed Committee over the scale of the memorial, in particular when compared to the Arc de Triomphe. At the same time, there was hope, for the objections were mainly due to its precise location. 'If the memorial were erected on such a site as that now suggested, he [Pontremoli] did not feel the same objection.' The plain meaning behind this amicable, delicate, diplomatic exchange between Robinson and the Frenchmen was all to do with geographical context. Placed within the environs of an ancient French city like St Quentin, or even over a major road running through the heart of the Somme battlefield, Lutyens's design was an overbearing statement of British grandeur. Shifted to the remote wilderness of the Thiepval ridge, it would soar over, but become one with, a backwater of the French landscape showing no sign of resurrection to its former self. Britain and its Empire could make their gesture, with its warm affirmation of Anglo-French unity, here without in any way impugning the dignity of France. The honour of the dead could sit enshrined majestically atop the long-buried corpse of a community. And no matter how splendid the memorial, Thiepval would remain miles from anywhere and on the road to nowhere.

Lutyens returned to France in October for a series of further meetings in which his designs were discussed, as was the amount of land required for the memorial site. The culmination was a conference at the British Embassy, during which the IWGC and French government agreed on the Thiepval plan. Ware now informed the Commissioners of the agreement. The main memorial on French soil was to be at Thiepval: 'The site chosen was an exceptionally good one and was, at the end of the Thiepval spur, the most conspicuous point on the Somme battlefield and visible from the Albert–Bapaume Road.' As for Lutyens's design, although shifted from St Quentin, it 'seemed eminently suitable for the new site at Thiepval'. In the end, it would take another couple of years before all details were formally agreed and signed off.

Among the remaining issues was Lutyens's persistent desire to place the memorial over a road, making it part of a thoroughfare like the Menin Gate. With the shift to Thiepval, he was considering the Authuille–Thiepval road, which was the way soldiers approached the front line during the war. Historically it was important; in terms of local communications it was extremely insignificant, being little better than a lane used mainly for local agricultural traffic. But this option, too, was ruled out, most likely because the French remained resistant to a roadway site, even when the route chosen was through a very sleepy backwater. The size also remained an issue, and Lutyens responded by further reductions in scale, bringing it below the crucial Arc de Triomphe standard (not including the podium!). The site was then fixed after a further survey of the ridge top. Lutyens's preferred choice was rejected by the IWGC's Finance Committee as requiring a lengthy access road and thus the purchase of a bigger plot of land. Eventually, it was agreed to place the memorial near the very summit, on the former parkland of the chateau.[73] With the de Bréda family uninterested in returning, the IWGC was at least dealing with a willing seller. If Thiepval's pre-war heart had been the chateau, then it was perhaps fitting that a massive war memorial was designated to sit on that precise spot.

Building the colossal memorial was a major project. The land was cleared, and foundations laid. A great range of construction machinery had to be transported to the site, as well as many tons of those most prized Lutyens materials: red bricks and Portland stone.

Every bit of kit, every piece of material had to be brought up to a location still not fully connected to the road network, and everything was exposed to the howling winds of a still largely treeless landscape. Work continued for over two years in these trying conditions.

Completion of the memorial also brought to a close a remarkable story about the genius of the place and its meaning, a story rich in intricate relationships and semiology, as if a strange prophecy had been fulfilled. The Thiepval memorial to the missing was the last of the great IWGC construction projects to be finished. Its unveiling was not, however, the propitious culmination of a truly amazing commemoration scheme. Instead, it came at just the moment when the short, sharp spell of disillusionment about the war was at its height, and questions regarding what the dead had actually achieved were in the air. Violent death even stalked the unveiling ceremony itself. The event was planned for May 1932, but President Doumer had been assassinated a few days before, and so it was postponed until August. President Lebrun, Doumer's successor, then led the unveiling ceremony, along with the Prince of Wales, in front of a large crowd.[74] However, the ceremony was not quite as effective as the Menin Gate unveiling of 1927, nor even the British Legion pilgrimage of 1928. Unlike the Menin Gate in the heart of cosy Ypres, no train station sat handily nearby to bring in pilgrims from the Channel ports, there was no host of nearby amenities essential for weary pilgrims, and Lutyens's memorial lacked the obvious embrace of Blomfield's refined classicism.

Nonetheless, the Thiepval memorial was (and is), quite literally, awesome, measuring 135 feet in length, 185 feet in width and 160 feet in height from the base of the podium to the top. Sixteen pillars carry a network of intersecting arches, with the largest central arch having a width of 35 feet. Among the inscriptions was one commemorating the Anglo-French alliance, but this seemed a mere postscript to the majestic but austere, laconic statement running across the string course coping of the front and back elevations: 'THE MISSING OF THE SOMME'. In Portland tablets set in Portland wreaths adorning the tops of the main arches were inscribed the names of the major actions of the 1916 Somme battle. Looming high above the floor level, simple Somme villages and woods ('Guillemont', 'High Wood', 'Le Transloy', and 'Bazentin' among others) were transformed into mystical incantations. Then, below, came the panels containing the 73,000 names of the missing, which cover each surface of the sixteen columns, Lutyens's ingenious solution to the challenge of creating space for such a number.

However, the overall effect of the memorial came from the combination of the undoubted genius of the architect with placement of the site. Thiepval was perfect for the memorial, and the memorial could never have achieved the same effect anywhere but Thiepval. Thiepval was not just a dead village, it too was totally missing, and no soldiers were more tragically dead than the missing. A symbiotic relationship was created, but it was so paradoxical, so hard to fathom that the memorial never gained the same status in the hearts and souls of the pilgrims as the Menin Gate in Ypres.

Even the comparison with the Ulster Tower created a difficulty. Built in a familiar Scots Baronial style, and of a decent, but not fearful height – only the most sensitive vertigo sufferer was likely to be startled by looking out from its roof – the Ulster Tower welcomed, not least because there was a genial Northern Irishman and his family happy to see visitors. By contrast, the Thiepval memorial's mix of weight, weightlessness, height, and labyrinth-like enclosure threw down an emotional and intellectual challenge. It meant few people were left unmoved by Thiepval, and few were neutral, but many were very probably bamboozled and not sure what to make of it.

Of course, its sheer scale and importance meant it became a fixture on the pilgrim trail, but the lack of deep, warm, reflective testimony on it compared with the myriad words spilt out over the Menin Gate makes an equally important statement. The Canadian veteran Will Bird visited Thiepval in 1931, when the memorial was well advanced. He wandered over to the site following an agreeable time with McMaster at the Ulster Tower. On concluding an examination of the construction work and design, he was told (erroneously) that the memorial was to have a viewing platform at the top. He was obviously not that impressed. He stated tartly, 'As a lookout tower it has no rival, and that is the best thing that can be said about it.'[75] Postcards aplenty could be filled with messages about the Menin Gate, rather fewer about the Thiepval memorial. It is no wonder the cafe in its shadow did well. Pilgrims probably felt very happy just to plonk themselves down at a table and take a cup of tea or glass of beer.

The Thiepval memorial's ability to carve out its own space, while being intimately of its surroundings, was emphasized by the strange appendage of the cemetery. The plan for the cemetery cloister raised in the earliest stages of discussion was quite lost once focus was concentrated on the memorial. It was not until 1930 that the suggestion returned, and then it took until January 1932, a few months in advance of the unveiling ceremony, before the idea was formalized. Three hundred French and an equal number of British soldiers, whose bodies had been recently recovered from the battlefields, were buried in a plot immediately behind the cemetery.[76] As the graves sat close to the imposing plinth wall of the memorial, there was no easy access, and anyone wishing to visit them had to retrace their steps to the main path and walk the entire length of the memorial site before reaching them. The French appeared to have realized this in advance, declaring the cemetery 'insignificant against, and dwarfed by, this large memorial'. Showing his usual concern for the effect on French pride, Ware asked whether it might be possible to extend the cemetery after the unveiling. In reply, Robinson said the space allowed for somewhere around 1,500 graves, but it seemed they were prepared to let the matter rest unless the French continued to press it.[77]

It was not until after the Second World War that steps were incorporated into the back wall of the memorial. Until that point, an ugly lattice metal staircase had been strung over the wall, which merely served to emphasize what an afterthought the cemetery was and made the graves a somewhat forlorn and pathetic pendant to the main memorial.

Another memorial left in the shadow of the silent giant was the 18 Division obelisk, sitting on the roadside leading to the site. H.A. Taylor predicted this would happen when visiting in 1928. By the same token, he offered the consolation that, though it would be totally overshadowed by the Thiepval memorial when it was completed, it would also serve to frame it: 'so that the eye, surveying the scene, will be led naturally to the simple pillar commemorating one of the finest fighting units in the British Expeditionary Force, and one which, for all its valiant battle-spirit, chose to leave wherever its memorials stand, that noble inscription "'This is my command, that ye love one another.'"[78]

The Thiepval memorial also managed to conjure up the world of the dead in a literal sense, as was forcibly brought home to a British veteran in 1936. Wishing to get a good photograph of the memorial, he wandered into a nearby field in order to gain a better view. Getting into a good position in what he took to be an old shell hole, he 'was surprised to find among the debris therein an old army boot containing human bones in perfect position'. Wanting to assist the IWGC, he produced a sketch map showing as best he could where he had come across the remains. An IWGC team followed up the information, and found the boot, identifying it as German. After a search of the surrounding area, especially after it had been so extensively covered as a preliminary to constructing the memorial, they were content there were no further remains to be found. The discovery was put down to local metal scavengers who had turned it up and discarded it.[79]

THE EMERGENCE OF A COMMUNITY

The construction, completion, and unveiling of the memorial created its own frontier society. Due to the influx of workers, and then visitors, the tiniest signs of a local economy were sparked into life. Charles Smith (an IWGC gardener) and his wife opened a cafe. This soon became a fixture for pilgrims anxious for a good cup of tea (and very probably the chance to buy some picture postcards of the area), and was described as 'a "port of call" for thousands of men and women touring the battlefields in the Somme valley'.[80] A small Britannic world mushroomed up in the shadows of the tower and the giant memorial, for Smith took on the custodianship of the tower, an occupation the *Larne Times* described as 'The Loneliest Ulster Government Job', on McMaster's retirement.[81] On leaving the tower, McMaster's wife seems to have taken on a job in the cafe, with T.H. Opie, another former IWGC gardener, joining the team.[82] Opie was a great asset to the institution capable of entertaining visitors with his 'many stories of the battles that raged in the area'.[83]

With the creation of this thin veneer of human inhabitation, Thiepval made the vaguest gesture towards being a living community, a fact recognized by the IWGC, which altered the note in the Mill Road Cemetery register to read: 'The existence of the commune, after a long period of uncertainty, was preserved.'[84] But it was hardly a bustling rural community. 'Still largely as I left it [in 1916]', was how a special correspondent for the *Liverpool Echo*, a King's Liverpool Regiment veteran, described Thiepval in the summer of 1936. Of the chateau there was no more sign than at the end of the battle, and only three or four houses had been built since the end of the war: 'They could not rebuild Thiepval, so razed was it'.[85]

When the second great war came Thiepval's way, it remained a place unable to escape its past. German soldiers eagerly climbed the Ulster Tower in the summer of 1940, and even signed their names in the visitor books, scribbling messages, one referring to the fighting in the vicinity in 1918.[86] Some carved graffiti, including swastikas, into the stonework along the viewing platform on the roof. Accessing the Thiepval memorial, they climbed up to its top turret and did the same.

The trench tentacles, never far from the surface, also continued to cause problems. In the early 1950s, the IWGC took the decision to lay flat the headstones in plot one of Mill Road Cemetery, across the site of a German dugout, as the degree of subsidence made it impossible to keep proper alignment.[87]

*

According to Taylor, Thiepval would never escape its past, nor its link with Britain, due to its symbolism as the combined anvil and altar of the British army: 'one feels that Thiepval has an invisible link with the hearts of our people, for nowhere did the qualities of our race endure a greater test or emerge more triumphantly from it'.[88] He was only echoing the famous war correspondent Sir Philip Gibbs, who, writing in the early days of the battle, predicted Thiepval's future: 'it is historic ground. A hundred years hence men of our blood will come here with reverence as to sacred soil.'[89] Whatever the pilgrims of the twenties and thirties thought about it, they would have noted the same quality it has today: its truly awful, verging on uncanny, silence. Perhaps it helps keep Edmund Blunden's Titania blissfully asleep, surrounded by her new crop of green Grenadiers.[90]

11

Postcards from Behind the Lines: Armentières, Bailleul, Béthune, Poperinghe ... and Around

Poperinghe, once an obscure little Belgian town (11,511 inhab.), lying unnoticed amongst its hop-fields, is now known all the world over.
Muirhead's Belgium and the Western Front (1920)

'Pop' was a city beloved by the Army; its fame will live, together with that of Bailleul, Bethune, Doullens and Amiens, as a hospitable and pleasant city.
Lieutenant-Colonel T.A. Lowe (1920)

Bailleul was also the chief detraining centre for the area, and was probably the most widely known town behind the lines to the average British soldier.
Reverend J.O. Coop (1920)

Throughout the twenties and thirties veterans held in high regard those towns behind the old front lines that had provided havens of safety, comfort, and some sort of normality for men returning from front line duty. It was for this reason that, aside from the earliest period of battlefield visiting, when the rear area towns offered most in terms of facilities, those behind the lines mainly attracted the attention of veterans and pilgrims. Imbued with happy memories, these towns were places where veterans liked to return to tramp familiar streets, line up the beers in well-remembered bars, and relive the very best bits of the old days on active service. For the families of veterans and the bereaved, it was the chance to see the place where the postcards came from and the shops where their souvenirs and presents were bought and sent home. By contrast, the tourist, especially the tourist in a hurry, thought these towns well off the beaten track, lacking the glamour and interest of the front line areas with the thickly clustered cemeteries, memorials, and wonders of the old Western Front. They ignored Lieutenant-Colonel T.A. Lowe's advice that they base themselves in what he called the 'key towns' behind the old lines, where the visitor would be able to imbibe the right kind of atmosphere as the accommodation, cafes, restaurants, and shops were those the soldiers used. Such an experience prepared the mind for the scenes to come, and in this way 'the "atmosphere" of the great campaign will soon be acquired, and the imagination stimulated'.[1] Instead, the key towns were left largely to the old soldiers, who happily reacquainted themselves and found their memories uncorked as the glasses clinked on the estaminet tables.

THE CHARMS OF POPERINGHE

Only two towns, Armentières and Poperinghe, managed to straddle the divide to any great extent, as both had particular qualities capable of luring a wide range of visitors. Some eight miles west of Ypres, Poperinghe owed its fame and interest to a combination of qualities starting with the utterly pragmatic. Undamaged, at least by front line standards, its charming main square, winding streets full of Flemish gables, cafes, and hotels made it an excellent base from which to visit the Ypres battlefields. This was especially true in those few years after the Armistice, as was made clear by Lowe in 1920:

> To Poperinghe, therefore, the visitor is advised to journey if his object be to make a study of the Ypres Salient. No fears need to be entertained about comfortable accommodation, for although the city was shelled frequently and heavily, it yet preserved its outline in a truly miraculous fashion. The inhabitants have returned, damaged houses have been repaired, new hotels, restaurants and boarding-houses have sprung up in every direction.[2]

That same year, *The Pilgrim's Guide to the Ypres Salient* offered the same general advice but declared itself in favour of one particular hotel. 'Ypres is a more convenient centre than Poperinghe for visiting the Salient,' it admitted, 'But those who look for a higher

Hotel Skindles
POPERINGHE

BEST HOTEL IN THE SALIENT
HOMELIKE :: ELECTRIC LIGHT
BATHROOMS, CENTRAL HEATING

Restaurant Skindles
YPRES
OPPOSITE THE RAILWAY STATION

V. M. BENTIN

PROPRIETOR

SKINDLES HOTEL

Figure 11.1 Skindles Hotel, Poperinghe, c 1920.

standard of comfort than can be found in its wooden buildings, would do well to take rooms at Skindles Hotel, Poperinghe.' Skindles was *the* hotel in which to stay. Immediately opposite the railway station, and run by Monsieur V.M. Bentin, it was the second incarnation of his business, which was originally closer to the centre of the town. Bentin's restaurant and bar had been given the name

'Skindles' by British officers in honour of their favourite watering hole at home, Skindles on the Thames at Maidenhead. The nickname stuck, and the business savvy Bentin quickly adopted it as the official name. In 1920 he was advertising his business as the 'best hotel in the Salient' complete with its electric light, central heating, and bathrooms.[3] Typical of Skindles' many guests were James Duffield and his wife, residents of Clapham, South London. Determined to visit their son's grave in Poperinghe Old British Military Cemetery as soon as travel restrictions were lifted in July 1919, they immediately contacted Thomas Cook, which booked them accommodation in the hotel at the cost of £2 per person for their two-night stay.[4]

Skindles was a major attraction for veterans not just for just old times' sake, but also for actual remnants of the past, such as Zoë, the bar manager throughout the war, who was a well-known face to countless officers. Visiting in 1927, H.A. Taylor said Skindles was the place to 'encounter friends of the trenches, for everyone who served here and who passes this way again, drops in to chat with Mademoiselle Zoë, who is a mine of information and of war reminiscence'.[5] Zoë presided over the 'bevy of bright, fresh-looking merry girls, with whom every subaltern in the army was in love', wrote Lowe in happy remembrance of Skindles.[6] The Canadian veteran Will Bird was delighted to find her 'still serving officers, but not "officers only"' as was the case during the war.[7] Even the cynicism of Henry Williamson was conquered by the siren call of Skindles, but he quickly reverted to type, having found a flyer for the Ypres branch of the business (established in a hut in the station square in 1919, but in much grander premises by the time of his visit). He could not resist quoting from the advertising copy, complete with its spelling mistakes: '"Most modern and up-to-date hotel in the salient (sic) Home Comforts, Reading Room Baths-Hot and Gold (sic) English speaking staff".' He at least stripped off the barbs by then introducing the most warm and homely of comedy comparisons, stating that 'Bairnsfather himself could not have limned so exquisite a Soldier's Dream ten years ago. Gold baths in Ypres!' This passing reference to Bairnsfather, the 'Old Bill' cartoonist so beloved by soldiers, was only a brief respite from Williamson's usual dyspepsia, for he went on to recount an alleged rip-off in the form of a deeply poor meal at an alarmingly inflated price in a small cafe in one of Poperinghe's side streets.[8]

Williamson's complaints aside, the town was home to many well-remembered delights for the returning veteran. Far more, indeed, than simply Zoë and Skindles. Poperinghe was also home to 'Ginger', the presiding female genius of the 'La Poupée' bar, and there was Cyril's restaurant, a spot all remembered as constantly crowded with diners.[9] During the war, the charms of such places were very much reserved for officers, as Bird pointed out. Fortunately for the other ranks, 'Pop' contained plenty of other shopkeepers, bartenders, and restaurateurs happy to engage with every class of customer, and it was a town every veteran could recall with happiness. Bird was obviously delighted to return in 1931, 'On to Poperinghe, good old "Pop," looking very much the same as in the mad days. The station seems exactly the same, and I could only see three or four new buildings.'[10]

Familiarity was bliss. In the summer of 1926, Harold Maybury, a veteran and author of *Years of Remembrance*, a book about his wartime service on the Western Front, wrote to his local newspaper about his recent visit to the battlefields. He found it a strange experience, for, although much was rebuilt, it inspired flashes of recognition combined with confusion over new contexts. The landscape threw up constant reminders, but nothing was quite as he knew it. Among that sea of emotions and memories, Poperinghe was a rock, as he walked past bars and cafes where the same old men with the same expressions played exactly the same card games as they had during the war.[11] But it was also something deeper than simple recognition. Poperinghe's pivotal position between experiences made the wartime town the neck of the hourglass: every soldier marched from it up to the front; every soldier came back to it after service at the front. Poperinghe was the gateway between experiences and sensations. After the war it was remembered as the Janus face of the Western Front, as Lowe realized:

> Poperinghe stood for everything that meant civilization to the British Army. The road from Poperinghe to Ypres was known to every soldier: to march eastwards on this road meant work, trenches, mud – everything unpleasant; to march westwards meant rest, a 'comfy' dinner in the town, and possibly an evening at the club. Needless to say, the latter comforts were only enjoyed occasionally and at long intervals, so 'Pop' was a city beloved by the Army.[12]

Williamson saw Poperinghe in exactly the same way, as a gateway to the front, and therefore dualistic in significance determined by direction: 'To-morrow, Poperinghe – "Pop" of ancient joy (on returning) and apprehension (on leaving).'[13] It was this unique status that dragged veterans towards it. Two King's Own Shropshire Light Infantry veterans who visited Ypres in 1930 felt compelled to visit 'Pop', with its wartime memories of egg and chips, ales, coffee, and a good bath. Although they felt it had lost 'its war-time glamour and prosperity', they enjoyed having a beer where one of them was billeted in 1916, and left inspired by the memory of the town as a 'haven of rest ... to troops, tired and jaded with the shock of battle and the constant bombardments of the Salient!'[14] For the men of the 85 Field Ambulance association too, Poperinghe was a 'haven of rest to many, both during and since the war', a place of 'sacred memory'. Poperinghe remained true to what it had been.

Poperinghe's strategic position – militarily, emotionally, and spiritually – made it *the* place to enter the Ypres salient and battlefields. The authors of *The Immortal Salient*, Beatrix Brice and Lieutenant-General Sir William Pulteney, had no doubt at all that it constituted the most fitting route into Ypres. For them the route formed a west–east axis, which also meant crossing the Channel at Folkestone or Dover for Boulogne or Calais. 'There is one great advantage to British pilgrims by this route, in that they enter the Salient through Poperinghe and along the road that was the main thoroughfare for our troops. It has the feeling of being the right avenue of approach to the Holy Ground of British Arms,' they informed would-be visitors.[15] Taylor felt precisely the same, but brought to it the memory of the veteran, rather than advice to the pilgrim going in the footsteps of others. He rejected outright all other routes to the Ypres battlefields as totally unsuitable because they were nothing to do with the paths followed by British soldiers: 'There was but one route, the old familiar road, via Poperinghe, for who would visit Ypres and miss Poperinghe? From "Pop" onwards you may go via Vlamertinghe or via Brielen, as you please, but memory will not permit you to avoid Poperinghe.'[16] The Whaley Bridge British Legion tour visited Poperinghe in this spirit. In 'Pop' they saw again the old 'rest billet, the Cinema, Divisional Baths, and home of Toc H [Talbot House]'. Then, having made the town a stop in its own right, it became the convenient staging post to the battlefields. Rejoining the road to Ypres opened memories through contrasts, as they covered the distance a lot more quickly than in the old days 'when we foot-slogged this never-ending stretch of pavé through Vlamertinghe, passed Goldfish Chateau and Asylum Corner before dispersing in small parties for various parts of the Salient'.[17]

TOC H: 'POP'S' GREATEST WARTIME RELIC

The Whaley Bridge party also mentioned the other essential draw of Poperinghe, the one capable of enticing the tourist looking for the recommended sites of the old Western Front, Talbot House. Taking over a house in the town centre during the war, the army chaplain, the Reverend P.B. ('Tubby') Clayton, turned it into a rest and recreation centre for all ranks, which he named Talbot House, in honour of Gilbert Talbot, brother of his close collaborator, the Reverend Neville Talbot. Known as 'Toc H' in army signals parlance, this particular formulation of the house's name quickly caught on and was commonly used by soldiers. After the war, Clayton transformed the aims and objectives of the centre into a Christian movement, which spread across the Empire, backed by the Prince of Wales as patron.[18] Clayton's mix of Christian faith, warm hospitality, and jolly irreverence, which was in itself an admixture of the spirits of Bairnsfather and the most famous trench journal of the war, the *Wipers Times*, helped ensure the success of Toc H as a movement, and in the process made its original home a star attraction: 'this our Bethlehem', as it was described in a Toc H publication.[19]

The sight most were interested in was at the top of the house, where Clayton converted the loft into a chapel using an old carpenter's bench for an altar. Many soldiers who never attended church at home took communion there before going up the line, and among those who participated were considerable numbers who were brought to faith by their experiences and the quiet, persistent missionary work of Clayton and his colleagues.

The problem for many visitors was that in 1919 the house reverted back to its original owners, becoming once again the home of the Camerlynck family. In turn, the Camerlyncks tried to cope with the interest in their house by occasionally allowing access to the chapel, if sometimes (understandably) reluctantly. Henry Williamson was one of those admitted, and it turned into a rare event on his pilgrimage, being a place where he felt the past and present in perfect harmony. Responding to his ring of the doorbell, the maid politely accommodated his request to see the chapel. She led the way up the steep stairs and held the door open for Williamson and his fellow veteran companion 'before leaving with a slight movement of her head, neither bow nor nod, but a gesture of sensibility and understanding'. Seeing the worn floorboards where thousands of soldiers had knelt in reverential prayer was a deeply moving moment for Williamson. As he stood at the window, he saw the sun emerge from behind the clouds, heard birdsong, and then suddenly a loud report. But this time it was not gunfire at Ypres, instead it was the detonation of old pillboxes at Brandhoek as the land was cleared for farming. It was a reminder of the realities of the war's horrors, but it was part of the spiritual experience of memory. And so, after resting in the former chapel awhile, 'We clumped down the narrow stairs for the second, and maybe the last, time, and went on our way, bearing a fresh layer in memory, of youthful charm, grave and impersonal, waiting on two unknown English pilgrims.'[20]

Clayton also managed to arrange entry for Toc H pilgrims, and for members of the organization access to this sacred site truly was an act of pilgrimage. After first seeking the permission of Mr Camerlynck, Clayton took over a large party in 1926. On their arrival in Poperinghe the pilgrims were greeted as welcome guests by Camerlynck, who encouraged everyone to explore the house freely. Some were coming to relive old days accompanied by friends or relatives, while others were there in memory of loved ones, and others still simply wanted to imbibe the holy aura of the place. Clayton set himself up in the garden, a much-loved spot for all wartime users, to recite his memories of every corner of the house, and thus became the oracle interpreting the mystical site on behalf of those with no direct experience.

Because the chapel room was in a loft space that was of no practical use to the inhabitants of the house, the veterans found everything within it exactly as it had been. Here was authenticity, and it was described using military nomenclature, for this was the original Toc H site, or the 'Chapel Mark I', made famous across the world through 'representations of it and its memory'. Visiting a genuine relic of the Western Front was clearly a prized and moving experience, and the chapel's very simple, yet immensely powerful, symbolism of an old carpenter's bench used for the altar made it one of 'the greatest shrines in Christendom'.

In groups of twenty the 1926 pilgrims ascended the stairs and visited the room, achieving something akin to a beatific experience, for on their return 'they had changed faces' and were 'men and women who had seen something and understood their vision'. For the author of the account in the *Toc H Magazine*, the sight also returned him to the war: 'One could not but hear again other footfalls – the Army boots of successive congregations of ten years ago.' War, memory, and God's purpose in the world were brought together in a quiet attic in West Flanders.[21]

Behind the scenes Clayton was anxious to secure possession of the house in the Rue de l'Hôpital and was worried by Camerlynck's allegedly brusque attitude towards some visitors. Of course, Clayton was looking at the issue from a purely British perspective. He simply did not grasp the disruption caused to family life by a constant stream of visitors knocking at the door and wishing to look around. In 1929 rumours began to circulate that the house was to be sold.[22] Fortunately for the Toc H fellowship a generous benefactor appeared in the form of Lord Wakefield of Hythe, who purchased the house and presented it to the organization. Wakefield then agreed to fund alterations to the property, allowing the fellowship to use it as hostel and centre for pilgrims visiting the salient. Deafening cheers greeted the announcement when it was made by the Prince of Wales at the annual Toc H gathering in the Royal Albert Hall in December 1929.[23] Toc H was coming home.

The purchase of the original Talbot House incorporated this special site into the British memorial landscape. But Talbot House was unique, as the Toc H organization realized, and it had to be inserted into that memorial landscape carefully. The management committee therefore carefully, politely, but firmly avoided the calls to return the building to a precise replica of its wartime condition down to the tiniest details. This would have made it nothing but a museum, eroded its status as a living institution interacting with pilgrims and visitors and, in fact, undermined the essential Christian, and wartime, message of the offer of fellowship and friendship to all.[24] Instead, it was perceived as an active site of worship, but the very particular worship space also made it a place of commemoration and remembrance. The chapel reflected this dual status in its decoration, being a mix of original furnishings and gifts donated in memory of lost loved ones or as thank-offerings for survival, and included a collection of original wooden crosses from the graves of unknown soldiers which had been donated by the IWGC after reinternment.[25]

The final result was a fine balance. The practical innovations and interventions made to the building allowed it to exude its original spirit. It was a place of worship and a hostel acting as home to pilgrims staying in its rooms. At the same time, it also opened its doors to those seeking advice and information about the old battlefields, the cemeteries, and memorials, and as a welcome stop for visitors who just wanted to pop in, have a look around, and take the chance to grab a cup of tea.[26]

Following the formal reopening service and ceremony at Easter 1931, when people came from across Britain to be present, the house prepared to receive its first pilgrims, who arrived a few months later.[27] On reaching Poperinghe they carefully explored the Old House, 'a place familiar already in our minds from much readings and imaginings', before lunching at Skindles, the other great 'Pop' landmark. During the afternoon they were taken on a battlefield tour, and the day closed with an evening service in the chapel.[28]

Reviewing its methods and practices in 1933, one of the subjects the Toc H organization examined was the value of its battlefield pilgrimages, asking members to give their opinions. One correspondent, too young to have served, mulled over the meaning of the Old House, having stayed there informally and met up with other Toc H members who, likewise, were not part of an organized pilgrimage. All agreed that the very specialness of the place imparted something that was not easily defined or explained but that was undoubtedly of spiritual and moral value. This led the membership to discuss whether those who visited the Old House should be marked out in some way, and they considered the appropriateness of a special badge, tie, ribbon, or other marker. Ultimately, they rejected these ideas, and turned back to the sense of beatification achieved by visiting the Old House: 'Any mark must come to be the outward sign of a real bestowal of honour in men's minds.'[29]

'G.C.', chronicler of the 1935 Public Schools' battlefields pilgrimage, certainly felt Talbot House left a deep impression on young minds. 'On Sunday morning, many of us availed ourselves of the opportunity to pay a visit to "Talbot House," Poperinghe, and I should say that none of those who made the trip will ever forget this wonderful place,' he noted, before concluding, 'The Chapel itself is certainly something to remember, and the "Old House" is most reverently kept in memory of those elder brethren of ours who partook of its hospitality during the trying years of the Great War.'[30] For the veteran D.S. Ryeland, 'to stay, in peace time as well as in war, was to gain inspiration'; thus did Talbot House in Poperinghe link tourist, veteran, and pilgrim.

THE SEARCH FOR THE MADEMOISELLE: ARMENTIÈRES

The only other rear area town with anything like the cachet of Poperinghe for all types of visitors was Armentières, or 'Armenteers' as British soldiers always called it. This small town close to the Franco-Belgian border was held in great affection by soldiers as a place of rest and recreation. 'Armentières is associated in the British mind with the lighter side of war, and not without reason, for there was sometimes good fun in Armentières,' declared H.A. Taylor.[31] This sense of slight qualification – 'sometimes good fun' – is also detectable in Coop's guide. For him the downside was the nondescript architecture, but he admitted this did not matter to soldiers eager to sample the town's 'good shops, excellent estaminets, and numerous places of entertainment'.[32]

A keen observer of Belgian and French Flanders, R.H. Mottram was also deeply interested in what, to him, was the peculiar phenomenon of Armentières. His was intrigued by a town gaining its prominence solely through its wartime status. Having none of the cultural glory and heritage of Ypres, Arras, or Rheims, and yet achieving fame, it 'was the freak town of the whole front ... Its prominence and its fate during 1914–1918 are a mystery.' He then went on to explain the conundrum to his readers. Armentières' reputation was partly based on memories of its role as a rest centre behind a generally quiet sector of the line, allowing veterans to recall it with affection. However, its fame among a broader audience was built 'upon the fact that the name of Armentières fitted the words put to an old folk-tune and sung by the British Army until it may be said to have entered into the language: "Mademoiselle of Armentières, parlez-vous?"'[33] And here Mottram identified the key to the town's fame among pilgrims and visitors alike. Veterans might have wanted to see old billets and well-remembered estaminets, but the others wanted to know about the famous 'Mademoiselle from Armenteers'.

Those great compilers and historians of the British army's songs and slang of the Great War, the officer-veterans John Brophy and Eric Partridge, identified 'Mademoiselle from Armenteers' as in circulation from 1915. It consisted of numerous verses, some absolutely filthy – and always much sanitized for home front consumption – about the activities of this particular young woman.[34] And her fame continued into the post-war years, when the character was turned into a highly successful romantic melodrama for both stage and screen.

No one ever knew whether the Mademoiselle was based on an actual woman; Mottram certainly did not think so, but knew some who proudly claimed that privilege, although it is less clear whether they wanted to be associated with the romantic or lewd characteristics for which she was famous. The explanation Mottram offered was based on local geography and culture. The farms in the rural districts were home to close-knit families who kept every member under a tight rein, especially the young women, making them seem shy or reluctant to respond to any soldier desiring a little friendly chatter. By contrast, the women of the towns were more outgoing, lively, and used to dealing with a wider circle of men. Armentières symbolized the latter world, according to Mottram, thus explaining its home to *the* composite woman of the British army's world on the Western Front.[35]

Seton Hutchison concurred with Mottram's interpretation. In his meditations on the battlefields, he wrote, 'No one ever dared to define the personality of Mademoiselle d' Armentières, who gave immortality to the song. Perhaps she was ubiquitous ... you would find her replicas throughout every village in this triangle at Merville, Estaires and St. Venant.'[36]

Where veterans trod carefully, the *Daily Mail* went boldly. Refusing to go along with such an explanation, in September 1930 it told its readers she had been identified as Laure Millanquet, who had recently been found dead in the house where she had lived alone for many years. In 1914, so the article claimed, she was a famed beauty, provoking the attention of every British soldier who passed through the town. Tragically, her beloved had been killed in the war, which turned her into a recluse.[37] She was not the only publicly named, and accepted, claimant, however. In 1928, the French Foreign Secretary, Aristide Briand, selected Jeanne Victorine Lescornez as the eponymous miss when approached by a South African veterans' group who wished to name her as patron.[38] Armentières and the Mademoiselle could not be divided, even if the recipients of the title were many and disputed.

For those pilgrims interested in Armentières with the intention of finding the model for the famous resident, her house, or the bar over which she presided, the realities of the town could find them agreeing with Coop's judgement on its lack of architectural delights. The veteran W.J. Baumgartner led a mixed group of veterans and pilgrims in 1928. Taken on a charabanc tour from Ypres, they joined the 'very straight and very uninteresting road which brought us to drab Armentières' which prompted mystification as to 'how anyone came to write a song about' it.[39]

For veterans, enquiries about the Mademoiselle were often an opportunity to indulge in a bit of light-hearted fun. When Major Charles Salvesen led the 236 Siege Battery veterans, their families, and friends on a tour in the summer of 1935, Armentières was included on the agenda. Finding all the women extremely curious as to her address, 'the ladies simply had to be shown ... Unfortunately, as every one asked separately, naturally the famous maiden was given several domiciles. However, that did not spoil the party.'[40] Veterans were usually far less bothered by this quest. Instead, for them it was the case of sniffing out well-remembered spots, which usually meant a snifter or two at the same time. The members of the 2 London Regiment association made Armentières an integral part of their pilgrimages. On arrival in the summer of 1926, they split up and went their own way for an hour and a half, with most paying 'a visit to our old billets, where we were received with open arms'. Four years later a similar itinerary was followed, allowing for an hour in the town where 'many visited the streets in which they were billeted in early 1915'.[41]

Unlike Baumgartner and his chums, when Will Bird arrived in 1931, he was delighted to find a charming town filled with happy people, which was a joyous contrast with the rather more dour atmosphere he perceived over the border in Flanders.[42] Some veterans were nowhere near as willing as Bird to accept change and found Armentières almost distressingly unrecognizable. Perhaps able to cope with the transformation of the front lines, they were less well able to deal with alteration to a world they considered safe and solid.

Disappointed at seeing so little he remembered in the town, one Liverpool Scottish veteran was cheered to find an estaminet he recalled, which acted as an anchor for his memories, but the details let him down, as he mourned the loss of the old 'Eggs and Chips' sign which used to sit in the front window. At another cafe he knew well, he found that the former owner was now married and living in Paris. Rather than delight at gaining some solid information about someone he once knew, it only provoked further regret at the absence of another link with the past. Wandering out of the town along what the soldiers called Brick-Stack Lane, he saw the rebuilt brickworks complete with a towering chimney. Again, this was no good; restoration alienated him. Only one place was precisely as he remembered it. 'Crown Prince House', on the road to Chapelle d'Armentières, ruined and pockmarked by shrapnel, was exactly as it was when he had marched past in 1918. Seeing that and visiting the graves in Erquinghem churchyard allowed him to hear again 'the singing of the Liverpool Scottish as in the days that are gone'.[43] The old world was there, somewhere, but it needed some finding.

Far more disillusioned and damning was 'J.B.M.'. In an August 1930 article for the *Belfast News-Letter* he offered a piece of blunt advice to his fellow veterans, unequivocally telling them not to return. 'To go back now, eleven years after, and revisit the battlefields is only to be disillusioned. Many ex-soldiers have made the pilgrimage, but a greater number have not. To those who haven't I would say – Keep your memories.' Everything disappointed him because it was not *exactly* as he remembered it. Like the Liverpool Scottish veteran, he was bewildered by the new brickworks outside Armentières. Worse still, in this, his kingdom, a worker had the cheek to ask whether he had permission to enter the site. In Armentières there was nothing of the old, ramshackle charm, as everything was new and uniform. Even though he found Suzette still serving in her cafe, he could hardly forgive her for growing older and enjoying the usual experiences of life: 'Suzette was there, but what a different Suzette. I scarcely recognised the buxom mother of five as the petite, vivacious brunette of war days.' Trying to interpret this veteran's thoughts precisely – if indeed he had any clarity – is extremely difficult. Did he believe he was still the exact same person as over a decade earlier, and so was confused at not finding Suzette and her town exactly the same? Or was it, as so often with veterans, that the disappointment that coloured their response to a battlefield pilgrimage actually made them blind to the evidence all around them? In the case of 'J.B.M.', no sooner had he recounted the shock of finding Suzette a mother of five, he recorded visiting another cafe he once frequented regularly. Here 'the room was little changed ... The room became crowded with memories,' but the owner would not let him go upstairs to see where he was once billeted, which left him feeling slighted.[44] Remnants of the past were not enough for this veteran, only an access-all-areas pass would do.

BÉTHUNE AND BAILLEUL: THE PLACES OF SUDDEN DESTRUCTION

Most tourists never ventured much further into the old rear areas than Poperinghe and Armentières, whereas veteran-led or -dominated groups often made a wider sweep, including the towns of Bailleul and Béthune in their itineraries. As with Poperinghe and Armentières, the reason for the visits was to see once again places fondly remembered for offering comfort in the midst of the hardships of war. Like Zoë in Poperinghe and Suzette in Armentières, Bailleul had its particular female of high renown, and she, too, was the queen of a bar room. Tina's was the officers' haunt of choice in Bailleul. As Atherton Fleming noted in his guide, the officer arriving soaked from the trenches 'could at least get dry outside and wet inside at "Tina's," a pleasant reversal of the usual conditions of the line'.[45]

However, it was not just the running theme of good company and good surroundings that shaped the recollections of Bailleul and Béthune. The two towns had a special status, and provoked a particular emotional response because of their eventual fate. With the British forced into precipitate retreat in the spring of 1918, these two towns, so long decently behind the lines, suddenly became vulnerable, and with equal suddenness they were destroyed as the fighting engulfed them. Having stood so long as beacons of normality, good times, and good cheer, the brutality unleashed on them so late in the war inspired deep pity in many veterans. Unlike Ypres, which so many soldiers knew only as a ruined city, as a place where the only differences across time were in the details of destruction, Bailleul was different: one moment it was there, and the next it was gone. The speed of the destruction, combined with its intensity and timing, haunted many veterans.

T.A. Lowe felt the brutality of such destruction acutely and allowed emotion to infuse his normally reserved and clipped manner when recounting the effect of the 1918 retreat: 'We never knew how much they meant to us until we lost them. "Pop" became a skeleton of its former self, from which its inhabitants were forced to flee; Bailleul is now a heap of ruins; Bethune is unrecognizable.'[46]

Like his fellow veteran, Wilfrid Ewart was deeply upset by the sight of Bailleul. Recalling it as the vibrant, bustling centre of everyday life where a soldier could leave behind the world of the trenches, he was left shocked by the devastation: 'Not one atom remains and not even stagnation.' Realizing that his sense of desolation might not translate to those with no direct knowledge of the town, Ewart explained to his non-veteran readers, 'The completeness of it [the devastation] will not strike home as had you known it before April, 1918.'[47] Such comments encapsulated the different gazes of battle-field visitors, one informed by direct experience and the other not.

Mottram described Bailleul's destruction as 'an event which stands for me among the major tragedies of the war ... Bailleul, happy and homely, and relatively spared, became suddenly involved in the worst and grimmest battle of the War. I remember standing in its empty and deserted market-place, with debris falling all around and great shells clanging on its solid stones ... In that square that had always been so busy and so handsome, was not a soul left.' On quitting the town, he learned the most appalling part of the story: the damage was actually caused by British artillery determined to stop the Germans occupying a useful base.[48] Coop put things with almost brutal sangfroid and succinctness: 'Bailleul, the one-time market town for troops ... is now flatter than Ypres, and all this was done in one short month in the spring of 1918. We did it. When we set to work to flatten out a town, we did it thoroughly, as the tourist will perceive when he visits Bailleul.'[49]

Arriving in 1920, Stephen Graham saw a town every bit as wrecked as Ypres but recovering at a far slower rate, identifying the reason as the lack of economic stimulation by battlefield visitors. 'It does not cater for war pilgrims or take the money of tourists, and so there are no prominent hotels and few estaminets,' he wrote, and he went on, 'Most of its houses are down, its ways choked with ruin, and in the evening nondescript squads of workmen shuffle through the streets to their homes in barracks and cellar.' Debris was still piled up in heaps higher than houses. Where there were traces of recognizable buildings and houses, they had concertinaed into each other. Everywhere were signs of life debased into brutality, epitomized by the awful sight of two young boys playing a game of bat and ball, except that the ball was a live frog being tossed towards the boy with the bat.[50] Perhaps a sense of guilt lay behind veteran reactions to these towns. At best, the British abandoned them to the advancing enemy; at worst they actively aided their destruction.

Eventually Bailleul recovered and restored some of its fine buildings, including the belfry and town hall, which impressed many on

Figure 11.2   Temporary railway platform, Bailleul, c. 1920.

their return. In a place where British soldiers had spent so much time, and so much money, veterans were also gratified to find the locals recalled them with equal fondness. When the members of the 236 Siege Battery association arrived in the town in the summer of 1935, they were very much surprised to learn that the mayor wanted to greet them in a formal reception, 'but somewhat to our relief, owing to electioneering activities, this did not take place'. Instead, the chief clerk was instructed to lay on a lunch for the party, following which he arranged for them to climb the belfry tower, where one of the veterans revealed his skills at playing the carillon, 'much to the astonishment of the local inhabitants'.[51] As was so often remarked on such occasions, the entente cordiale was alive and well.

The last of the small towns capable of exerting a hold on veterans was Béthune. As with its sisters, it inspired affectionate memories, regardless of the realities of the place, because it had catered for Tommy's every need during the war, as Taylor recognized:

> The Bethune of those days was a rather grubby little town by comparison with British places of the same size, but, to the soldier, it was a kind of little Paris ... There were in Bethune many establishments for the recreation of the

soldier, as well as cafés where a palatable glass of beer and a good dinner might be obtained. There were little shops well stocked with a varied array of souvenirs and post cards, always full of soldiers in search of something to send to friends, or to take home should the anticipated period of leave be forthcoming before a hostile shell or bullet made leave a matter of no interest.[52]

The special status of Béthune was also clear to Mottram. 'Men could be thankful in Bailleul, comfortable in Poperinghe, but in Béthune everyone was happy.'[53] Fleming was equally prepared to lavish accolades and testimonials on the town, the 'oasis in the desert [for] the mud-soaked man down the line for a rest', a place of restaurants and bars, and in the Hôtel de France, the provider of meals 'that could only be equalled in Paris'.[54]

Tragically, Béthune was another town that had been smashed in the spring of 1918. Writing soon after the end of the war, Coop said the main square 'was reduced to a pathetic state of ruins'. As with Ewart's careful explanation of Bailleul's condition to the non-veteran, Coop felt the need to tell the visitor how difficult it was to envisage the state of the town at the war's end. The rubble having been swept and neatly piled, 'the traveller cannot have, nor can he form, any adequate picture of its desolate condition during the spring and summer of 1918. It was just a complete heap of blocks of stone and torn timber.'[55]

Given this mess, Taylor was very pleasantly surprised by the immense contrast he experienced on returning in 1928. New buildings, public and private, pleasant shops, and good roads inspired a rarity – a veteran who saw, accepted, and appreciated a distinct improvement on what he once knew. 'The reconstruction has been carried out with much more taste and skill than in any town of comparable size which I have visited. Indeed, I would say of Bethune that it is the most improved town in the old war zone,' he noted.[56] Will Bird was prepared to join him in singing the praises of the new Béthune. It was a place where 'all the buildings [were] gay with colors, banks and stores most prosperous looking, the shop windows displaying first-class goods', all topped off by a fine park.[57] It was to this Béthune that Edmund Blunden brought a fictionalized version of himself, Duncan, in the 1933 novel he co-authored with his wife, *We'll Shift Our Ground*. Arriving late at night, the veteran

Duncan spends a few moments 'placing myself gently back into *my* Béthune. Can't hurry over that,' before he and his wife discover a fine restaurant still willing to serve at such a late hour.[58]

THE BEHIND-THE-LINES MEMORIALS

The post-war status of these towns was not simply as a place where veterans rediscovered happy days, or as a model for tasteful and effective reconstruction, as in the case of Béthune. They were also invested with major significance by the IWGC. When starting its plans to combine battle exploits memorials with those to commemorate the missing, Béthune and Armentières were identified as appropriate sites for major memorials. Armentières was to commemorate 15,000 missing and the operations around Ploegsteert and on the Lys, while Béthune was to mark the battles of Neuve Chapelle and Festubert, as well as other operations in the neighbourhood, along with some 20,000 names of the missing. Armentières was, however, swiftly rejected after initial investigations could find no site suitable for a substantial memorial.[59] As Armentières was a relatively small town, it is also possible there were some French concerns that a major British memorial might seem too dominating, and therefore overshadow the French contribution to victory.

Planning proceeded smoothly over Béthune, with the mayor enthusiastic. He offered two sites for consideration, both in the heart of the town and connected by the newly renamed Boulevard Kitchener. With things progressing nicely, an architectural competition for the memorial design was launched, and was completed by 1925. The Western Front veteran and IWGC junior architect Captain J.R. Truelove was the successful candidate.[60] Knowing precisely what Truelove had in mind is difficult to determine, as the French authorities then rejected his design in the same package that included Lutyens's original plans for Arras and St Quentin.[61] With the IWGC now forced to scale down the memorials and include them in cemetery sites, the hunt was on for suitable alternatives. Eventually, Truelove was asked to design memorials for the Vis-en-Artois cemetery on the Arras–Cambrai road and at Le Touret, Richebourg-l'Avoué, a small village near Béthune.

The extent to which Truelove altered his designs rather than starting from scratch is not clear, but it seems likely he revised his original plans for Béthune. He certainly delved into his knowledge of the classical orders and their related allegorical languages. At Vis-en-Artois, the result was two graceful colonnades linked by a central screen wall holding a bas relief of St George slaying the dragon, carved by Ernest Gillick. Emerging from either side of the linking screen were cenotaph-like pylons, echoing Lutyens's entrance to the Étaples cemetery, as well as his original designs for the Arras memorial. Digging deeper still into classical iconography, this veteran of industrial warfare decorated the Ionic colonnades with spears and shields. Truelove's inclusion of the St George sculpture added a disturbing element to the seemingly cool and dignified classical memorial design. A nude St George rides bareback driving his lance into the dragon, while the monster digs its talons into the horse's belly. Although allegorical, Gillick's work emphasized the delicacy and vulnerability of human and animal flesh, as the rents caused by the dragon's claws are palpable. At first glance the sculpture appears a neat, if obvious, decorative flourish, but when contemplated for any length of time it takes the memorial beyond cool to chilling.

Truelove did something similar at Le Touret. The entire memorial is framed by a Florentine loggia surrounding a rectangular court. Running along the eastern side of the court is a colonnade forming a long gallery walk from the entrance. For the centre of the court, Truelove designed a small column resembling a classical altar or augury plinth, suggesting sacrifice and destiny. Included among the motifs was the IWGC lamp of remembrance, and in the entablature of the court were inscribed the names of the battles commemorated by the memorial, while the colonnades contain the panels with the names of the missing. Calm, peace, repose, and dignity were the watchwords of Truelove's design; the veteran used his knowledge and experience to give his comrades the qualities they deserved. And yet, as with Vis-en-Artois, Truelove left an edge, in the altar with its implications of blood and, indeed, guts.[62] Perhaps towns associated with great joy and relief were not quite the right hosts for these memorials after all.

THE BRITISH LEAGUE OF HELP AND THE TOWNS AND VILLAGES OF FOND MEMORY

There was a final bond between these rear area towns and British visitors in the form of the local associations forged by the British League of Help. As well as the links made between local units and particular places in France, usually because they attacked or defended them, questions of civic pride came into play when contemplating an adoption. Large and important towns and cities in Britain felt their dignity and status demanded somewhere of roughly equivalent size and prestige in France, which is why Bristol and Bradford adopted Béthune and Bailleul, respectively (also ensuring neat alliteration). Bristol could then take pride in rebuilding 'a town which recalls pleasant memories in the minds of many a Bristol soldier', linking civic status to local memory. For J.H. Palin, mayor of Bradford and veteran of the Western Front, there was the personal connection through his own wartime experiences. On an official visit to Bailleul in 1925 to celebrate the new relationship between the two towns, he recalled the friendly welcome he received whenever he managed to snatch a little time away from the trenches.[63] Palin's direct knowledge of Bailleul gave greater emotional depth to the adoption. Here was a veteran who actually knew what wartime Bailleul was like.

The south coast seaside towns of Bexhill-on-Sea, Eastbourne, Hastings, and Rye were motivated by the associations between their local unit, the Sussex Regiment, and in particular 5th battalion, with a series of villages behind the Somme front lines where the men had been based for a considerable period between 1915 and the start of the Somme battle. Bexhill adopted Bayencourt; Eastbourne, Bray-sur-Somme; Hastings, Sailly-au-Bois, and Rye, Coigneux. Although nothing like the scale of Bailleul or Béthune, all four villages shared with them the same wartime experience: behind the lines for much of the conflict, they were suddenly engulfed in 1918. However, when the tide of war broke over these villages, it did so long after the Sussex men had come and gone. Therefore, what these seaside towns did was adopt places where their men had once found escape from the dangers and monotonies of trench duty. They were behind the lines, only just, and because of that by no means entirely safe, but nonetheless they were away from the front.

When Sussex veterans returned to the villages their towns had adopted, they were treated to especially warm welcomes. In 1934 a party from Hastings arrived complete with a contingent of 5th battalion veterans led by Captain W.F. Brown, chairman of the old comrades association. Replying to the mayor's address, Brown 'expressed the pleasure of the members in revisiting old billets, and some found the actual people with whom they stayed during the war'.[64] The donors also had the gratifying opportunity of seeing for themselves the difference their contributions had made to the adopted community.

Bexhill-on-Sea raised enough money for a new water tower to be installed in Bayencourt, and a large delegation made its way over the Channel for the formal unveiling ceremony of this new civic amenity in June 1924. Local dignitaries expressed their deep gratitude, causing the Mayor of Bexhill to swell with pride, and he was able to tell the local newspaper that the town had done a 'real sound, pukka, job'.[65] He obviously thought the argot of the Tommy was just the right note on such an occasion.

THE SPECIAL PLACES OF WOMEN VETERANS

Towns and cities behind the lines were not solely remembered and visited as places of peace, joy, and respite; for some they were the location of their war work and contribution to victory. Among the many veterans visiting France and Belgium were those of the women's services stationed across the Channel. Unlike former infantry and artillery men, battlefield sites held no particular interest for many of these women, who were instead far more focused on the lines of communication and the army's main bases: for these veterans, the remains of trenches, dugouts, and the seeming recovery of the landscape from the devastation of the war was of less concern; instead, the world they remembered was one of huts, camps, supply dumps, and the army's massive infrastructural requirements. When they visited, often it was the finding of open space where there was once a maze of temporary buildings that provoked the greatest surprise.

Along with some friends from the local branch of her old comrades association, in the summer of 1922 Liverpool veteran of Queen Mary's Women's Army Auxiliary Corps (WAAC) Dorothy Holder visited Étaples, where all that remained of their old camp was a few concrete slabs and strands of wire. They wandered around trying to work out the footprint and location of each hut, but 'everything is so overgrown that it is hard to place the different hospitals', turning the whole thing into 'a sad pilgrimage to old "Camp One"'. On a happier note, she did find local people she remembered, and enjoyed catching up with them, including 'Anna [who] is always pleased to see the "Old Waacs"', and she could report, 'The little chubby tram conductor is now quite a man, and has a moustache.'

Edith Winter went to Rouen in the same year, 'revisiting the various places dear to the heart of the Waacs'. Her search for fragments of the past was equally lacking in dramatic discoveries, but she did come across an old army signboard, which cheered her up. Unlike the disappeared camps, Rouen was just as she remembered it; the only thing missing was the throng of soldiers clad in khaki.

Ada Strange went over to Calais with some fellow WAAC veterans to find their one-time base and have a good look round. A second motivation was to see old acquaintances, for 'during the war I met some very nice French people' who had kept in touch and invited her over on many occasions. On arrival, she and her friends were recognized instantly and warmly welcomed by the French couple, who ran a chip stall near her old camp. All then merrily 'drank the Entente and the Bon Sénté [sic]' together. Moreover, the couple 'enquired after heaps of the girls' and told them how much they missed all their old friends from Britain. Further investigations and explorations in and around Calais saw the WAAC veterans celebrate a reunion with their former laundress, and they were delighted to find the woman who had worked in the YWCA and had since got married.[66]

As with male veterans visiting behind the lines haunts, the wonder of stumbling across familiar faces and sights allowed the years to disappear and brought the old days to life. Behind the lines was a world as rich as the former battle zones, but not all visitors could grasp it to the same degree as the veteran.

*

Now seldom visited, but then a fixture for veterans and pilgrims, the 'behind the lines' towns offered an alternative vision of the war. Far from the world of trenches, though not entirely immune to gunfire and aerial bombing, these havens held a central place in veterans' memories. They were the flip side of the front line, being the home of vitality and normality. Recollections of the trenches were sharpened and reinforced by the contrasting ones of beers in Armentières estaminets. Indeed, at times, the recall of drinks, eats, and laughs might even override the memory of the front. As Siegfried Sassoon put it in his poem 'To One Who Was With Me in the War', 'Remembering, we forget much that was monstrous.'

L'Envoi

My husband and myself thank you very much for helping us to go to France, and visit the grave of our dear son in Etaples Cemetery. We shall never forget the kindness shown to us on that journey; everything was done to make that sad mission as happy and pleasant as possible, and when we saw the beautiful cemetery where our dear son lays [*sic*], and heard the loving words that were spoken of them, our hearts were filled with pride that we had our dear one resting there. It has always been our wish to go over there, and it has made us very glad.

A pilgrim from New Cross, London, who joined the St Barnabas Hostels' pilgrimage (1923)

After lunch our time was our own and Frank Miller asked me if I would guide him to find the Artillery Wood Cemetery, somewhere near Boesinghe. We set off, passed the other Essex Farm and its cemetery and many others too, which we inspected. There were so many little cemeteries that we were a long time finding the one we wanted. A Belgian directed us to it, and Frank found the grave we had set out to find, the resting place of the son of two of his friends who greatly desired a photo of the grave. We took snaps of it. We called at Essex Farm for a lemonade and the lady living there told us that she remembered the approach of the Germans in 1914 and fled to Poperinghe. By the time we reached the hotel, we had truly been in the past. After an excellent meal we sat at the little tables outside the hotel, silent for a good while until I suggested an itinerary for the week.

Gerald Dennis, veteran of 21 King's Royal Rifle Corps (1928)

For the most part the north-eastern section of France is only to be regarded from the utilitarian point of view; it is a passage, in fact, to more interesting regions for which the long-distance tourist may be bound ... It may be urged truly, enough, however, that the north-east is the area of the battle-fields, and it would be natural to assume that it possesses a special interest for the tourist on that ground alone.

Charles L. Freeston, *France for the Motorist* (1927)

Pilgrim, veteran, tourist, all were drawn to the battlefields. Each came looking for something a little different. Often suspicious of the tourist, the pilgrim and veteran had more in common with the sightseer than they liked to admit, for all shared a common fascination and desire. They were drawn to the places where the cataclysm of war had been expressed most fully. The Western Front battlefields were one of the great anvils of the war, on which the guns hammered out death and destruction, causing untold misery. It was thus hardly surprising that people were drawn to a land that was weird, wonderful, and, above all, sanctified through the sacrifice of life on an immense scale.

During the conflict the British people consumed a diet of stories about the fighting fronts, with the Western Front taking up the vast majority of newspaper, cinema, and art and photographic display space. This made the front both quotidian and mysterious. People got glimpses of it, and in those glimpses they realized that it was horrific and unlike anything anyone had ever seen before. Exhibitions of model trenches, captured guns, tanks trundling around city centres, and war photograph exhibitions displaying supersize images of the work of photographers such as Frank Hurley attracted huge crowds. In those images they saw mud, shattered trees, men helping the wounded, men struggling along tracks weaving through the pulped landscape, tanks, guns, explosions, craters. The combined effect was mesmerizing and inspired a longing to see the real thing.

As soon as people started to cross the Channel with the intention of satiating their curiosity, arguments about the morality of gawping at the horrors of war began to ring out loudly. Indeed, sardonic prophesy came from the pen of Philip Johnstone (the pseudonym of Lieutenant J.S. Purvis) even before the war was over. In tune with the sentiments of Bruce Bairnsfather and C.E. Montague, his poem 'High Wood' was published in the *Nation* in February 1918 (and is quoted at the start of chapter 7). In it he foresaw the wood as a lucrative tourist trap expertly exploited for all its worth by the owners. Tourists are told that the authenticity of the site is unquestionable, and they are reminded to keep to the paths, not damage company property, and to patronize the hospitality facilities. Seeing, wondering, and consuming had been identified as the essential tourist habits. Those undertaking the journey responded by declaring their mission one of understanding and empathy. They were on the battlefields to comprehend fully how the war had been fought and see for themselves the sites of so much loss. Visiting was thus presented as an act of homage to the epic endeavours of those who had defended them. And doubtless that was the motivation of many tourists. Unfortunately for them, it was one few public voices of the time were prepared to consider sympathetically. Instead, it was far easier to condemn visiting as nothing more than ghoulish self-indulgence.

In a completely different moral category, a sanctified one, were the bereaved. Under the shawl of religious allusion and imagery, they were deemed pilgrims to a land made sacred through the spilling of blood. Steeped in rituals, individual and collective, the bereaved trod their way through this new holy land. Their world was far from one in which the forces of secularization were victorious. High Church rhetoric met Bunyan's Protestant *Pilgrim's Progress* on the battlefields.

So deep an impression did this group make on the public imagination that they gained their own mythology and mystique, as encapsulated by Rudyard Kipling in his short story *The Gardener*, first published by *McCall's Magazine* in the spring of 1925. This poignant tale tells the story of Helen Turrell, a spinster who had devotedly cared for her illegitimate son, Michael, having originally told him he was adopted in order to avoid scandal for them both. On the outbreak of war Michael volunteers for military service and is killed not long after joining his unit in France. Declared missing by the authorities, it is not until after the armistice that the heartbroken woman is informed that his body has been discovered, identified, and reinterred in the British military cemetery of 'Hagenzeele', a name with the ring of Flanders authenticity in its echoes of Hazebrouck, Dadizeele, and Hooge.

Helen immediately sets out to visit the cemetery. Once in France she meets a Lancashire woman who descends into hysterics at not being able to find any trace of her son's grave. Another woman is encountered looking for the grave of a secret lover. Unable to find a way of communicating properly with others suffering the agony of loss, Helen's inarticulate stumblings inadvertently offend the woman. On the following day Helen finally reaches the cemetery. Seeing a man planting flowers, she asks for help. Bidding her to follow him, he takes her to the grave, marked with a simple wooden cross. What happened at the graveside is left untold by Kipling. Instead, he concluded the story with: 'When Helen left the Cemetery, she turned back for a last look. In the distance she saw the man bending over his young plants; and she went away, supposing him to be the gardener.'[1]

A bereaved father, Kipling brought all his personal knowledge, thoughts, and feelings to the story. As a senior IWGC committee member who had been exposed to the voluminous correspondence it received, he also infused the work with the experiences of so many veterans, tourists, pilgrims, and most especially the female bereaved. Kipling's portrayal of the women carries within it every sentiment of those letters asking for information, requesting photographs or wreaths to be laid. Each of the three women has been made fanatical through the intensity of their emotions, but despite the solidarity arising from their grief, so intense is their sense of individual loss they find it hard to engage with each other in a meaningful way. As each woman knows, only when they stand at the grave they so long to see will catharsis occur, and in Helen's case the instrument is an IWGC gardener, a veteran who has taken on this work of homage to his dead comrades. So powerful is his quiet, gentle, but authoritative presence that he seems Christ-like; the ending of Kipling's story echoes the resurrection morning in the Garden of Gethsemane, where Mary Magdalen confuses the risen Jesus for the gardener.

Kipling's story reflected the public perception of the pilgrim as not just a bereaved woman, but a bereaved mother. Although fathers, sons, and brothers were among the many pilgrims, time and again emphasis was placed on the woman. The idea of the pathetic, weeping mother finally achieving consolation by standing at the graveside was one expressed in countless accounts of pilgrimages. It created a polarity. On the one side was the pilgrim, representing the true soul of the old battlefields, and on the other was the supposedly garish tourist, utterly lacking humility or self-awareness.

However, the poor, lost mother was not simply a press construction but a reality, as the Reverend Matthew Mullineux knew. Seeing for himself the awful sight of women drifting around France desperately seeking the final resting place of a lost loved one led him to establish the St Barnabas Hostels organization. Bringing consolation to the broken-hearted was Mullineux's mission, as it was for the war graves visitation services of the YMCA, the Salvation Army, and the Church Army. These organizations devoted themselves to the care of the bereaved pilgrims, whether they arrived in groups, large or small, or as individuals. By arranging every detail of the trip, these services ensured the bereaved, many of whom came from humble backgrounds, were not distracted from their focus by any other concerns. They were brought to the grave or memorial and provided with their special moment of communion.

With the bereaved pilgrim awarded a greater moral status than the tourist, there was the potential for the landscape to be perceived as their possession. Their myriad tears made them far more rightful owners of it than the visitor. Interrupting and complicating that perception was the veteran. Outside of the specialist old comrades associations, this figure was always a male and was always perceived to be male. Moreover, he was also labelled a soldier of the trenches or gun-pits and not a former stores clerk based behind the lines in Rouen. Despite this simplistic view, the veteran filled a liminal position, defying easy categorization. In one sense he was a revenant still wandering across the landscape he once owned as he descended into his former self and ways. At the same time, he was a joyous and boisterous actor busily engaged in fulfilling Sassoon's dictum that in 'remembering, we forget much that was monstrous' as the drinks flowed along with all the old jokes and all the old songs. Yet he was also a quasi-pilgrim, having intimate knowledge of the dead and sharing the pilgrim's desire to find moments in which to grieve, commemorate, and remember the lost.

The continual presence of veterans on the battlefields meant that remembrance of the conflict in this landscape was 'entirely different from that expressed on Armistice Day across Britain. By the mid-twenties the needs and demands of the bereaved dominated the observations. The Armistice Day festivities seen in the early twenties collapsed under the weight of grief poured out by the bereaved. This served to push the complex ex-service memory of the war to the periphery of public expression.[2] On the former battlefields, the veteran was back in his territory, a warrior returned to King Pellam's Launde, as David Jones had it in his epic, autobiographical war reminiscence poem *In Parenthesis*.

This landscape, this kingdom of death, underwent huge changes during the twenties. In 1919 the condition of the former battlefields was, as so many visitors noted, indescribable. Devastation was the term everyone used so freely and continually that it was almost cheapened as a term; it lost its power to convey the awful reality. Thanks to the amazing zeal and dedication of those who returned, the battlefields were transformed. In some places the speed of the restoration was dramatic, while in others, whether because of remoteness or the intensity of the fighting or both, the pace was much slower. Visitors, especially veterans, could be disoriented, dismayed even, by this restoration of the landscape. Some veterans felt as if their possession had been stolen from them while their back was turned, and its new owner had altered it out of all recognition. Such protestations ignored the fact that the battlefields never fully shook off the scars of war. As the careful and observant visitor found, with just a little bit of close examination the landscape revealed its recent history clearly. Even the most disillusioned of veterans uncovered traces of their old world and reconnected with their former selves.

CEMETERIES AND MEMORIALS: THE PERMANENT REMINDERS OF THE WAR

Not least among the reminders were the ubiquitous war memorials and cemeteries. On no other former fighting front did the Imperial War Graves Commission oversee such intensity and scale of activity. By 1937 the IWGC had completed work on 2,300 cemeteries and burial sites containing the graves of 725,559 service and auxiliary personnel. Twenty-seven memorials to the missing had been completed, and the last one (Villers-Bretonneux) was in the final stages of construction, commemorating 317,040 men.[3] Alongside them sat countless memorials to individuals, battalions, regiments, divisions, and corps. Together they formed a chain of architectural interventions in the landscape snaking across northern France and western Belgium. Moreover, before the towns, villages, and farms were rebuilt and the newly planted saplings matured, the cemeteries and memorials were the dominant features in the landscape. The neat hedges, low walls, and Crosses of Sacrifice sticking up like masts in a sea of mud drew all eyes their way. The Reverend T.B. Stewart Thomson used an alternative image, declaring the crosses standing tall in the cemeteries the equivalent of the caravan waymarkers across the desert.[4] As the attentive also realized, the front line cemeteries followed and marked the ebb and flow of battle: in their preservation of trench names and features of the battlefield, the IWGC's architects made the conflict a permanent feature of the landscape in more ways than one. The legacy of the battlefield continued in these cities of the dead.

Caring for these cemeteries and memorials was another army, the staff of the IWGC, which consisted almost entirely of demobilized soldiers: the veteran remained a permanent presence, willing to greet and assist all kinds of visitors. To the tourist and mourner, they were guide and comforter, to the veteran, an old comrade. The IWGC's staff lived alongside the communities of Belgian and French pioneers working desperately hard to rebuild their livelihoods, homes, and culture, often in remote and inaccessible places. Resolute in this intention, the people of the devastated zones sometimes seemed at first sight to be detached from or indifferent to British visitors. As was often realized, this was a misunderstanding. Having suffered so much devastation and having so much work to do left many with little time or energy to spare. They lived with the effects of the war every single day. Their silence was not the result of ignorance; quite the opposite, it was a surfeit of knowledge.

THE PEOPLE OF THE BATTLEFIELDS

For some, whether French, Belgian, or British expatriate, the war was the source of their income. With amazing rapidity, a tourist infrastructure grew up across the old Western Front. Starting with wooden hut hotels and cafes before being replaced with permanent buildings, they were the symbols, and practical expressions of, all those who saw the great potential of the hospitality industry. In the plethora of hotels Britannique and Tommy cafes, tea, British tobacco, British beers, chips, and sandwiches were served to thousands of visitors. Tour operators invested in charabancs and cars. Cluttered itineraries were then offered, ensuring the visitor got value for money in terms of quantity even if they never lingered at any one site for longer than a quarter of an hour. Others took the opportunity to preserve fragments of the battlefield on their land or knock up a scratch 'museum' of the artefacts they had salvaged; farmers' wives and children sold shell nose-caps, shrapnel, and other items as souvenirs. In the main towns souvenir shops flourished, fuelled by the insatiable appetite of visitors for a genuine relic to display on the mantelpiece or sideboard. And, of course, there were the postcards. Thousands of them, depicting everything from ruined tanks to pristine new cemeteries and memorials. The former battlefields were sites of memory, sites of mourning, and sites of tourist delight at one and the same time, and often there was no hard and fast distinction between those who did the remembering, the mourning, and the wondering. The preconceptions and previous experiences of visitors were both challenged and shaped by the truly amazing world they found on the former battlefields. The self, the space, and the place became intertwined as tightly as a remnant of barbed wire clinging to an old 'piquet' in some Flanders drainage ditch.

Visiting the battlefields as an official guest in late 1917, F.S. Oliver decided to record his experiences in a diary. For him it would create a legacy that 'may prove to be of some interest to my grandchildren and great-grandchildren a hundred years hence'.[5] The grandchildren and great-grandchildren of that generation have not just read those accounts, they have also set out, like him, to see those places for themselves. The landscapes of the Western Front continue to haunt; people continue to be drawn towards this final witness to the devastation caused by the First World War.

Notes

INTRODUCTION

1. Full details are provided for quotations. For all other material referenced in this chapter, please see the Bibliography.
2. Lowenthal, *The Past is a Foreign Country*, 383.
3. Winter, *Remembering War*, 276–7.
4. Rowlands, 'The Role of Memory', 144–5.
5. Tarlow, 'An Archaeology of Remembering', 151.
6. Margry (ed.), *Shrines and Pilgrimage*, 32.
7. Reader and Walter (eds.), *Pilgrimage*, 8.
8. Edensor, 'Staging Tourism', 324.
9. Macfarlane, *The Old Ways*, 17.
10. Ibid., 22.
11. Daly, Salvante, and Wilcox (eds.), *Landscapes*, 5.
12. Stewart and Strathern (eds.), *Landscape, Memory, and History*, 6.
13. Foster, 'Creating a *Temenos*', 259–90.
14. Godden, 'Designing Memory'.
15. Bennett, 'Art, Affect, and the "Bad Death"'.

CHAPTER ONE

1. See *Manchester Courier*, 28 September 1914; *Sunderland Daily Echo*, 1 October 1914; *Daily Mirror*, 8 December 1914; *Birmingham Gazette*, 24 December 1914; *Dorking and Leatherhead Advertiser*, 5 December 1914; *Daily Mail*, 8 December 1914.
2. *Cheltenham Chronicle*, 23 January 1915.
3. Craster, 'Fifteen Rounds a Minute', 61–2.
4. *Bath Chronicle*, 25 March 1916 and 28 July 1917.
5. St John, *A Journey*.
6. Gwynn (ed.), *The Anvil of War*, 214, 220, 224, 225.
7. Hankey, *The Supreme Command*, vol. I, 349–50.

8 Hankey, *The Supreme Command*, vol. II, 757.
9 Ibid., 759–60.
10 Cadbury Research Library, University of Birmingham (hereafter CRL), NC 1/27/1-43, Austen Chamberlain letter, 19 November 1917.
11 Gwynn (ed.), *The Anvil of War*, 274.
12 'A Souvenir for Visitors to the Front', *Times*, 27 December 1918.
13 Elton, *Montague*, 132–3, 146, 148.
14 Gwynn (ed.), *The Anvil of War*, 216.
15 CRL, NC 1/27/1-43, Item 13, letter from Austen Chamberlain to Neville Chamberlain, 19 November 1917.
16 *Western Morning News*, 28 January 1919.
17 *Sussex Mirror*, 16 February 1915.
18 Gwynn (ed.), *The Anvil of War*, 240, 242, 252.
19 St John, *Journey*, 128–132.
20 Gwynn (ed.), *The Anvil of War*, 221, 227–32, 288.
21 Masefield, *The Old Front Line*, 12.
22 *Drogheda Argus*, 9 December 1915.
23 Gwynn (ed.), *The Anvil of War*, 228.
24 *Drogheda Argus*, 9 December 1915.
25 Gwynn (ed.), *The Anvil of War*, 235.
26 *Liverpool Daily Post*, 11 September 1916.
27 Orpen, *An Onlooker*, 36.
28 Masefield, *The Old Front Line*, 11.
29 *Western Mail*, 5 February 1918.
30 *Coventry Evening Telegraph*, 7 January 1915.
31 *Runcorn Guardian*, 15 February 1918.
32 Gwynn (ed.), *The Anvil of War*, 251.
33 *Warwick and Warwickshire Advertiser*, 17 April 1915.
34 *Drogheda Argus*, 9 December 1915.
35 *Western Mail*, 5 February 1918.
36 *Runcorn Guardian*, 15 February 1918.
37 *Huddersfield Daily Examiner*, 31 January 1918.
38 St John, *Journey*, 122–3.
39 Gwynn (ed.), *The Anvil of War*, 216–18.
40 *Runcorn Guardian*, 15 February 1918.
41 *Drogheda Argus*, 9 December 1915.
42 *Dublin Daily Express*, 16 September 1915.
43 *Sussex Mirror*, 16 February 1915.
44 *Bath Chronicle*, 28 July 1917.
45 *The Western Front: Drawings by Muirhead Bone*, 2.
46 Gwynn (ed.), *The Anvil of War*, 234, 267.
47 Masefield, *The Old Front Line*, 34, 55.

48 Ibid., 31, 34, 55.
49 Lauder, *A Minstrel*, 294–5.
50 St John, *Journey*, 189–92.
51 Gwynn (ed.), *The Anvil of War*, 287.
52 Tucker, *Johnny Get Your Gun*, 158–9.
53 Ibid.
54 Brown, *The Somme*, 302.
55 Quoted in Brown, *The Somme*, 296.
56 Brown, *The Somme*, 333–6.
57 *Manchester Courier*, 5 April 1915.
58 *Birmingham Daily Press*, 13 June 1917.
59 Elton, *Montague*, 144, 149.
60 See *Sketch*, 28 April 1915; *Illustrated London News*, 3 February 1917.
61 Bairnsfather, *Mud to Mufti*, 128.
62 *Yorkshire Evening Post*, 21 August 1916.
63 *Answers*, 7 September 1918.
64 Chasseaud, *Rats Alley*, 86.
65 *Michelin Guides: The Marne 1914*, foreword.

CHAPTER TWO

1 Clout, *After the Ruins*, 28–30.
2 *Dundee Evening Telegraph and Post*, 11 November 1918.
3 Hansard, House of Commons debates, 12 November 1918, vol. 110, c. 2,477.
4 CWGC/1/1/7/B/42, Exhumation – France and Belgium, General File, Part 1, Note from Major Stopford, DGRE, 19 November 1918.
5 *Times*, 18 November 1918.
6 *Daily Mail*, 25 November 1918.
7 CWGC/2/2/1/6, Commission meeting no. 06, 19 November 1918.
8 TNA CAB 24/103/70, Cabinet papers, Travelling concessions for relatives visiting graves of the fallen in France and Belgium, 12 April 1920.
9 See, for example, the announcement in the *Times*, 16 May 1919, which was also published by many other national and local newspapers.
10 Muirhead (ed.), *Muirhead's Belgium*, v–vi.
11 *Answers*, 31 May 1919.
12 *Daily Mail*, 15 May 1919.
13 *Daily Express*, 29 May 1919.
14 Church Army, Annual Report, 1918–1919, 10–11.
15 *Evening News*, 10 June 1919 (see comment on the article in *Dundee Evening Telegraph*, 11 June 1919).
16 *Nottingham Journal*, 20 September 1919.

17 *Aberdeen Press and Journal*, 12 June 1919.
18 Hansard, House of Commons debates, 1 July 1919, vol. 117, c. 751.
19 *Pall Mall Gazette*, 5 July 1919; Hansard, House of Commons debates, 12 August 1919, vol. 119, cc. 1,077.
20 CWGC/2/2/1/17, Commission meeting no. 17, 18 November 1919.
21 See *Dover Express*, 18 April, 8 July, 3 October, 17 October, 24 October 1919; 20 February, 20 March, 13 August, 29 October, 5 November, 31 December 1920; *Folkestone, Hythe, Sandgate and Cheriton Herald*, 22 February 1919; *Diss Express*, 31 January 1919.
22 Belgian State Railways advertisement in *The Pilgrim's Guide*, xx.
23 *Railway Gazette*, 4, 11 June 1920.
24 *Hull Daily Mail*, 12 May 1920.
25 *Burnley News*, 22 May 1920.
26 *Manchester Evening News*, 21 May 1920.
27 *Dundee Courier*, 24 May 1920.
28 *Chelmsford Chronicle*, 8 July 1921; see also, *Country Life*, 27 August 1921.
29 *Tamworth Herald*, 13 April 1922.
30 S.E.&C.R. advertisement in *The Pilgrim's Guide*, x.
31 *Gloucester Echo*, 26 August 1919.
32 Frame, *My Life*, 78.
33 *Answers*, 31 May 1919.
34 *Aberdeen Press and Journal*, 20 October 1919.
35 Dafoe, *Over the Canadian Battlefields*, 15.
36 Story, *The Battlefields, Ypres*, 3.
37 Thomas Cook and Son, *Battlefields and Cities of Belgium*.
38 *Aberdeen Press and Journal*, 20 October 1919.
39 *Sunday Times*, 1 February 1920.
40 Captain R.S.P. Poyntz and Franco-British Travel Bureau advertisements, *The Pilgrim's Guide*, xxi, xxxii.
41 Thomas Cook and Son, *Battlefields and Cities of Belgium*; Thomas Cook and Son, *The Battlefields of Belgium and France*.
42 *Pall Mall Gazette*, 5 July 1919.
43 *Pall Mall Gazette*, 5 July 1919.
44 *Gloucester Echo*, 26 August 1919.
45 See advertisement in the *Bystander*, 21 July 1920.
46 Battlefields Bureau advertisement in *The Pilgrim's Guide*, v.
47 Queen Mary's Army Auxiliary Corps, *Old Comrades' Association Gazette* 2, no. 5 (1921): 7.
48 CWGC/2/2/1/8, Commission meeting no. 08, 21 January 1919.
49 CWGC/2/2/1/26, Commission meeting no. 26, 21 September 1920.
50 CWGC/2/2/1/8, Commission meeting no. 08, 21 January 1919.
51 CWGC/2/2/1/9, Commission meeting no. 09, 18 February 1919.
52 *Dundee Evening Telegraph*, 11 June 1919.

53 *Folkestone, Hythe, Sandgate, and Cheriton Herald*, 5 March 1921.
54 *The Bulletin: The Official Organ of the National Federation of Discharged and Demobilised Sailors and Soldiers*, 19 February 1920.
55 TNA CAB 24/103/70 and CAB 24/106/45, 12 April, 1 June 1920.
56 Hansard, House of Commons debates, 13 February 1919, vol. 112, cc. 272–372; 18 March 1919, vol. 113, cc. 1,935, 2,083; 12 August 1919, vol. 113, c. 1,077.
57 TNA CAB 24/104/5 and CAB 23/21/18, 12, 21 April, 11 June 1920.
58 TNA CAB 24/103/70 and CAB 24/106/45, 1 June 1920.
59 TNA ADM 1/8611/154, Visits to War Graves in France and Belgium by Relatives of Deceased Officers and 19 Men of the Armed Forces, Correspondence, 15 July, 19 August, 10 September 1920.
60 TNA T 161/23/6, Imperial War Graves Commission: Powers to provide transport and passes for visits of relations to War Graves, Correspondence, 16 June, 29 June, 30 June; 1 July, 2 July, 8 July 1920; 20 August 1920. See also report in *Lincolnshire Echo*, 15 March 1921.
61 TNA ADM 1/8611/154, Correspondence, 28 July, 12 August, 31 August, 12 September, 5 December 1921; File notes, 2 February, 11 February, and 20 February 1920.
62 TNA ADM 1/8611/154, Correspondence, 25 April 1921.
63 TNA FO 737/13/10, War Grave Passes (War of 1914–1918), War Graves Pass, May 1939; ADM 1/8611/154, Correspondence, 25 April, 29 July 1921.
64 *The Bulletin*, 5 August 1920.
65 TNA ADM 1/8611/154, Correspondence, 15 July 1920; TNA HO 45/11080/419854, War Graves Passes for relatives of late members of H.M. Forces buried in France and Belgium, Correspondence between Foreign Office and War Office, 20 May, 17 June 1921.
66 TNA ADM 1/8611/154, File note, 17 June 1921.
67 TNA HO 45/11080/419854, Correspondence, 21 September 1921, 13 March 1922, 13 February 1922; TNA FO 737/13/10, War Grave Passes (War of 1914–1918), War Graves Pass, May 1939.
68 Church Army, Annual Report, 1918–1919, 10–11; Church Army advertisement in *The Pilgrim's Guide*, ix.
69 Church Army advertisement in *The Pilgrim's Guide*, ix.
70 *Graphic*, 19 April 1919.
71 *The Bulletin*, 5 August 1920.
72 The St Barnabas Hostels, France [Reverend Matthew Mullineux], *How to Reach 'The Hallowed Acres' in France and Belgium* (imprint and date unknown, but likely to be 1924), 4–5.
73 *Times*, 29 March 1920.
74 YMCA advertisement in *The Pilgrim's Guide*, xi.
75 *Red Triangle Magazine* 3, no. 6, February 1920, 213.
76 *Red Triangle Magazine* 5, no. 1, September 1921, 16.

77 Salvation Army International Heritage Centre, London (hereafter SAIHC): Annual Reports, 1923, 34; 1924, 34; 1925, 52; 1929, 43.
78 CWGC/1/1/16/20, Prince of Wales Visit with British Legion, 1928. See also SAIHC, Extract from *Year Book, 1930*, 2; *Coventry Herald*, 10 February 1922.
79 *Coventry Herald*, 10 February 1922.
80 Church Army, Annual Report, 1923–1924, 60.
81 Church Army, Annual Report, 1919–1920, 31; CWGC/1/1/16/20, Prince of Wales Visit with British Legion, 1928.
82 IWM Box no. 86/48/1, Private papers of W.G. Marlborough, Church Army War Graves Department leaflet, 1927.
83 CWGC/1/1/16/20, Prince of Wales Visit with British Legion, 1928.
84 *Ypres Times* 1, no. 8 (July 1923): 24; St Barnabas Hostels, *Hallowed Acres*, 6–7.
85 *Ypres Times* 3, no. 3 (July 1926): 77; 3, no. 6 (April 1927): 57.
86 *St. Andrew's Citizen*, 16 July 1938.
87 CWGC/1/1/16/20, Prince of Wales Visit with British Legion, 1928.
88 *Red Triangle Magazine* 3, no. 11 (1920): 416–18.
89 CWGC/1/1/16/20, Prince of Wales Visit with British Legion, 1928, Letter by Ware to General Trotter, 10 August 1928.
90 St Barnabas War Graves Pilgrimages appeal, *Spectator*, 29 March 1924.
91 St Barnabas Hostels, *Hallowed Acres*, 3.
92 *Banbury Advertiser*, 16 February 1922.
93 *Yorkshire Evening Post*, 5 May 1920.
94 For an example of a typical Mullineux talk see *West Sussex Gazette*, 18 March 1920.
95 St Barnabas Hostels, *Hallowed Acres*, 4, 5.
96 *Banbury Advertiser*, 16 February 1922. For an example of its publication elsewhere see *Motherwell Times*, 3 March 1922; *Hendon and Finchley Times*, 10 September 1920.
97 *Lincolnshire Echo*, 15 March 1921.
98 *Country Life*, 4 September 1920.
99 CWGC/2/2/1/17, Commission meeting no. 17, 18 November 1919.
100 *Times*, 4 September 1920.
101 SAIHC, Salvation Army Year Book, 1922, 8; Salvation Army Annual Report, 1921, 23.
102 *Graphic*, 28 August 1920.
103 *Country Life*, 4 September 1920.
104 *Graphic*, 28 August 1920.
105 Church Army, Annual Report, 1919–1920, 31.
106 *Belfast News-Letter*, 31 March 1923.
107 *Burnley Express*, 20 May 1922.

108 *Red Triangle Magazine* 5, no. 1 (1921): 15.
109 *Burnley Express*, 20 May 1922; *Country Life*, 4 September 1920.
110 *Belfast News-Letter*, 13 July 1925.
111 Church Army, Annual Report, 1923–1924, 60–1.
112 Frame, *My Life*, 78, 80.
113 *East Kent Gazette*, 15 October 1921. Her brother was Christopher Wicks, buried at St Pierre Cemetery, Amiens.
114 Nellie Burrin, 'Western Front Pilgrimage', 10–12.
115 Watson, *Fighting Different Wars*.
116 YMCA advertisement in *The Pilgrim's Guide*, xi.
117 *Graphic*, 28 August 1920.
118 Vance, *Maple Leaf Empire*, 88–9, 93–4.
119 CWGC/2/2/1/6, Commission meeting no. 06, 19 November 1918.
120 *Graphic*, 28 August 1920.
121 *Graphic*, 28 August 1920.
122 *Sphere*, 14 January 1928.
123 *Graphic*, 28 August 1920.
124 *Country Life*, 4 September 1920.
125 SAIHC, *War Cry*, April 1952, obituary of General Mrs c. Higgins.
126 *Daily Mail*, 28 February 1920.
127 *Daily Mail*, 28 February 1920. See also similar coverage in *Edinburgh Evening News* and *Manchester Evening News*, 28 February 1920.
128 TNA BT 31/25486/163088, Fields of Honour Society Ltd.
129 CWGC/1/1/17/2, The King's Pilgrimage, Letter from Mullineux to IWGC, 1 May 1923.
130 See advertisement in the *Scotsman*, 1 July 1922.
131 CWGC/1/1/17/2, The King's Pilgrimage, Letter from IWGC to Mullineux, 1 June 1923.
132 Story, *Present Day Paris*, 12.
133 West, *Three Days' Journey*, 9.
134 *Pall Mall Gazette*, 12 July 1919.
135 Coop, *A Short Guide*, 12.
136 Ibid.
137 Frame, *My Life*, 79.
138 Graham, *The Challenge*, 33.
139 *The Pilgrim's Guide*, 50.
140 Graham, *The Challenge*, 36.
141 *The Pilgrim's Guide*, 52.
142 *Cheltenham and County Looker-On*, 3 January 1920.
143 *Gloucestershire Echo*, 26 August 1919; Braithwaite Buckle, *A Kingly Grave*, 18–19.
144 *Cheltenham and County Looker-On*, 3 January 1920.

145 IWM Documents 140, Private papers of Miss O. Edis, Diary, 3, 4–5, 12, 43, 51, 56, 64.
146 *Ypres Times* 2, no. 3 (July 1924): 68.
147 *Ypres Times* 5, no. 4 (October 1930): 114.
148 *Ypres Times* 4, no. 3 (July 1928): 81.
149 *Ypres Times* 5, no. 3 (July 1930): 86.
150 *Berwick Advertiser*, 16 August 1928; *Ypres Times* 3, no. 8 (October 1927): 230.
151 *Ypres Times* 5, no. 4 (October 1930): 114; 5, no. 8 (October 1931): 235.
152 *St. Andrew's Citizen*, 23 July 1938, 20 August 1938.

CHAPTER THREE

1 St Barnabas Hostels, *Hallowed Acres*, 6–7.
2 Public Records Office of Northern Ireland (hereafter PRONI) PM7/4/9, Provision of radio set for caretaker, Ulster Tower. See correspondence from Cockerell.
3 *Ypres Times* 3, no. 3 (July 1926): 12.
4 *Ypres Times* 6, no. 7 (July 1933): 218.
5 Taylor, *Good-Bye*, 25.
6 Coop, *A Short Guide*, 9.
7 *Pilgrim's Guide*, 49; *Muirhead's Belgium*, lxxiv.
8 Laurentz, *Milestones*; see also *Pall Mall Gazette*, 12 November 1912.
9 For press coverage of her career see *Hearth and Home*, 15 November 1900; *Daily Mail*, 14 May 1907; *Sunday Times*, 21 July 1907.
10 *Cheltenham and County Looker-On*, 27 December 1919, 3 January 1920.
11 IWM Documents 140, Edis diary, 5, 19.
12 *Cheltenham and County Looker-On*, 27 December 1919, 3 January 1920.
13 Fleming, *Battlefields*, 88
14 Coop, *A Short Guide*, 81.
15 IWM Documents 140, Edis diary, 70.
16 *Aberdeen Daily Journal*, 12 June 1919.
17 *East Kent Gazette*, 15 October 1921.
18 Wilfrid Ewart, 'Autumn on the Somme', *Country Life* 46, no. 1,192, 8 November 1919, 605.
19 Coop, *A Short Guide*, 13.
20 IWM Documents 140, Edis diary, 80.
21 CRL, YMCA/ACC69/2/12, Letter sent by J. Hastings Eastwood to his wife, Jessie, 10 February 1919.
22 IWM Documents 140, Edis diary, 6, 8, 9, 15, 31, 48, 53, 61, 66, 80, 81.
23 Story, *Ypres*, vol. 2 of *The Battlefields*, 6–7.
24 *Cheltenham and County Looker-On*, 27 December 1919, 3 January 1920.
25 *Pilgrim's Guide*, xii–xv.

26 *Cheltenham and County Looker-On*, 27 December 1919; 3 January 1920.
27 *Pilgrim's Guide*, 63.
28 Fleming, *Battlefields*, 88.
29 Coop, *A Short Guide*, 13, 32, 68.
30 *Cheltenham and County Looker-On*, 3 January 1920.
31 CRL, YMCA MSS 843/2/2, Letter sent by Alice Knight to her parents, 4 August 1919.
32 *Yorkshire Post*, 23 August 1921.
33 *Scotsman*, 14 May 1923.
34 *Ypres Times* 4, no. 3, July (1928): 82.
35 *Hull Daily Mail*, 25 August 1922.
36 *Thetford and Watton Times*, 25 August 1928.
37 Tameside Local Studies and Archives Centre (hereafter TLSAC), MR2/26/79, Account of a battlefields tour of France, 1936.
38 *Cheltenham On-Looker*, 27 December 1919.
39 See, for example, *Pilgrims Guide*, 49; *Muirhead's Belgium*, lxxiv–lxxvi.
40 *Gloucester Citizen*, 10 September 1925.
41 *Toc H Magazine* 7, no. 11 (November 1929): 363–4.
42 *Ypres Times* 7, no. 8 (October 1935): 245–6.
43 *Scotsman*, 6 July 1926.
44 Williamson, *The Wet Flanders Plain*, 106–8.
45 Graham, *The Challenge*, 39–40.
46 Coop, *A Short Guide*, 95.
47 Fleming, *Battlefields*, 2–3.
48 Story, *Ypres*, vol. 2 of *The Battlefields*, 3, 6.
49 Lowe, *The Western Battlefields*, 2.
50 *Graphic*, 27 September 1919.
51 Coop, *A Short Guide*, 60, 68.
52 Burrin, 'Western Front Pilgrimage', 10–12.
53 CRL, YMCA MSS 843/2/2, Alice Knight, letter to parents, 5 November 1919.
54 *Sphere*, 18 August 1928.
55 See *St. Andrew's Citizen*, 16, 23, 30 July; 6, 13, 20, 27 August; 3, 10, 17 September 1938.
56 *Ypres Times* 3, no. 4 (October 1926): 104.
57 *Toc H Magazine* 7, no. 11 (November 1929): 362.
58 *Ypres Times* 4, no. 3 (July 1928): 81.
59 *Scotsman*, 22 August 1923.
60 Lowe, *The Western Battlefields*, 2.
61 *Sphere*, 9 July 1927.
62 Fleming, *Battlefields*, 11.
63 Ibid., 36.
64 *The Story of an Epic Pilgrimage*, 34.

65 Somerset Heritage Centre (hereafter SHC), DD/X/SIM/4, Personal account of Mrs E.A. Smith's attendance at the battlefields pilgrimage, 1928.
66 *Toc H Magazine* VIII, no. 11 (November 1929): 362.
67 *Aberdeen Daily Journal*, 13 January 1919.
68 IWM Documents 13273, Private papers of C.R. Jones, Account of battlefield pilgrimage, 1920, made by his father, Charles Jones; see Fleming, *Battlefields*, 45.
69 IWM Documents 13273, Jones battlefield pilgrimage.
70 Brown, *The Somme*, 2.
71 Carrington, *Soldier*, 82.
72 *The Western Front: Drawings by Muirhead Bone*, 2.
73 Townroe, *A Pilgrim*, 26.
74 See Stone, 'The Far Right and the Back to the Land Movement', 182–98.
75 This is probably a misprint for 'Tar Heels', a nickname for soldiers from North Carolina. The US 30 Division fought in the Ypres salient in 1918 and included units from North and South Carolina.
76 *Ypres Times* 3, no. 6, (April 1927): 154.
77 *Ypres Times* 4, no. 4 (October 1928): 118.
78 *Belfast News-Letter*, 4 August 1930.
79 *Ypres Times* 5, no. 2 (April 1930): 41–2.
80 *Northampton Mercury*, 6 August 1937.
81 *Ypres Times* 7, no. 4 (October 1934): 116.
82 *Sphere*, 18 August 1928.
83 Coop, *A Short Guide*, 13.
84 Lowe, *Western Battlefields*, 3.
85 Coop, *A Short Guide*, 13.
86 *Bucks Examiner*, 21 September 1934.
87 IWM Documents 140, Edis diary, 26, 17–18.
88 Lowe, *Western Battlefields*, 3.
89 Ewart, 'Autumn on the Somme', 605.
90 Lowe, *Western Battlefields*, 1.
91 *Dundee Courier*, 27 May 1920.
92 *Answers*, 31 May 1919.
93 *Gloucester Echo*, 26 August 1919.
94 Coop, *A Short Guide*, 16.
95 CWGC/2/2/1/26, Commission meeting no. 26, 21 September 1920.
96 CWGC/1/1/7/B/68, Direction Boards, correspondence between Goodland and Ware, 14, 17, 28 November 1923.
97 CWGC/1/1/7/B/68, Letter from K.T. Gemmell, 29 April 1924; Letter from Hon. Mrs Rosalind Lyle, n.d., c. 1924; Letter from F.J. Kirby, 25 August 1926; Letter from Alfred Whittle, 30 August 1926; Letter from C.H.W. Cook, 27 August 1930.

98   See Olson, 'Maps for a New Kind of Tourist', 205–20.
99   CWGC/1/1/7/B/68, Goodland to Ware, 17 November 1923.
100 *Ypres Times* 2, no. 2 (April 1924): 53.
101 Ibid.
102 See Connelly, *Celluloid War Memorials*.
103 *Ypres Times* 2, no. 2 (April 1924): 53.
104 *Ypres Times* 2, no. 4 (October 1924): 112.
105 *Ypres Times* 2, no. 2 (July 1924): 68.
106 *Ypres Times* 2, no. 7 (July 1925): 191; see also 3, no. 1 (January 1926): 17.
107 Pulteney and Brice, *The Immortal Salient*, 22–3.
108 CWGC/1/1/10/G/10, Preservation of Belgian Front at Certain Points, Letter from War Office to Ingpen, 30 July 1920.
109 CWGC/1/1/3/5 FX C&FA, Correspondence, Inspection of Cemeteries in Contract 1923 P, 27 October 1924.
110 CWGC/1/1/7/B/68, Ware memorandum, 1 May 1925; Eighth Annual Report of the Imperial War Graves Commission Annual Report, 1928, 32.
111 Fourth Annual Report of the Imperial War Graves Commission, 7.
112 CWGC/1/1/7/B/68, Correspondence, 12 July 1926; Memorandum Director of Records to Director of Works, 28 March 1927.
113 CWGC/1/1/7/B/68, Kenyon to Lieutenant-Colonel H.F. Robinson, Deputy Director of Works, 20 December 1923.
114 Hurst, *The Silent Cities*, ix.
115 CWGC/1/1/7/B/68, Undated sketch designs.
116 Hurst, *The Silent Cities*, ix.
117 CWGC/1/1/7/B/68, Goodland to Ware, 2 June 1927.
118 Hurst, *The Silent Cities*, ix.
119 CWGC/1/1/7/B/68, Goodland to Ware, 24 June 1927; Ware to Robinson, 23 January 1928.
120 *Ypres Times* 2, no. 7 (July 1925): 190.
121 *Ypres Times* 4, no. 6 (April 1929): 167.
122 *Ypres Times* 3, no. 6 (April 1927): 154.
123 *Ypres Times* 5, no. 8 (October 1931): 249.
124 *Ypres Times* 3, no. 4 (October 1926): 88.
125 *Ypres Times* 5, no. 6 (April 1931): 170.
126 *Ypres Times* 3, no. 2 (April 1926): 36.
127 *Athenaeum*, 12 November 1920, 653.
128 CWGC/1/1/7/B/68, Letters from J.S. Dows, 15, 17 October 1929.
129 The history of the demarcation stones can be pieced together for a range of sources. See *L'Auto*, 25 February, 9 November 1921, 5 June 1922; *Excelsior*, 9 November 1921, 6 March 1922; *Figaro*, 5 March, 23 August 1922; *La Revue du Touring-club de France* 391 (June 1927): 109; *La Révil du Nord*, 12 November 1921; *Le Soir* (Brussels), 29 January 1922; *Ypres Times* 1, no. 7 (April 1923): 180–3; 1, no. 9 (October 1923): 266–7; Coombs, *Before*

Endeavours Fade, 6; Holt, *The Ypres Salient*, 122. See also https://sites.google.com/site/wraros/demarcationstonesww1; http://www.greatwar.co.uk/article/demarcation-stone.htm; http://memorialdormans.free.fr/bornesVauthierHistoire.htm (accessed 16 July 2019).
130 CWGC/1/1/10/G/10, Correspondence, 18 February, 7 March 1922.
131 *Ypres Times* 1, no. 7 (April 1923): 180–3; 1, no. 9 (October 1923): 266–7. CWGC/1/1/10/G/10, Ingpen to PAS, 5 August 1924.
132 *Punch*, 2 October 1918, 213.
133 *Scotsman*, 8 August 1923.
134 See *Times*, 11 October 1928; *Yorkshire Post*, 24 October 1929; *Ypres League* 4, no. 6 (April 1929): 184.
135 *Surrey Mirror*, 17 August 1928.
136 Taylor, *Good-Bye*, 106.
137 Bird, *Thirteen Years After*, 7. CWGC/1/1/10/G/10, British Embassy, Brussels to Ware, 13 February 1922.

CHAPTER FOUR

1 *Banbury Advertiser*, 13 October 1921.
2 *Berwick Advertiser*, 3 December 1920.
3 *Aberdeen Press and Journal*, 20 October 1919.
4 Longworth, *The Unending Vigil*, 129.
5 CRL, YMCA MSS 843/2/2, Alice Knight, letter to parents, 6 June 1919.
6 Frame, *My Life*, 79.
7 Braithwaite Buckle, *A Kingly Grave*, 17–19.
8 *St. Andrew's Citizen*, 18 October 1919.
9 *Pall Mall Gazette*, 18 July 1919.
10 Dafoe, *Over the Canadian Battlefields*, 32.
11 Claeys, 'World War I and the Reconstruction of Rural Landscapes', 108–28. See 108 for the headline figures. Also, Clout, *After the Ruins*, 19–58.
12 See Claeys, 'World War 1 and the Reconstruction of Rural Landscapes'; Clout, *After the Ruins*.
13 IWM Documents 140, Edis diary, 55, 66.
14 *Berwick Advertiser*, 3 December 1920.
15 *Northern Constitution*, 10 November 1923.
16 Braithwaite Buckle, *A Kingly Grave*, 21.
17 Graham, *The Challenge*, 18.
18 *Red Triangle Magazine* 3, no. 11 (July 1920): 417.
19 CRL, YMCA MSS 843/2/2, Alice Knight, letter to parents, 28 August 1919.
20 IWM Documents 13273, Jones battlefield pilgrimage.
21 *St. Andrew's Citizen*, 18 October 1919.
22 *Red Triangle Magazine* 3, no. I (September 1919): 1.

23 CRL, NC 18/1/196-237, Item 224, Letter from Neville Chamberlain to Hilda Chamberlain, 23 August 1919.
24 *North-East Lanark Gazette*, 17 October 1919.
25 Dafoe, *Over the Canadian Battlefields*, 31, 33.
26 Graham, *The Challenge*, 89.
27 IWM Documents 140, Edis diary, 12.
28 Frame, *My Life*, 79.
29 *Red Triangle Magazine* 3, no. 6 (February 1920): 213; 3, no. 11 (July 1920): 416; *Scotsman*, 21 June 1919.
30 Dafoe, *Over the Canadian Battlefields*, 32.
31 *Scotsman*, 21 June 1919.
32 *Pall Mall Gazette*, 18 July 1919; CRL, NC 18/1/196-237, Item 224, Letter from Neville Chamberlain to Hilda Chamberlain, 23 August 1919.
33 *Banbury Advertiser*, 13 October 1921; *Scotsman*, 29 November 1919; *Aberdeen Weekly Journal*, 14 November 1919; *Chelmsford Chronicle*, 20 August 1920; *Birmingham Daily Gazette*, 3 November 1921.
34 Lowe, *The Western Battlefields*, 9.
35 *Ypres Times* 1, no. 4 (July 1922): 95.
36 IWM Documents 13273, Jones battlefield pilgrimage.
37 CRL, NC 18/1/196-237, Item 224, Letter from Neville Chamberlain to Hilda Chamberlain, 23 August 1919.
38 CRL, YMCA/ACC69/2/12, Letter sent by J. Hastings Eastwood to his wife, Jessie, 10 February 1919.
39 *Banbury Advertiser*, 13 October 1921.
40 *Hull Daily Mail*, 25 August 1922.
41 CRL, YMCA MSS 843/2/2, Alice Knight, letter to parents, 28 August 1919.
42 *East Kent Gazette*, 15 October 1921.
43 CRL, YMCA/ACC69/2/12, Letter sent by J. Hastings Eastwood to his wife, Jessie, 10 February 1919.
44 Ewart, 'Autumn on the Somme', 605.
45 CWGC/1/1/7/B/36, Land Acquisition Belgium, Correspondence, 21 July, 4 August 1920.
46 CWGC/1/1/3/5. See correspondence 20, 27, 30 September 1924; 17, 21, 30 October 1924.
47 *Daily Mail*, 23 September 1919.
48 Graham, *The Challenge*, 36.
49 CRL, NC 18/1/196-237, Item 224, Letter from Neville Chamberlain to Hilda Chamberlain, 23 August 1919.
50 *North-East Lanark Gazette*, 17, 24 October 1919.
51 See Dendooven, *De vergeten soldaten van de Eerste Wereldoorlog*, based on his PhD thesis, 'Asia in Flanders Fields'; James, *The Chinese Labour Corps*; Xu, *Strangers on the Western Front*.
52 *Daily Mail*, 15 September 1919.

53 West, *Three Days' Journey*, 19.
54 Ibid., 9, 11, 12, 19.
55 IWM Documents 140, Edis diary, 17, 78.
56 CRL, YMCA MSS 843/2/2, Alice Knight, letter to parents, 23 September 1919.
57 CRL, YMCA MSS 843/2/2, Alice Knight, letter to parents, 5 November 1919.
58 Fleming, *Battlefields*, 74.
59 CRL, YMCA MSS 843/2/2, Alice Knight, letter to parents, 5 November 1919.
60 *Aberdeen Press and Journal*, 20 October 1919.
61 IWM Documents 140, Edis diary, 16.
62 Ewart, 'Autumn on the Somme', 605.
63 Coop, *A Short Guide*, 46; *Red Triangle Magazine* 3, no. 11 (July 1920): 417.
64 IWM Documents 140, Edis diary, 40, 52.
65 *Aberdeen Press and Journal*, 13 January 1919.
66 D.E.F., 'Ahead of the Army', 281.
67 *Sheffield Daily Telegraph*, 19 February 1919.
68 Dafoe, *Over the Canadian Battlefields*, 38, 40.
69 CRL, NC 18/1/196-237, Item 224, Letter from Neville Chamberlain to Hilda Chamberlain, 23 August 1919.
70 CRL, YMCA MSS 843/2/2, Alice Knight, letter to parents, 23 September 1919.
71 Frame, *My Life*, 84–5.
72 *Scotsman*, 5 July 1919.
73 *Yorkshire Post*, 23 August 1921.
74 *Memories*, 37.
75 *Red Triangle Magazine* 3, no. 11 (July 1920): 416.
76 IWM Documents 140, Edis diary, 8.
77 *Cheltenham On-Looker*, 27 December 1919.
78 Dafoe, *Over the Canadian Battlefields*, 50–1.
79 CRL, NC 18/1/196-237, Item 224, Letter from Neville Chamberlain to Hilda Chamberlain, 23 August 1919.
80 *Memories* 2, no. 12 (Spring 1923): 145.
81 *Scotsman*, 20 May, 18 June 1919.
82 IWM Documents 140, Edis diary, 15, 54.
83 Lowe, *Western Battlefields*, introduction.
84 Williamson, *The Wet Flanders Plain*, 85–9.
85 Graham, *The Challenge*, 19.
86 *Kirkintilloch Herald*, 6 August 1919.
87 CRL, YMCA/ACC69/2/12, Letter sent by J. Hastings Eastwood to his wife, Jessie, 10 February 1919.
88 Graham, *The Challenge*, 73.

89   *Pall Mall Gazette*, 18 July 1919.
90   *Aberdeen Press and Journal*, 20 June 1919.
91   *North-East Lanark Gazette*, 17 October 1919.
92   CRL, NC 18/1/196-237, Item 224, Letter from Neville Chamberlain to Hilda Chamberlain, 23 August 1919.
93   *North-East Lanark Gazette*, 24 October 1919.
94   Ewart, 'Autumn on the Somme', 605.
95   *Sheffield Weekly Telegraph*, 27 September 1919.
96   *Red Triangle Magazine* 5, no. 1 (September 1921): 16.
97   *Red Triangle Magazine* 6, no. 3 (March 1922): 54–5.
98   *Red Triangle Magazine* 7, no. 4 (April 1923): 88.
99   Bird, *Thirteen Years After*, 127.
100 Bird, *Thirteen Years After*, 127.
101 Graham, *The Challenge*, 94.
102 *St. Andrew's Citizen*, 18 October 1919.
103 Ewart, 'Autumn on the Somme', 605.
104 *Daily Mail*, 13 September 1919.
105 Bird, *Thirteen Years After*, 127–9.
106 Swinton (ed.), *Twenty Years After*, 804.
107 Ewart, 'Autumn on the Somme', 605.
108 *Daily Mail*, 13 September 1919.
109 *Red Triangle Magazine* 3, no. 11 (July 1920): 416.
110 *Yorkshire Post*, 23 August 1921.
111 Graham, *The Challenge*, 90, 92–3, 106.
112 *Daily Mail*, 16 September 1919.
113 *Red Triangle Magazine* 3, no. 6 (February 1920): 214.
114 *Red Triangle Magazine* 3, no. 11 (July 1920): 418.
115 Graham, *The Challenge*, 93.
116 *Ypres Times* 3, no. 8 (October 1927): 231.
117 Jones, *In Parenthesis*, 185.
118 *Manchester Evening News*, 21 May 1920; *Burnley News*, 22 May 1920.
119 *Red Triangle Magazine* 7, no. 4 (April 1923): 88.
120 *Bunbury Advertiser*, 13 October 1921.
121 *Yorkshire Post*, 23 August 1921; *Thetford and Watton Times*, 25 August 1928.
122 *Belfast Telegraph*, 5 December 1921.
123 Graham, *The Challenge*, 90, 95.
124 *Belfast News-Letter*, 4 August 1930.
125 *Ypres Times* 5, no. 8 (October 1931): 232.
126 Coop, *A Short Guide*, 4–75.
127 *Dundee Courier*, 1 July 1925.
128 *Ypres Times* 3, no. 5 (January 1927): 128.
129 *Ypres Times* 3, no. 4 (October 1926): 92.
130 Williamson, *The Wet Flanders Plain*, 139–44.

131 *Biggleswade Chronicle and Bedfordshire Gazette*, 15 August 1930.
132 Mottram, *A Journey*, 238.
133 Seton Hutchinson, *Pilgrimage*, 55–7.
134 Coop, *A Short Guide*, 78.
135 Fleming, *Battlefields*, 45.
136 Taylor, *Good-Bye*, 44.
137 The literature on the Delville Wood memorial is extensive. For the best overviews see Nasson, 'Delville Wood'; Nasson, *Springboks on the Somme*, 205–42. See also TNA WO 32/5869, War Memorials, Union of South Africa; CWGC/1/1/10/A/12 & /13, /14, /15, Memorials – South Africans – Delville Wood.
138 Dennis, *A Kitchener Man's Bit*, 236.
139 Taylor, *Good-Bye*, 44; *Ypres Times* 3, no. 5 (January 1927): 127.
140 For a detailed history of Ploegsteert and the wood in the war see Spagnoly and Smith, *A Walk Around Plugstreet*.
141 For more details see Godden, 'Designing Memory'.
142 *LRB Record* 3, no. 113 (August 1927): 141.
143 Williamson, *The Wet Flanders Plain*, 105.
144 Bird, *Thirteen Years After*, 35.
145 *Leeds Mercury*, 22 April 1935.

CHAPTER FIVE

1 See poem XL in A.E. Housman's collection *A Shropshire Lad*.
2 Mottram, *Journey*, 6.
3 From 'To One Who Was with Me in the War'. See Hart-Davis (ed.), *Siegfried Sassoon*, 151–2.
4 *Ypres Times* 7, no. 7 (July 1935): 217.
5 Nottinghamshire County Archives (hereafter NCA), DD/1332/198, Papers of Neville Talbot, 'Pilgrimage of Padres to Talbot House, Poperinghe and Ypres, 27 April–1 May 1930'. Incomplete manuscript.
6 *Ypres Times* 5, no. 4 (October 1930): 112.
7 *Ypres Times* 8, no. 1 (January 1936): 10.
8 Behrend, *As From Kemmel Hill*, 148–9.
9 Taylor, *Good-Bye*, 15.
10 *Yorkshire Post*, 20 August 1921.
11 *Ypres Times* 3, no. 6 (April 1927): 152, 154.
12 *Memories* 4, no. 29 (Summer 1927): 164.
13 *Ypres Times* 6, no. 4 (October 1932): 110.
14 *Yorkshire Post*, 26 July 1930.
15 *Ypres Times* 6, no. 7 (July 1933): 211.
16 *Ypres Times* 5, no. 4 (October 1930): 112.

17 *Ypres Times* 7, no. 8 (October 1935): 236, 239.
18 SHC, DD/X/SIM/4, Smith battlefields pilgrimage, 1928.
19 *Yorkshire Post*, 30 March 1929.
20 *Ypres Times* 8, no. 1 (January 1936): 11.
21 *Ypres Times* 5, no. 3 (July 1930): 77.
22 TLSAC, MR2/3/1/60, 'Tour of the Battlefield: Battle of the Aisne', April 1934.
23 *Sphere*, 23 August 1924.
24 *Ypres Times* 7, no. 8 (October 1935): 236.
25 *Nottingham Evening Post*, 28 May 1928.
26 *Ypres Times* 5, no. 2 (April 1930): 40–5.
27 Ibid., 40.
28 *Ypres Times* 5, no. 8 (October 1931): 238–9.
29 *Ypres Times* 3, no. 6 (April 1927): 152.
30 *Memories* 1, no. 2 (September 1920): 34.
31 *Memories* 1, no. 7 (December 1921): 184.
32 *Memories* 1, no. 2 (September 1920): 34, 43; 1, no. 3 (December 1920): 74.
33 *Memories* 4, no. 29 (Summer 1927): 160–1.
34 *Memories* 1, no. 2 (September 1920): 37
35 TLSAC, MR2/3/1/60, 'Tour of the Battlefield: Battle of the Aisne', April 1934. See also Caddick-Adams, 'Footsteps Across Time'.
36 Wylly (ed.), *Sherwood Foresters*, 238.
37 *A Souvenir of the Battlefields Pilgrimage*, 143.
38 *Scotsman*, 6 July 1926.
39 Behrend, *As from Kemmel Hill*, 154.
40 *Sphere*, 9 July 1927.
41 *Ypres Times* 1, no. 4 (July 1922): 95.
42 *Ypres Times* 6, no. 3 (July 1932): 79.
43 *Ypres Times* 7, no. 7 (July 1935): 215.
44 *Ypres Times* 4, no. 3 (July 1928): 81.
45 *Ypres Times* 6, no. 4 (October 1932): 107, 109.
46 *Leeds Mercury*, 2 August 1932.
47 *Yorkshire Post*, 20 August 1921.
48 *Ypres Times* 5, no. 2 (April 1930): 42–3; no. 3 (July 1930): 78; 7, no. 3 (July 1934): 80; *Memories* 3, no. 20 (Spring 1925): 146.
49 *Memories* 1, no. 2 (September 1920): 37.
50 Dennis, *A Kitchener Man's Bit*, 235.
51 Wylly (ed.), *Sherwood Foresters Regimental Annual, 1926*, 238–9.
52 *Sheffield Daily Independent*, 3 August 1928.
53 *Ypres Times* 3, no. 4 (October 1926): 93.
54 *Yorkshire Evening Post*, 29 August 1919.
55 *Belfast News-Letter*, 4 August 1930.

56 *Ypres Times* 5, no. 2 (April 1930): 45.
57 *Ypres Times* 5, no. 8 (October 1931): 19–21.
58 *Berwick Advertiser*, 16 August 1928.
59 *St. Andrew's Citizen*, 23 July 1938.
60 *Memories* 1, no. 3 (December 1920): 74.
61 *Memories* 3, no. 20 (Spring 1925): 146.
62 *Sphere*, 9 July 1927.
63 *Belfast News-Letter*, 4 August 1930.
64 *Scotsman*, 6 July 1926.
65 NCA, DD/1332/198, Talbot pilgrimage.
66 *Church Times*, 28 April 1922.
67 Behrend, *As From Kemmel Hill*, 155.
68 *Ypres Times* 3, no. 4 (October 1926): 104–5.
69 *Sphere*, 9 July 1927.
70 *Ypres Times* 3, no. 4 (October 1926): 92.
71 *Ypres Times* 3, no. 6 (April 1927): 153.
72 Swinton (ed.), *Twenty Years After*, vol. II, 885.
73 *Ypres Times* 4, no. 5 (January 1929): 134.
74 *Ypres Times* 7, no. 7 (July 1935): 215–6.
75 *Sphere*, 23 July 1927.
76 *Thetford and Watton Times*, 25 August 1928.
77 *Surrey Mirror*, 17 August 1928.
78 *Ypres Times* 1, no. 4 (July 1922): 95.
79 *Ypres Times* 3, no. 4, (October 1926): 88–9, 92.
80 *Ypres Times* 3, no. 6 (April 1927): 152.
81 *Leeds Mercury*, 22 April 1935.
82 *Ypres Times* 6, no. 1 (January 1932); 23–4.
83 *Ypres Times* 6, no. 4 (October 1932): 107.
84 *Ypres Times* 7, no. 4 (October 1934): 114–15.
85 *Ypres Times* 7, no. 8 (October 1935): 236–8.
86 *Ypres Times* 8, no. 1, (January 1936): 12.
87 Ibid.
88 *Ypres Times* 5, no. 3 (July 1930): 76
89 *Ypres Times* 5, no. 7 (July 1931): 215–16.
90 *Ypres Times* 6, No. 3 (July 1932): 74–6.
91 *Ypres Times* 3, no. 3 (July 1926): 74.
92 *Ypres Times* 4, no. 3 (July 1928): 82.
93 TLSAC, MR2/3/1/60, 'Tour of the Battlefield: Battle of the Aisne', April 1934.
94 *With the Leyland Tigers to Wipers, 1929*, 32.
95 *Ypres Times* 7, no. 8 (October 1935): 236–8.
96 *Ypres Times* 3, no. 8 (October 1927): 232.

97 *Church Times*, 2 September 1932.
98 *Ypres Times* 6, no. 3 (July 1932): 89.
99 *Ypres Times* 5, no. 2 (April 1930): 41.
100 *Ypres Times* 5, no. 2 (April 1930): 40–5; 7, no. 3 (July 1934): 80.
101 *Ypres Times* 4, no. 3 (July 1928): 82.
102 *Memories* 2, no. 12 (Spring 1923): 144.
103 Taylor, *Good-Bye*, 97.
104 *Thetford and Watton Times*, 25 August 1928.

CHAPTER SIX

1 *Ypres Times* 2, no. 8 (October 1925): 18.
2 *Scotsman*, 13 October 1925. See also Seton Hutchison, *Pilgrimage*, 1–3.
3 *Nottingham Evening Post*, 28 May 1928.
4 *Falkirk Herald*, 15 April 1939.
5 *Memories* 4, no. 29 (Summer 1927): 162.
6 *Belfast Telegraph*, 5 December 1921. See also *Larne Times*, 10 December 1921.
7 *Ypres Times* 3, no. 2 (April 1926): 36.
8 *Church Times*, 2 September 1932.
9 See Wilkinson, *The Church of England*; Connelly, *The Great War*; Goebel, *The Great War and Medieval Memory*. For France, see Dupront, *Du sacré*.
10 *Ypres Times* 2, no. 2 (April 1924): 39; 2, no. 4 (October 1924): 111–2.
11 TNA FO 612/306, War Grave Passes. The Passport Office then oversaw the operation until the system was abolished in the 1960s.
12 London Metropolitan Archives, Acc. 1297, MET 10/591 Metropolitan Railway Old Comrades' Association, Correspondence with Southern Railway, 12 August 1932.
13 IWGC Annual Report 1939–1940, 12.
14 *Ypres Times* 7, no. 4 (October 1934): 116.
15 *Hull Daily Mail*, 11 September 1936.
16 *Ypres Times* 7, no. 7 (July 1935): 218.
17 *Yorkshire Post*, 5 August 1937.
18 *Great Western Railway Magazine*, August 1937, 398–9.
19 *Falkirk Herald*, 16 April 1938.
20 See IWGC Annual Reports, 1926–1940.
21 *Sphere*, 9 July 1927.
22 *Story of an Epic Pilgrimage*, 21; *Times*, 13 August 1928. See also SHC, DD/X/SIM/4, Smith battlefields pilgrimage, 1928.
23 *Sphere*, 18 August 1928.
24 *Ypres Times* 7, no. 1 (January 1934): 26.
25 *Ypres Times* 2, no. 8 (October 1925): 218.

26 *Ypres Times* 3, no. 2 (April 1926): 50.
27 *Ypres Times* 3, no. 4 (October 1926): 104.
28 *Ypres Times* 4, no. 3 (July 1928): 84.
29 Newfoundland was at that time was a Dominion of the British Empire, and not yet part of Canada.
30 *Toc H Magazine* 7, no. 11 (November 1929): 358–76.
31 *Ypres Times* 3, no. 3 (July 1926): 74; 4, no. 3 (July 1928): 83; 4, no. 4 (October 1928): 118; 4, no. 3 (July 1928): 83.
32 *Toc H Magazine* 7, no. 11 (November 1929): 359.
33 *Ypres Times* 6, no. 3 (July 1932): 78.
34 *Ypres Times* 3, no. 8 (October 1927): 233.
35 *Ypres Times* 4, no. 6 (April 1929): 172.
36 *St Barnabas Pilgrimages, Ypres, Somme, 1923*, 6; CWGC/1/1/1/9/B/16, Memorials to the Missing – Menin Gate unveiling, Letter from Mullineux to Ware, 19 July 1927.
37 CWGC/1/1/1/9/B/16, Mullineux to Ware, 19 July 1927.
38 *Ypres Times* 3, no. 4 (October 1926): 101.
39 Ibid., 101, 105.
40 *Ypres Times* 5, no. 8 (October 1931): 232.
41 *Ypres Times* 4, no. 4 (October 1928): 119.
42 *Ypres Times* 8, no. 1 (January 1936): 10–11.
43 *Church Army Gazette*, 24 March 1923.
44 *Berwickshire News and General Advertiser*, 8 July 1924.
45 Braithwaite Buckle, *A Kingly Grave*, 12, 13–14.
46 See, for example, the article by 'Ex-Staff Officer' in the *Scotsman*, 29 December 1919.
47 *The Bulletin*, 5 August 1920; CWGC/2/2/1/26, Commission meeting no. 26, 21 September 1920; *Army and Navy Gazette*, 11 December 1920.
48 *St. Andrew's Citizen*, 18 October 1919.
49 CRL, NC 18/1/196-237, Item 224, Letter from Neville Chamberlain to Hilda Chamberlain, 23 August 1919.
50 Command Paper 565, 'Memorandum of the Secretary of State for War Relating to the Army Estimates for 1920–21', 15 [173].
51 IWGC Annual Report 1921–1922, 4.
52 *Falkirk Herald*, 16 April 1938. See also IWGC Annual Report 1937–1938, 24.
53 *Ypres Times* 1, no. 3 (March 1922): 72; 1, no. 7 (April 1923): 205; 2, no. 8 (October 1925): 217, 3, no. 1 (January 1926): 18.
54 *Coventry Herald*, 10 February 1922.
55 IWM Documents 140, Edis diary, 16.
56 West, *Three Days' Journey*, 7.
57 Graham, *The Challenge*, 19–21, 28.
58 *Red Triangle Magazine* 3, no. 6 (February 1920): 214.

59 Dafoe, *Over the Canadian Battlefields*, 45.
60 *Scotsman*, 21 June 1919.
61 IWM Documents 140, Edis diary, 20.
62 West, *Three Days' Journey*, 7.
63 CRL, YMCA MSS 843/2/2, Alice Knight, letter to parents, 5 November 1919.
64 Longworth, *The Unending Vigil*, 129.
65 *Times*, 2 September 1920.
66 *Graphic*, 28 August 1920.
67 *Scotsman*, 21 June 1919.
68 *Berwickshire News and Advertiser*, 21 November 1922.
69 Ware, *The Immortal Heritage*, 56.
70 *Ypres Times* 5, no. 3 (July 1930): 78.
71 *Old Comrades' Association Gazette* 3, no. 4 (October 1922); *Country Life*, 4 September 1920.
72 *Sphere*, 9 July 1927.
73 *East Kent Gazette*, 15 October 1921.
74 *Hull Daily Mail*, 25 August 1922.
75 *Ypres Times* 3, no. 2 (April 1926): 37.
76 *Red Triangle Magazine* 5, no. 1 (September 1921): 15.
77 *Gloucester Citizen*, 10 September 1925.
78 *Dundee Evening Telegraph*, 4 August 1931.
79 *East Kent Gazette*, 15 October 1921.
80 *Ypres Times* 3, no. 2 (April 1926): 37.
81 *Ypres Times* 3, no. 8 (October 1927): 233.
82 *Ypres Times* 6, no. 3 (July 1932): 78.
83 *Church Times*, 10 September 1926.
84 *St. Andrew's Citizen*, 6 August 1938.
85 Taylor, *Good-Bye*, 27.
86 *Ypres Times* 6, no. 3 (July 1932): 84.
87 Taylor, *Good-Bye*, 148; *With the Leyland Tigers*, 27.
88 *Ypres Times* 7, no. 8 (October 1935): 237.
89 *Church Times*, 10 September 1926.
90 For a study of the IWGC's approach to architecture see Godden, 'Designing Memory'.
91 'Geometry of sleep' is how the poet and Second World War veteran Charles Causley described an IWGC cemetery. See his poem, 'At the British War Cemetery, Bayeux' in his collection *Union Street*.
92 NCA, DD/1332/198, Talbot pilgrimage.
93 Braithwaite Buckle, *A Kingly Grave*, 27–8.
94 Burrin, 'Western Front Pilgrimage', 10–12.
95 *Ypres Times* 6, no. 4 (October 1932): 111.

96 See Henry Benson articles in *Cornish Guardian*, 17 November 1922; *Perthshire Advertiser*, 15 August 1928; *West Sussex Gazette*, 16 August 1928.
97 *Belfast Telegraph*, 7 December 1921.
98 *Red Triangle Magazine* 3, no. 6 (February 1920): 213.
99 *Belfast News-Letter*, 13 July 1925.
100 *Coventry Herald*, 10 February 1922.
101 *Ypres Times* 5, no. 8 (October 1931): 231–2.
102 *Belfast Telegraph*, 7 December 1921; 7 April 1923.
103 *Belfast News-Letter*, 21 September 1923.
104 *Ypres Times* 1, no. 9 (October 1923): 271.
105 *Ypres Times* 3, no. 8 (October 1927): 233, 234.
106 *Ypres Times* 7, no. 4 (October 1934): 116.
107 Church Army Annual Report 1919–1920, 31.
108 *Red Triangle Magazine* 7, no. 4 (April 1923): 88.
109 *Church Army Gazette*, 24 March 1923.
110 *Church Army Gazette*, 10 November 1923.
111 *Ypres Times* 3, no. 4 (October 1926): 89.
112 *Scotsman*, 20 October 1921.
113 *Aberdeen People's Journal*, 2 September 1939.
114 *Yorkshire Evening Post*, 16 July 1937.
115 SHC, DD/X/SIM/4, Smith battlefields pilgrimage, 1928.
116 IWM Documents 13273, Jones battlefield pilgrimage.
117 *Ypres Times* 5, no. 4 (October 1930): 114.
118 *Berwickshire News and General Advertiser*, 20 November 1923.
119 *Memories* 2, no. 10 (September 1922): 67.
120 Frame, *My Life*, 82–3.
121 *Ypres Times* 6, no. 7 (July 1933): 212.
122 *East Kent Gazette*, 15 October 1921.
123 See, for example, *Yarmouth Independent*, 1 January 1921; *Central Somerset Gazette*, 18 February 1921; *Hendon and Finchley Times*, 22 January 1921; *Rugby Advertiser*, 21 October 1921.
124 See, for example, *Biggleswade Chronicle*, 30 June 1922; *Central Somerset Gazette*, 21 July 1922; *Framlingham Weekly News*, 14 October 1922; *Market Harborough Advertiser*, 27 February 1923.
125 *Market Harborough Advertiser*, 27 February 1923.
126 *Toc H magazine* 4, no. 10 (October 1926): 340–5.
127 *Toc H Magazine* 5, nos. 8 & 9 (August 1927): 347.
128 *Toc H magazine* 4, no. 10 (October 1926): 343.
129 SHC, DD/X/SIM/4, Smith battlefields pilgrimage, 1928.
130 *Church Times*, 6 April 1923.
131 *Daily Express*, 13 April 1936.
132 Sheffield City Archives (hereafter SCA), MD 7753/1, Sheffield Pilgrimage to the French and Belgian Battlefields, 8–13 July 1938.

133 *Berwickshire News and General Advertiser*, 7 October 1919.
134 *Scotsman*, 6 July 1926.
135 *Ypres Times* 2, no. 8 (October 1925): 199–201.
136 *Ypres Times* 4, no. 6 (April 1929): 165–6.
137 *Ypres Times* 7, no. 3 (July 1934): 72.
138 *Britannia* 1, no. 13 (December 1928): 1106.
139 *Taunton Courier*, 15 August 1923.
140 See for examples of press coverage *Dover Express*, 19 September 1924; *Liverpool Echo*, 12 September 1930; *Uxbridge and West Drayton Gazette*, 13 May 1938.
141 *Ypres Times* (Pilgrimage special issue, August 1922), 26–7.
142 *Hull Daily Mail*, 25 August 1922.
143 *With the Leyland Tigers*, 13.
144 *Berwick Advertiser*, 16 August 1928.
145 For a history of the British League of Help see Lewis, 'Adoptive Kinship'.
146 *Sheffield Daily Telegraph*, 2 June 1930.
147 *Ypres Times* 7, no. 3 (July 1934): 81.
148 *Motherwell Times*, 15 July 1927.
149 *Belfast News-Letter*, 4 August 1930.

CHAPTER SEVEN

1 *Answers*, 31 May 1919.
2 *Daily Mail*, 15 September 1919.
3 *Country Life*, 14 June 1919.
4 *The Trouble Buster: Journal of the USA General Hospital*, no. 2, 27 March 1920.
5 *The Watch on the Rhine*, 23 May 1919.
6 *Daily Mirror*, 16 April 1919.
7 *Aberdeen Daily Journal*, 20 October 1919.
8 *The Home Sector: A Weekly for the New Civilian*, 31 January 1920.
9 *Michelin Ypres*, 69, 76–81.
10 *Country Life*, 4 September 1920.
11 *Athenaeum*, 12 November 1920, 653.
12 *Graphic*, 27 September 1919.
13 *Tatler*, 1 October 1919.
14 Story, *Ypres*, vol. 2 of *The Battlefields*, 4–5.
15 *Daily Express*, 28 March 1919.
16 *Graphic*, 31 May 1919.
17 *Manchester Guardian*, 6 September 1919.
18 *Daily Mirror*, 16 April 1919.
19 *Nottingham Evening Post*, 4 February 1920.
20 CWGC/2/2/1/17, IWGC meeting, no. 17, 18 November 1919.

21 CWGC/1/1/9/B/23, Ypres – General File, Copy of final draft, 3 December 1919.
22 See, for example, *Cambridge Daily News*, 4 December 1919; *Kent and Sussex Courier*, 12 December 1919; *Londonderry Sentinel*, 13 December 1919; *Newcastle Evening Chronicle*, 5 December 1919; *Nottingham Journal*, 6 December 1919.
23 *Ypres Times* 3, no. 4 (October 1926): 105.
24 *Ypres Times* 3, no. 6 (April 1927): 153.
25 *Sphere*, 9 July 1927.
26 CRL, NC 18/1/196-237, Item 224, Letter from Neville to Hilda Chamberlain, 23 August 1919; NC1/26/162-188, Item 184, Postcard from Neville to Anne Chamberlain, 20 August 1919.
27 *Sphere*, 9 July 1927.
28 Ibid.
29 *Graphic*, 18 June 1921.
30 *Country Life*, 22 July 1922.
31 *Aberdeen Daily Journal*, 20 October 1919.
32 Burrin, 'Western Front Pilgrimage', 11.
33 *Berwick Advertiser*, 16 August 1928.
34 CWGC 2/2/1/17, Commission meeting no. 17, 18 November 1919.
35 *The Watch on the Rhine*, 16 October 1920.
36 *The Watch on the Rhine*, 2 May 1919.
37 *Scotsman*, 6 July 1926.
38 Williamson, *The Wet Flanders Plain*, 80–1.
39 *Morpeth Herald*, 7 January 1938.
40 *Morpeth Herald*, 3 June 1938.
41 *Nash's Pall Mall Magazine*, March 1924, 60–73.
42 Lowe, *Western Battlefields*, 3.
43 Coop, *A Short Guide*, 13–14.
44 *Cheltenham and County Looker-On*, 3 January 1920.
45 *Illustrated London News*, 14 June 1919.
46 Coop, *A Short Guide*, 15; Lowe, *Western Battlefields*, 3.
47 Fleming, *How to See the Battlefields*, 88.
48 CRL, YMCA MSS 843/2/2, Alice Knight, letter to parents, 5 November 1919. See also letters to parents, 4 August, 23 September 1919.
49 Coop, *A Short Guide*, 15; Fleming, *How to See the Battlefields*, 88.
50 *Dundee Courier*, 25 May 1920.
51 *North-East Lanark Gazette*, 17 October 1919.
52 *Aberdeen Daily Journal*, 20 October 1919.
53 *Dundee Evening Telegraph*, 20 August 1920.
54 *Daily Mail*, 14 August 1923; *Dover Express*, 28 June 1929.
55 *Ypres Times* 6, no. 4 (October 1932): 107

56 *West London Observer*, 12 August 1921.
57 *Manchester Evening News*, 21 May 1920; *Burnley News*, 22 May 1920; *Banbury Advertiser*, 13 October 1921.
58 *Banbury Advertiser*, 13 October 1921.
59 *Northern Constitution*, 3 November 1923.
60 *East Kent Gazette*, 15 October 1921.
61 *Gloucester Citizen*, 10 September 1925.
62 Townroe, *A Pilgrim*, 193–4.
63 *Thetford and Watton Times*, 25 August 1928.
64 *Dundee Courier*, 7 July 1919. For the tank as propaganda symbol see Wright, *Tank*, 81–133.
65 Ibid.
66 Coop, *A Short Guide*, 32; *The Pilgrim's Guide*, 61.
67 *Muirhead's Belgium*, 49; Ward, Lock and Co., *Handbook to Belgium*, 55.
68 *Edinburgh Evening News*, 11 September 1920.
69 *Banbury Advertiser*, 13 October 1921.
70 *Cheltenham Chronicle*, 28 February 1920.
71 Lowe, *Western Battlefields*, 21.
72 Bowman Dodd, *Up the Seine*, 379.
73 *Scotsman*, 9 June 1919.
74 Fleming, *How to See the Battlefields*, 40.
75 *Kirkintilloch Herald*, 24 September 1919.
76 *Ballymena Observer*, 31 January 1919.
77 Kennedy, *Ypres to Verdun*, 12 and plates XI and XII.
78 *Aberdeen Daily Journal*, 20 October 1919.
79 Pulteney and Brice, *The Immortal Salient*, 31.
80 *Todmorden and District News*, 10 August 1928.
81 *Sheffield Independent*, 5 November 1927.
82 In Flanders Fields Museum Research Centre (hereafter IFFMRC), Joan Simmonds diary, 5–15 July 1930.
83 Bird, *Thirteen Years After*, 7.
84 Ibid., 15; *Memories* 4, no. 29 (Summer 1927): 164.
85 Swinton (ed.), *Twenty Years After*, 101–2.
86 *Daily Mail*, 10 May 1920.
87 *Leicester Daily Post*, 20 March 1920.
88 *Ypres Times* 3, no. 1 (April 1922): 80.
89 Coombs, *Before Endeavours Fade*, 64. Also see revised 1983 edition, 67.
90 *Lancashire Daily Post*, 3 August 1935.
91 *Ypres Times* 3, no. 8 (October 1927): 232.
92 *Ypres Times* 7, no. 8 (October 1935): 244.
93 *Dundee Courier*, 2 July 1925.
94 *Ypres Times* 4, no. 5 (January 1929): 134.

95 *Ypres Times* 3, no. 3 (July 1926): 75.
96 Taylor, *Good-Bye*, 115.
97 *Staffordshire Advertiser*, 14 October 1922.
98 See TNA WO 32/5877, 34 Division memorial.
99 See TNA WO 32/5888, 9 Scottish Division memorial.
100 See Australian War Memorial, AWM 27 623/4, 1 Division Memorial; CWGC 1/1/1/10/B/58, Battle Exploit Memorials – Australian.
101 See Fox, *The King's Pilgrimage*, 5-6.
102 For examples see, *Hull Daily Mail*, 12 August 1927 (exposed concrete); *West Sussex Gazette*, 28 December 1922 (pillbox as Cross of Sacrifice base).
103 For further details on the complex evolution of Tyne Cot Cemetery see Bostyn *et al.*, *Passchendaele 1917*; Baker, *Architect and Personalities*, 91; Godden, 'Memory, Landscape and the Architecture of the Imperial War Graves Commission'; Godden, 'Designing Memory'.
104 Taylor, *Good-Bye*, 154.
105 *Lancashire Daily Post*, 21 July 1924. The story had originally appeared in Australian newspaper the *Sydney Guardian*.
106 *Lancashire Daily Post*, 21 July 1924.
107 *Daily Telegraph*, 20 September 1932.
108 Thurlow, *Pill-Boxes*, 9.
109 *Ypres Times* 3, no. 8 (October 1927): 234.
110 Burrin, 'Western Front Pilgrimage', 11.
111 *Dundee Courier*, 1 July 1925.
112 *Times*, 16, 19 September, 7 October, 11 November 1929.
113 Taylor, *Good-Bye*, 115.
114 For details see Christie, *The Newfoundlanders in the Great War*.
115 For a discussion of the memorial space see Paul Gough, 'Sites in the Imagination'.
116 *Ypres Times* 6, no. 3 (July 1932): 77.
117 Bird, *Thirteen Years After*, 134.
118 IFFMRC, Joan Simmonds diary, 5–15 July 1930.
119 *Ypres Times* 7, no. 7 (July 1935): 206.
120 *Lichfield Mercury*, 16 August 1929.
121 Brice, *The Battle Book*, 99.
122 For the discussions and memorial plans for Hill 60 see TNA WO 32/3138, Memorials and Graves: War Memorials, Hill 60, J.J. Calder's presentation to the nation for memorial; WO 32/5890, Memorials and Graves: 9th Queen Victoria's Rifles, London Regiment, at Hill 60, Ypres, Belgium; CWGC/1/1/10/B/29, Battle Exploits Memorial 1/22 London Regiment (The Queen's); 1/1/10/B/67, Battle Exploits Memorial 1st Australian Tunnelling Company – Hill 60; 1/1/15/31, Gifts – Hill 60 – Press cuttings only.
123 *Northern Whig*, 6 September 1923.

124 *Ypres Times* 4, no. 6 (April 1929): 167.
125 *Perthshire Advertiser*, 30 June 1928.
126 *Hull Daily Mail*, 25 August 1922.
127 *Ypres Times* 6, no. 7 (July 1933): 211.
128 *Ypres Times* 4, no. 3 (July 1928): 82.
129 *Toc H Magazine* 5, nos. 8 & 9 (August 1927): 360.
130 *Scotsman*, 22 August 1923.
131 *Nelson Leader*, 28 August 1929.
132 *Sunderland Echo*, 18 November 1935.
133 *Ypres Times* 7, no. 8 (October 1935): 245.
134 *Coventry Evening Times*, 8 February 1934.
135 Williamson, *The Wet Flanders Plain*, 93–5; *Daily Mail*, 17, 19, 30 September 1930.
136 Brice, *Ypres: Outpost of the Channel Ports*, 3.
137 Author's collection. It is possible that Mrs Moon was the wife of either E.A.V. or Reginald Moon, both of whom were IWGC gardeners and worked in the area.
138 *Scotsman*, 6 August 1928.
139 *Sheffield Daily Telegraph*, 20 July 1931.
140 *Ypres Times* 4, no. 4 (October 1928): 119.
141 *With the Leyland Tigers*, 24.
142 *Sheffield Daily Telegraph*, 22 June 1925.
143 *Ypres Times* 3, no. 3 (July 1926): 74
144 *Daily Herald*, 12 September 1930.
145 *Daily Mail*, 17 September 1930.
146 *Daily Mail*, 17, 20 September 1930.
147 TNA WO 32/3138, Hill 60; CWGC1/1/15/31, Gifts – Hill 60.
148 CWGC/2/2/1/132, Commission meeting no. 132, 23 July 1930.
149 *Liverpool Echo*, 29 July 1937.
150 *Daily Mail*, 29 August 1919.
151 Coop, *A Short Guide*, 18.
152 Lowe, *Western Battlefields*, 3–4.
153 Williamson, *The Wet Flanders Plain*, 56, 69–71, 94–5.
154 *Ypres Times* 5, no. 8 (October 1931): 232.
155 *Ypres Times* 3, no. 4 (October 1926): 91; IFFMRC, Joan Simmonds diary, 5–15 July 1930.
156 *Gloucester Citizen*, 23 June 1926.

CHAPTER EIGHT

1 *Ypres Times* 7, no. 7 (July 1935): 215.
2 *Aberdeen Press and Journal*, 7 February 1923.
3 See, for example, *Times*, 8 June 1917.

4   *Colour Magazine* 9, no. 2 (September 1919), 'Special Canadian War Memorials Number': 39–40.
5   *Ballymena Observer*, 31 August 1923.
6   *Daily Mail*, 23 September 1919.
7   'The Ruin in the War Area', *Architectural Review* 46, no. 277 (December 1919): 172.
8   Price, 'The Reconstruction of Belgium'.
9   For studies of the debates surrounding the rebuilding of Ypres see Claeys, 'World War I'; Derez, 'A Belgian Salient'; Dendooven, *Ypres as Holy Ground*; Jaspers, 'Huib Hoste and the Reconstruction of Zonnebeke'; Lauwers, 'Le Saillant d'Ypres'. See also Woolley, 'National Congress of Belgian Architects'.
10 CWGC2/2/1/8, Commission meeting no. 8, 21 January 1919.
11 The formal title was Town Major. In very rough terms, this was something like the military equivalent of a mayor.
12 TNA WO 32/5569, Memorials and Graves: War memorials: suggestions for preservation of battlefield at Ypres as war memorial, Letter from Beckles Willson to Charles de Broqueville, 5 July 1919.
13 *Graphic*, 15 November 1919.
14 Reproduced in the *Chelsea Gazette* 1, no. 4 (22 November 1919).
15 *Times*, 3 October 1919.
16 Graham, *The Challenge*, 36. See also 18–37.
17 *Larne Times*, 7 January 1922.
18 *Memories*, Vol. 1, No. 3 (December 1920): 74.
19 Williamson, *The Wet Flanders Plain*, 70, 91. 101–3.
20 Bird, *Thirteen Years After*, 13, 31–2.
21 *Sphere*, 23 July 1927.
22 *Dundee Courier*, 20 July 1921.
23 *Het Ypersche*, 8, 29 September, 13, 20 October 1928.
24 *Ypres Times* 3, no. 1 (January 1926): 18.
25 *Ypres Times* 2, no. 3 (July 1924): 82; 3, no. 1 (January 1926): 23.
26 *Ypres Times* 3, no. 5 (January 1927): 145.
27 Surrey History Centre, QRWS/30/MURP/2, L.N. Murphy papers, Correspondence between Murphy's son and the Regimental Museum, 1997–1998.
28 Buckinx, *Ypres et ses environs*.
29 *Sphere*, 23 July 1927.
30 See TNA WO 339/120758, Lieutenant-Colonel Edward Percy Cawston; CWGC/8/1/4/2/54, Private G.C. Hopkins – Schoonselhof Cemetery.
31 TNA WO 32/3138, Memorials and Graves: War Memorials, Hill 60, J.J. Calder's presentation to the nation for memorial; WO 32/5890, Memorials and Graves: 9th Queen Victoria's Rifles, London Regiment,

at Hill 60, Ypres, Belgium; CWGC/1/1/10/B/29, Battle Exploits Memorial 1/22 London Regiment (The Queen's); 1/1/15/31, Gifts – Hill 60 – Press cuttings only; *Ypres Times* 1, no. 1 (October 1921): 20; *Daily Mirror*, 28 August 1927; *Westminster Gazette*, 3, 7 September, 15 October 1927.

32 TNA ED 121/30, British School at Ypres, Report by H.J.R. Murray, 20 March 1929; CWGC/2/2/1/118, Commission meeting no. 118, 13 March 1929.
33 For a history of the school see Fox and Elliott, *The Children Who Fought Hitler*.
34 See *Ypres Times* 4, no. 1 (January 1928) 22; 4, no. 4 (October 1928): 118.
35 See *Church Times*, 17, 24 October 1919; 21 August 1925.
36 *Ypres Times* 5, no. 2 (April 1930): 42.
37 *Ypres Times* 2, no. 5 (January 1925): 129.
38 See Blomfield, *Memoirs of an Architect*, 174–97; Fellows, *Reginald Blomfield*, 103–14; Dendooven, *Ypres as Holy Ground*.
39 *Ypres Times* 3, no. 8 (October 1927): 222.
40 *St. Barnabas, Menin Gate Pilgrimage, 1927*, 31–2.
41 *Daily Express*, 25 July 1927.
42 NCA, DD/1332/198, Talbot pilgrimage.
43 *St. Barnabas, Menin Gate Pilgrimage, 1927*, 7–8.
44 *Graphic*, 23 July 1927.
45 *St. Barnabas, Menin Gate Pilgrimage, 1927*, 2, 8–9
46 *Ypres Times* 5, no. 2 (April 1930): 53.
47 *Sunday Express*, 9 December 1934.
48 *Ypres Times* 8, no. 1 (January 1936): 13.
49 *Ypres Times* 5, no. 8 (October 1931): 232.
50 Bird, *Thirteen Years After*, 1–2.
51 *Ypres Times* 7, no. 7 (July 1935): 215.
52 *Ypres Times* 5, no. 8 (October 1931): 229–30.
53 *Hull Daily Mail*, 25 August 1922.
54 *Sunderland Echo*, 11 November 1935.
55 Seton Hutchison, *Pilgrimage*, 179–81.
56 Bird, *Thirteen Years After*, 15.
57 Fleming, *Battlefields*, 4.
58 Williamson, *The Wet Flanders Plain*, 92.
59 Taylor, *Good-Bye*, 152.
60 Fleming, *Battlefields*, 6.
61 Williamson, *The Wet Flanders Plain*, 97–8.
62 Quoted in *Canadian Battlefield Memorials*, 21.
63 Taylor, *Good-Bye*, 151.
64 Bird, *Thirteen Years After*, 8.

65 Williamson, *The Wet Flanders Plain*, 62–7, 85–9.
66 *Ypres Times* 6, no. 8 (October 1933): 243.
67 *Ypres Times* 6, no. 7 (July 1933): 210.
68 *Ypres Times* 6, no. 8 (October 1933): 245.

CHAPTER NINE

1 An excellent summary of the battles of Arras can be found in *Michelin Guides to Arras*, 1–20.
2 During the spring and summer of 1915, the French launched two ferocious assaults with the intention of capturing the ridge. Despite making some advances, success eluded them on both occasions at the cost of 150,000 casualties. See Greenhalgh, *The French Army*, 95, 116.
3 *Michelin Guide to Arras*, 26.
4 For a military, social, and cultural history of Canada's relationship with Vimy see Cook, *Vimy*. For a study of the battle see Nicholls, *Cheerful Sacrifice*.
5 Taylor, *Good-Bye*, 76.
6 Fleming, *Battlefields*, 31–2.
7 Townroe, *A Pilgrim*, 132.
8 Coop, *A Short Guide*, 60.
9 CRL, YMCA/ACC69/2/12, Letter sent by J. Hastings Eastwood to his wife, Jessie, 10 February 1919.
10 *Red Triangle Magazine* 4, No. 1 (September 1920), 19–20.
11 Dafoe, *Over the Canadian Battlefields*, 37.
12 IWM Documents 13273, Jones battlefield pilgrimage.
13 *Aberdeen Daily Journal*, 28 November 1922. See also Benson's article in the *Belfast News-Letter*, 7 April 1923.
14 West, *Three Days' Journey*, 16.
15 *Ypres Times* 3, no. 5 (January 1927): 130.
16 *Ypres Times* 5, no. 3 (July 1930): 85–6.
17 Seton Hutchison, *Pilgrimage*, 233
18 Dafoe, *Over the Canadian Battlefields*, 37.
19 Mottram, *Journey*, 177.
20 Taylor, *Good-Bye*, 75–8.
21 Townroe, *A Pilgrim*, 132.
22 Seton Hutchinson, *Pilgrimage*, 233; Townroe, *A Pilgrim*, 130.
23 Mottram, *A Journey*, 178–9.
24 Taylor, *Good-Bye*, 75.
25 Seton Hutchinson, *Pilgrimage*, 233; Townroe, *A Pilgrim*, 132; Taylor, *Good-Bye*, 75.
26 Graham, *The Challenge*, 115–6.

27 *Dundee Courier*, 25 May 1920.
28 West, *Three Days' Journey*, 16–19.
29 *Red Triangle Magazine* 4, no. 1 (September 1920): 20.
30 *Daily Mail*, 16 September 1919.
31 *Belfast News-Letter*, 7 April 1923.
32 Dafoe, *Over the Canadian Battlefields*, 31.
33 Chemin de Fer du Nord, *Lens, Arras, Albert*, 34.
34 IWM Documents 140, Edis diary, 74.
35 Taylor, *Good-Bye*, 95–6.
36 Swinton (ed.), *Twenty Years After*, vol. II, 841.
37 See *The Western Front: Drawings by Muirhead Bone*, 10.
38 Dafoe, *Over the Canadian Battlefields*, 34.
39 Story, *Present Day Paris*, 156–7.
40 *Shipley Times and Express*, 28 March 1919.
41 Graham, *The Challenge*, 113.
42 Coop, *A Short Guide*, 49, 69.
43 West, *Three Days' Journey*, 10.
44 Dafoe, *Over the Canadian Battlefields*, 36.
45 *Red Triangle Magazine* 3, no. 1 (September 1919): 2; *Red Triangle Magazine* 3, no. 6 (February 1920): 214.
46 *Belfast News-Letter*, 7 April 1923.
47 Mottram, *A Journey to the Western Front*, 180.
48 Williamson, *The Wet Flanders Plain*, 109–12.
49 Swinton (ed.), *Twenty Years After*, vol. II, 879–93.
50 See Lewis, 'Adoptive Kinship', 210–11.
51 *Newcastle Daily Chronicle*, 11 September 1920.
52 Taylor, *Good-Bye*, 96.
53 *Hull Daily Mail*, 14 January 1919, 4 May 1923. For examples of annual coverage, see 5 May 1924, 4 May 1925, 3 May 1926, 2 May 1927, 17 October 1927.
54 CWGC/1/1/9/A/2, Memorials to the Missing Pt. 2, Letter from Ingpen to Ware, 16 March 1922; Report of an interview with M. Cousel, Préfet du Pas de Calais, 7 April 1922.
55 CWGC/1/1/10/B, Battle Exploits Memorial Committee minutes of meetings, 5, 6, 7 August, 30 September 1919; Report of the Committee, 14 February 1921.
56 CWGC/1/1/9/B/23, Ypres – General File, Correspondence between Ware and Kenyon, 8–12 August 1919.
57 CWGC/1/1/9/B, Memorials to the Missing, Ingpen to PAS, 28 April 1926; Report on visit to the Somme and Arras, Robinson to Ware, 19 June 1926; Ware to Kenyon, 22 December 1926; CWGC/2/2/1/105, Commission Meeting No. 105, 14 December 1927.

58 For a discussion of the cemetery and memorial see Skelton and Gliddon, *Lutyens and the Great War*, 132–5; Geurst, *Cemeteries of the Great War*, 289–91
59 Hay, *9th (Scottish) Division Memorial*, 18. See also TNA WO 32/5888, 9 Scottish Division memorial.
60 For details of the memorial see WO 32/5859, 12 Division war memorial.
61 CWGC 1/1/10/B/WG 869, Pt. 1, 37 Division memorial.
62 *Ypres Times* 5, no. 3 (July 1930): 86.
63 IWM Documents 140, Edis diary, 72.
64 Coop, *A Short Guide*, 60.
65 Story, *Present Day Paris*, 156.
66 Newman, *Cycling in France*, 95.
67 Lowe, *Western Battlefields*, 37.
68 *Ypres Times* 1, no. 4 (July 1922): 94.
69 Fleming, *Battlefields*, 26.
70 Graham, *Challenge*, 113.
71 Williamson, *The Wet Flanders Plain*, 112–3.
72 Bird, *Thirteen Years After*, 59, 69.
73 Clout, *After the Ruins*, 156, 265–66.
74 The literature on the Vimy Ridge memorial is extensive. For an excellent overview of the history of every aspect of the design, planning and construction of the memorial see Hucker and Smith, *Vimy*.
75 *Ypres Times* 3, no. 4 (October 1926): 93.
76 *Daily Express*, 17 October 1927. N.B., the official guide to the Canadian memorials which reprints this article inadvertently states it was published in October 1928. See *Canadian Battlefield Memorials*, 39.
77 *Daily Express*, 17 October 1927.
78 *Sheffield Daily Telegraph*, 7 August 1928.
79 SHC, DD/X/SIM/4, Smith battlefields pilgrimage, 1928.
80 Taylor, *Good-Bye*, 87–93.
81 *Ypres Times* 3, no. 5 (January 1927): 130.
82 *Berwick Advertiser*, 16 August 1928.
83 SHC, DD/X/SIM/4, Smith battlefields pilgrimage, 1928.
84 *Ypres Times* 6, no. 3 (July 1932): 78.
85 Swinton (ed.), *Twenty Years After*, vol. II, 884–5.
86 Quoted in Hucker and Smith, *Vimy*, 53.
87 For a record of the Canadian pilgrimage for the unveiling see Murray (ed.), *The Epic of Vimy*. For British reactions to the memorial see *Scotsman*, 27 July 1936; *Hull Daily Mail*, 4 May 1937; *Gloucester Journal*, 14 August 1937; *Bedfordshire Times and Independent*, 10 December 1937; *Hendon and Finchley Times*, 24 June 1938; *St. Andrew's Citizen*, 6 August 1938; *Shields Daily News*, 8 August 1938; *Sheffield Daily Telegraph*, 13 July 1939.
88 Mottram, *A Journey to the Western Front*, 188–90.
89 Swinton (ed.), *Twenty Years After*, vol. II, 885.

90 British Legion, *Souvenir of the Battlefields Pilgrimage*, 34; CWGC/1/1/10/G/13, Windows in Memorial Chapel Notre Dame de Lorette.
91 *Ypres Times* 6, no. 4 (October 1932): 109.
92 Taylor, *Good-Bye*, 52.
93 British Legion, *Souvenir of the Battlefields Pilgrimage*, 34.
94 Bird, *Thirteen Years After*, 56.

CHAPTER TEN

1 For a detailed study of 1 July 1916 see Middlebrook, *The First Day on the Somme*.
2 Prior and Wilson, *The Somme*, 252.
3 See Gibson, *Behind the Lines*, 216–19.
4 Masefield, *The Old Front Line*, 72.
5 *Michelin Guides to The Somme*, vol. I, 11.
6 Chemin de Fer du Nord, *Lens, Arras, Albert*, 16.
7 Taylor, *Good-Bye*, 53; *Kent and Sussex Courier*, 22 April 1921.
8 *Sevenoaks Chronicle and Kentish Advertiser*, 8, 15 April 1921.
9 *Yorkshire Post*, 20 August 1921.
10 *Hull Daily Mail*, 25 August 1922.
11 *Northern Constitution*, 3 November 1923.
12 Williamson, *The Wet Flanders Plain*, 147
13 *Ypres Times* 3, no. 5 (January 1927): 129.
14 *Sphere*, 9 July 1927.
15 *A Souvenir of the Battlefields Pilgrimage*, 98.
16 *Berwick Advertiser*, 16 August 1928.
17 *Derby Daily Telegraph*, 27 August 1928; Taylor, *Good-Bye*, 58.
18 Taylor, *Good-Bye*, 59.
19 Quoted in Middlebrook, *The Somme Battlefields*, 111.
20 Quoted in Clout, *After the Ruins*, 25, 265.
21 *Yorkshire Evening Post*, 11 June 1928.
22 Mottram, *A Journey*, 240–1.
23 For details of the construction of the memorial see TNA WO 32/5868, War memorials: 36 Ulster Division at Thiepval, France; CWGC/1/1/10/B/17, Battle Exploit Memorials – 36th Ulster Division; Switzer, *Ulster, Ireland and the Somme*, 80–108; Switzer and Graham, "Ulster's Love in Letter'd Gold".
24 Taylor, *Good-Bye*, 54.
25 *Larne Times*, 31 December 1921.
26 *Belfast News-Letter*, 21 November 1921.
27 TNA WO 32/5868, War Memorials: 36 Ulster Division at Thiepval, France, Letter from Sir James Craig to BEMC, 20 October 1919.

28 Taylor, *Good-Bye*, 54; *Hull Daily Mail*, 25 August 1922.
29 *Northern Whig*, 21 November 1921.
30 *Northern Constitution*, 3 November 1923.
31 *Northern Constitution*, 10 November 1923.
32 Taylor, *Good-Bye*, 54.
33 Williamson, *The Wet Flanders Plain*, 138.
34 PRONI WGC/1/1 and WG/1/3, Thiepval Memorial Visitors Books, Mary Halloran, 14 July 1924; Mr and Mrs McLean, 11 June 1925; Mr and Mrs T. Martin, 15 July 1925; H.A. Wilson, 16 July 1924. J. Allan Mulholland, 30 August 1925; G.O. Mallows, 1 July 1926; J.B. Cormac, 20 August 1931.
35 *Yorkshire Post*, 20 August 1921.
36 *Larne Times*, 31 December 1921
37 *Northern Constitution*, 3 November 1923.
38 *Berwick Advertiser*, 16 August 1928.
39 *Ypres Times*, Vol. 3, No. 5 (January 1927): 130.
40 TNA WO 32/5868, Correspondence between Stuart Oswald and Northern Ireland government, 7 September 1925; Northern Ireland government to BEMC and reply, 2, 7 November 1925.
41 PRONI FIN 18/9/4, Thiepval Memorial Committee proposed acquisition of additional land containing trench and dugout, adjoining the Tower, Correspondence 19, 21, 29 January, 28 August 1929.
42 *Belfast Telegraph*, 31 July 1928.
43 Mottram, *A Journey*, 243.
44 CWGC/7/3/1/1/M79/1, Sixteenth Scale Working Drawing Sheet No. 1 – Mill Road Cemetery, Thiepval.
45 Mottram, *A Journey*, 242.
46 *Belfast News-Letter*, 4 August 1930.
47 *Yorkshire Post*, 20 August 1921.
48 *Ypres Times* 3, no. 8 (October 1927): 233.
49 PRONI PM7/4/9, Correspondence 17 September, 25, 27, 28, 31 October 1924.
50 The radio station 2LO was named for the transmitter of the fledgling British Broadcasting Company, located in the Strand.
51 Taylor, *Good-Bye*, 55–6, 59.
52 Switzer, *Ulster, Ireland and the Somme*, 100.
53 Quoted in Switzer, *Ulster, Ireland and the Somme*, 100.
54 PRONI FIN 18/8/46, Ulster War Memorial, Thiepval, Letter from Lord Craigavon to H.M. Pollock, Finance Minister, Northern Ireland Government, 4 August 1928.
55 *Sphere*, 9 July 1927.
56 PRONI FIN 18/8/46, Correspondence, 11 June 26 September 1934; Report by Major Charles Boyle on visit, 27 June 1934; CWGC/2/2/1/184,

Commission meeting no. 184, 18 July 1934; CWGC/1/1/10/B/17 & B/18, Battle Exploits Memorials – 36 Division; CWGC/6/4/1/2/4237, Staff card for William McMaster.
57 *Kent and Sussex Courier*, 22 April 1921.
58 *Liverpool Echo*, 30 June 1936.
59 *Yorkshire Evening Post*, 11 June 1928.
60 *Kent and Sussex Courier*, 4 June 1926. See also 13 May 1921.
61 *Yorkshire Evening Post*, 11 June 1928.
62 Taylor, *Good-Bye*, 57.
63 *Edinburgh Evening News*, 30 July 1932.
64 *A Souvenir of the Battlefields Pilgrimage*, 128.
65 CWGC/1/1/10/B/2, Battle Exploits Memorial Committee – Minutes of Meetings No. 5, 7 August 1919; Meeting No. 7, 23 January 1920.
66 Ibid.
67 CWGC/1/1/9/A/11, Battlefield Memorials – Memorandum by the Secretary of State for War, 8 July 1921; CWGC/1/1/9/A/13, Memorials – Report of the National Battlefields Memorial Committee; Letters from the Chairman, 7 June 1921.
68 CWGC/1/1/9/A/19, Minutes of the Advisory Committee appointed to advise the IWGC on sites and memorials, Battle Memorials III St. Quentin, July 1923.
69 CWGC/1/1/9/B/ACON 56, Correspondence, 28 April, 29 June 1926.
70 CWGC/1/1/9/A/19, Minutes of the Advisory Committee appointed to advise the IWGC on sites and memorials, Meeting No. 6, 2 July 1925.
71 CWGC/1/1/9/B/ACON 56, Memorials to the Missing, Robinson to Ware, 19 June 1926.
72 *Ypres Times* 3, no. 5 (January 1927): 128.
73 CWGC/1/1/9/B/ACON 56, Memorials to the Missing, Correspondence, 29 June, 9, 20, 29 July, 12 October 1926; CWGC/1/1/9/A/19, Minutes of the Advisory Committee appointed to advise the IWGC on sites and memorials, Meeting No. 7, 4 May 1928. For the detailed planning and construction of the memorial see Stamp, *The Memorial*.
74 For an account of the unveiling ceremony, see *Times*, 2 August 1932.
75 Bird, *Thirteen Years After*, 196.
76 CWGC/1/1/7/B/11, Anglo-French Mixed Committee, 7 January 1932; CWGC/1/1/9/B/41, Anglo-French Cemetery at Thiepval, 11 November 1930.
77 CWGC/2/2/1/147, Commission meeting no. 147, 13 January 1932.
78 Taylor, *Good-Bye*, 58.
79 CWGC/1/1/7/B/46, Exhumations Belgium and France General File Part 5, Letter from E.S. Haynes to IWGC, 20 April 1936; Registration Officer to Ware, 30 April 1936.

80 *Sheffield Daily Telegraph*, 24 January 1939.
81 *Larne Times*, 9 December 1939.
82 *Surrey Mirror*, 2 July 1937; *Northern Ireland Daily Mail*, 6 November 1937; CWGC/6/4/1/2/6154, Staff card, Charles Smith; CWGC/6/4/1/2/4978, Staff card, T.H. Opie.
83 Hammerton, *I Was There*, 830. It seems that the cafe changed hands again on the eve of war, for the *Sheffield Daily Telegraph* reported on 28 September 1939 that Mrs Bower, a local woman, had received a telegram from her sister, Mrs Loughran, the proprietor of the Thiepval cafe, reassuring her about conditions in France.
84 Middlebrook, *Somme Battlefields*, 111.
85 *Liverpool Echo*, 30 June 1936.
86 PRONI WG/1/3, Thiepval Memorial Visitors Books.
87 CWGC/9/2/2/4640_001, Mill Road Cemetery Thiepval, France; as this photograph, taken between 1 June 1954 and 30 June 1955, shows, the work was complete by this point.
88 Taylor, *Good-Bye*, 59.
89 *Daily Chronicle*, 7 July 1916.
90 See Edmund Blunden's poem 'Premature Rejoicing' in Gardner, *Up the Line to Death*, 91.

CHAPTER ELEVEN

1 Lowe, *Western Battlefields*, 9.
2 Ibid., 12.
3 *Pilgrim's Guide*, xvi, 52. For more on the legendary status of Sklindles, see Williamson, *The Wet Flanders Plain*, 39–42; see also Swinton, *Twenty Years After*, vol. I, 93–5.
4 IWM Documents 16791, Private papers of J.W.F. Duffield, Correspondence between J. Duffield and Thomas Cook, 1, 12 August 1919.
5 *Sheffield Independent*, 1 August 1927. See also *Good-Bye*, 122.
6 Lowe, *Western Battlefields*, 12.
7 Bird, *Thirteen Years After*, 43.
8 Williamson, *The Wet Flanders Plain*, 41–5.
9 *The Pilgrim's Guide*, 78.
10 Bird, *Thirteen Years After*, 43.
11 *Liverpool Echo*, 19 August 1926.
12 Lowe, *Western Battlefields*, 11.
13 Williamson, *The Wet Flanders Plain*, 33.
14 *Ypres Times* 5, no. 2 (April 1930): 44.
15 Pulteney and Brice, *The Immortal Salient*, 80.
16 Taylor, *Good-Bye*, 122.

17 *Ypres Times* 5, no. 4 (October 1930): 113.
18 For details see Clayton, *Tales of Talbot House,* and Lever, *Clayton of Toc H.*
19 *Coming of Age: A Record of Events at the Coming-of-Age Festival of Toc H in June and July 1936; A supplement to the Toc H Journal* (August 1936). Published in *Toc H Magazine* 14, no. 1 (January 1936): 10.
20 Williamson, *The Wet Flanders Plain,* 38–9.
21 *Toc H Magazine* 4, no. 10 (October 1926): 341–2.
22 Quoted in Lever, *Clayton of Toc H,* 147.
23 See report in *Western Daily Press,* 9 December 1929.
24 *Toc H Magazine* 9, no. 7 (August 1931): 314.
25 *Toc H Magazine* 9, no. 5 (May 1931): 195.
26 *Ypres Times* 5, no. 8 (October 1931): 244–45.
27 *Toc H Magazine* 9, no. 5 (May 1931): 194–7.
28 *Toc H Magazine* 9, no. 7 (August 1931): 312–14.
29 *Toc H Magazine* 11, no. 11 (December 1933): 394–5.
30 *Ypres Times* 7, no. 7 (July 1935): 206.
31 Taylor, *Good-Bye,* 217.
32 Coop, *A Short Guide,* 40.
33 Mottram, *A Journey,* 100–1.
34 Brophy and Partridge, *The Long Trail,* 16.
35 Mottram, *A Journey,* 102.
36 Seton Hutchison, *Pilgrimage,* 218.
37 *Daily Mail,* 6 September 1930.
38 Seton Hutchinson, *Pilgrimage,* 218.
39 *Ypres Times* 4, no. 3 (July 1928): 82.
40 *Ypres Times* 7, no. 8 (October 1935): 237.
41 *Ypres Times* 3, no. 3 (July 1926): 74; 5, no. 3 (July 1930): 86.
42 Bird, *Thirteen Years After,* 47.
43 *Liverpool Echo,* 19 August 1926.
44 *Belfast News-Letter,* 4 August 1930.
45 Fleming, *Battlefields,* 13–14.
46 Lowe, *Western Battlefields,* 12.
47 *Ypres Times* 1, no. 4 (July 1922): 94.
48 Mottram, *Journey,* 46–7.
49 Coop, *A Short Guide,* 38.
50 Graham, *The Challenge,* 41.
51 *Ypres Times* 7, no. 8 (October 1935): 237.
52 Taylor, *Good-Bye,* 209.
53 Mottram, *A Journey,* 87.
54 Fleming, *Battlefields,* 21.
55 Coop, *A Short Guide,* 43–4.
56 Taylor, *Good-Bye,* 210.

57 Bird, *Thirteen Years After*, 49.
58 Blunden and Norman, *We'll Shift Our Ground*, 109.
59 CWGC 1/1/9/A/19, Minutes of the Cabinet Advisory Committee appointed to advise the IWGC on sites and memorials, second meeting, 13 February 1922; fourth and fifth meetings, 26 March, 2 December 1924.
60 CWGC 2/2/1/78, IWGC committee meeting no. 75, 10 June 1925. See also CWGC 1/1/9/B, Memorandum on memorials to the missing, 29 June 1926.
61 CWGC 1/1/9/B, Memorandum by Ingpen on memorials to the missing, 28 April 1926.
62 For a discussion of the designs see Barker and Atterbury, *The North of France*, 48–50, 61.
63 *Western Daily Press*, 24 March 1925; *Yorkshire Post*, 7 September 1925.
64 *Hastings and St. Leonard's Observer*, 15 September 1934.
65 *Bexhill-on-Sea Observer*, 7 June 1924.
66 *Old Comrades' Association Gazette* 3, no. 4 (October 1922): 6–7.

L'ENVOI

1 Rutherford, *Rudyard Kipling*, 320.
2 See Gregory, *The Silence of Memory*, 51–92.
3 See Ware, *The Immortal Heritage*, 63, 66–7.
4 *Motherwell Times*, 15 July 1927.
5 Gwynn, *The Anvil of War*, 287.

Bibliography

PRIMARY UNPUBLISHED SOURCES

Australian War Memorials

27 623/4, 1 Division memorial.
27 623/5, 2 Australian Division memorial.
27 623/6, 3 Australian Division memorial.
27 623/7, 4 Australian Division memorial.
27 623/8, 5 Australian Division memorial.
27 623/3, Memorials – Australian Corps and Misc.

Cadbury Research Library, University of Birmingham

Chamberlain archives: Neville Chamberlain correspondence.
NC 1/26/162-188.
NC 1/27/1-43.
NC 18/1/196-237.
YMCA/ACC69/2/12, J. Hastings Eastwood correspondence.
YMCA MSS 843/2/2, Alice Knight correspondence.

Commonwealth War Graves Commission

Annual Reports of the Imperial War Graves Commission, 1919–42 (London: HMSO).
Cemetery registers, 1921–39.
CWGC/1/1/1/9/B/16, Memorials to the Missing – Menin Gate unveiling.
CWGC/1/1/3/5 FX, C&FA Correspondence.
File series CWGC/1/1/7/B/55; CWGC/1/1/7/B/56; CWGC/1/1/7/B/57;
 CWGC/1/1/7/B/58; CWGC/1/1/7/B/59; CWGC/1/1/7/B/60;
 CWGC/1/1/7/B/61; CWGC/1/1/7/B/62; CWGC/1/1/7/B/63;
 CWGC/1/1/7/B/64, Cemetery estimates and designs.

CWGC/1/1/7/B/42, Exhumation – France and Belgium, General File, Part 1.
CWGC/1/1/7/B/68, Direction Boards – France and Belgium.
CWGC/1/1/7/B/36, Land Acquisition Belgium.
CWGC/1/1/9/B/23, Ypres – General File.
CWGC/1/1/9/A/2, Memorials to the Missing Part 2.
CWGC/1/1/9/A/11, Battlefield Memorials – Memorandum by the Secretary of State for War.
CWGC/1/1/9/A/13, Memorials – Report of the National Battlefields Memorial Committee.
CWGC/1/1/9/A/19, Minutes of the Advisory Committee appointed to advise the IWGC on sites and memorials, Battle Memorials.
CWGC 1/1/9/B, Memorandum on Memorials to the Missing.
CWGC/1/1/9/B/3, Memorials to the Missing.
CWGC/1/1/7/B/11, Anglo-French Mixed Committee, 7 January 1932.
CWGC/1/1/9/B/41, Anglo-French Cemetery at Thiepval, 11 November 1930.
CWGC/1/1/7/B/46, Exhumations Belgium and France General File Part 5.
CWGC/1/1/9/B, Memorials to the Missing.
CWGC/1/1/10/A/12 & /13, /14, /15, Memorials – South Africans – Delville Wood.
CWGC/1/1/10/B, Battle Exploits Memorial Committee – Minutes of Meetings.
CWGC/1/1/10/B/2, Battle Exploits Memorial Committee – Minutes of Meetings.
CWGC/1/1/10/B/17 & B/18, Battle Exploits Memorials – 36 Division.
CWGC/1/1/10/B/29, Battle Exploits Memorial 1/22 London Regiment (The Queen's).
CWGC 1/1/1/10/B/58, Battle Exploit Memorials – Australian.
CWGC/1/1/10/B/67, Battle Exploits Memorial 1st Australian Tunnelling Company – Hill 60.
CWGC/1/1/10/G/10, Preservation of Belgian Front at Certain Points.
CWGC/1/1/10/G/13, Windows in Memorial Chapel Notre Dame de Lorette.
CWGC/1/1/15/31, Gifts – Hill 60 – Press cuttings only.
CWGC/1/1/16/20, Prince of Wales Visit with British Legion, 1928.
CWGC/1/1/17/2, The King's Pilgrimage.
CWGC/2/2/1/1-/231, Minutes of IWGC Commission meetings nos. 1–231 (1917–1939).
CWGC/6/4/1/2/4237, Staff card for William McMaster.
CWGC/6/4/1/2/6154, Staff card for Charles Smith.
CWGC/6/4/1/2/4978, Staff card for T.H. Opie.
CWGC/7/3/1/1/M79/1, Sixteenth Scale Working Drawing Sheet no. 1 – Mill Road Cemetery, Thiepval.
CWGC/8/1/4/2/54, Private G.C. Hopkins – Schoonselhof Cemetery.

File series CWGC/8/1/4-300, Enquiry files containing correspondence between the IWGC and next-of-kin of a First World War casualty commemorated by them.
CWGC/9/2/2/4640_001, Mill Road Cemetery Thiepval, France.

Imperial War Museum

IWM Box no. 86/48/1, Private papers of W.G. Marlborough, Church Army War Graves Department leaflet, 1927.
IWM Documents 140, Private papers of Miss O. Edis. Diary.
IWM Documents 13273, Private papers of C.R. Jones, Account of battlefield pilgrimage, 1920, made by his father, Charles Jones.
IWM Documents 16791, Private papers of J.W.F. Duffield.

In Flanders Fields Museum Research Centre, Ieper

Joan Simmonds diary, 5–15 July 1930.

Library and Archives Canada

RG 25 Vol. 330, IWGC Proceedings.
RG 25 Vol. 336 W 18/26, St Julien memorial – commemorations of 2nd Battle of Ypres
RG 38 Vol. 419, Meetings of the Canadian Battle Monuments Commission.
M-19-8 Battle Exploits Memorial, Newfoundland.

London Metropolitan Archives

Acc. 1297, MET 10/591, Metropolitan Railway Old Comrades Association.

The National Archives, UK

ADM 1/8611/154, Visits to War Graves in France and Belgium by Relatives of Deceased Officers and Men of the Armed Forces.
BT 31/25486/163088, Fields of Honour Society Ltd.
CAB 24/103/70, Cabinet papers.
CAB 24/106/45, Cabinet papers,
CAB 24/104/5, CAB 23/21/18, Cabinet papers
ED 121/30, British School at Ypres.
FO 737/13/10, War Grave Passes (War of 1914–1918), War Graves Pass, May 1939.

FO 612/306, War Grave Passes.
HO 45/11080/419854, Letter to Brussels Embassy, 21 September 1921, 13 March 1922; Home Office letter to Chief Constables, 13 February 1922.
T 161/23/6, Imperial War Graves Commission: Powers to provide transport and passes for visits of relations to War Graves, Treasury memorandum, 16 June 1920.
WO 32/3138, Memorials and Graves: War Memorials, Hill 60.
WO 32/5569, Memorials and Graves: War Memorials: suggestions for preservation of battlefield at Ypres as war memorial.
WO 32/5863, Memorials and Graves: Tank Corps memorial at Pozieres.
WO 32/5868, War Memorials: 36 Ulster Division at Thiepval, France.
WO 32/5869, War Memorials, Union of South Africa
WO 32/5877, Memorials and Graves: 34 Division memorial.
WO 32/5888, Memorials and Graves: 9 Scottish Division memorial.
WO 32/5890, Memorials and Graves: 9th Queen Victoria's Rifles, London Regiment, at Hill 60.
WO 339/120758, Lieutenant-Colonel Edward Percy Cawston.

Nottinghamshire County Archives

DD/1332/198, Papers of Neville Talbot, 'Pilgrimage of Padres to Talbot House, Poperinghe and Ypres, 27 April–1 May 1930'. Incomplete manuscript.

Public Records Office of Northern Ireland

PRONI FIN 18/8/46, Ulster War Memorial, Thiepval.
PRONI FIN 18/9/4, Thiepval Memorial Committee proposed acquisition of additional land containing trench and dugout, adjoining the Tower.
PRONI PM7/4/9, Correspondence concerning the provision of a radio set for the caretaker of the Thiepval War Memorial in France.
PRONI WGC/1/1 and WG/1/3, Thiepval Memorial Visitors Books.

Salvation Army International Heritage Centre

Salvation Army Year Books, 1918–31.
Salvation Army Annual Reports, 1919–31.
Obituary notices for General Higgins.

Sheffield City Archives

MD 7753/1, Sheffield Pilgrimage to the French and Belgian Battlefields, 8–13 July 1938.

Somerset Heritage Centre

DD/X/SIM/4, Personal account of Mrs E.A. Smith's attendance at the battlefields pilgrimage, 1928.

Surrey History Centre

QRWS/30/MURP/2, L.N. Murphy papers. Correspondence between Murphy's son and the Regimental Museum, 1997–1998.

Tameside Local Studies and Archives Centre

MR2/26/79, Account of a battlefields tour of France, 1936.

W. Sussex Records Office

RSR/MSS/7/73, Correspondence regarding the 12 Division memorial maintenance.

NEWSPAPERS AND JOURNALS

National newspapers

Answers; Athenaeum; Britannia; Bystander; Country Life; Daily Express; Daily Mirror; Daily Telegraph; Graphic; Illustrated London News; Manchester Guardian; Observer; Pall Mall Gazette; Scotsman; Sketch; Spectator; Sphere; Sunday Times; Tatler; Times; Westminster Gazette.

Regional newspapers

Aberdeen Press and Journal; Ballymena Observer; Banbury Advertiser; Bath Chronicle; Berwick Advertiser; Berwickshire News and General Advertiser; Bedfordshire Times and Independent; Bexhill-on-Sea Observer; Biggleswade Chronicle and Bedfordshire Gazette; Belfast News-Letter; Birmingham Daily Press; Birmingham Gazette; Burnley Express; Burnley News; Cambridge Daily News; Chelmsford Chronicle; Central Somerset Gazette; Cheltenham Chronicle; Cheltenham and County Looker-On; Cornish Guardian; Coventry Evening Telegraph; Coventry Herald; Diss Express; Dorking and Leatherhead Advertiser; Drogheda Argus; Dublin Daily Express; Dundee Courier; Dundee Evening Telegraph and Post; Dover Express; East Kent Gazette; Evening News; Edinburgh Evening News; Folkestone; Hythe; Sandgate and Cheriton Herald; Framlingham Weekly News; Gloucester Echo; Hastings and St. Leonard's Observer; Hendon and

Finchley Times; Huddersfield Daily Examiner; Hull Daily Mail; Kent and Sussex Courier; Kirkintilloch Herald; Lancashire Daily Post; Larne Times; Leeds Mercury; Lichfield Mercury; Lincolnshire Echo; Liverpool Daily Post; Londonderry Sentinel; Manchester Courier; Manchester Evening News; Market Harborough Advertiser; Morpeth Herald; Motherwell Times; Newcastle Evening Chronicle; Northampton Mercury; North-East Lanark Gazette; Nottingham Evening Post; Nottingham Journal; Nelson Leader; Northern Ireland Daily Mail; Northern Whig; Perthshire Advertiser; Rugby Advertiser; Runcorn Guardian; St. Andrew's Citizen; Sheffield Daily Independent; Shields Daily News; Shipley Times and Express; Staffordshire Advertiser; Sunderland Daily Echo; Surrey Mirror; Sussex Mirror; Tamworth Herald; Thetford and Watton Times; Uxbridge and West Drayton Gazette; Warwick and Warwickshire Advertiser; West London Observer; Western Daily Press; Western Mail; Western Morning News; West Sussex Gazette; Yarmouth Independent; Yorkshire Evening Post.

Regimental and battalion journals

The Bulletin: The Official Organ of the National Federation of Discharged and Demobilised Sailors and Soldiers; Chelsea Gazette: The Official Organ of the Chelsea and South Kensington Branch of the National Federation of Discharged and Demobilised Sailors and Soldiers; The Home Sector: A Weekly for the New Civilian; LRB Record; Memories: The Magazine of the 19th London Old Comrades Association; Old Comrades' Association Gazette; Old Contemptible: Journal of the Old Contemptibles Association; The Trouble Buster: Journal of the USA General Hospital; The Watch on the Rhine.

Trade and other specialist journals

Anglo-Belgian Notes: Official Organ of the Anglo-Belgian Union; Architectural Review; Church Army Gazette; Church Times; Great Western Railway Magazine; Railway Gazette; Red Triangle Magazine; Toc H Magazine; Ypres Times: Journal of the Ypres League.

Belgian and French newspapers

L'Auto Excelsior; Figaro; La Revue du Touring-club de France; La Révil du Nord; Le Soir (Brussels); *Het Ypersche.*

Hansard and Parliamentary papers

Command Paper 565, 'Memorandum of the Secretary of State for War Relating to the Army Estimates for 1920–21'. London: HMSO, 1920.
Hansard. House of Commons debates, House of Lords debates, 1915–39.

Maps

Belgian State Railways. *Map of the British Cemeteries in Flanders*. Brussels: Belgian State Railway Press and Publicity Service, n.d., c. 1925.

CATA Tours, Bruges. *Champ de Bataille de l'Yser : The Yser Battlefield*. Brussels: J. De Grieve and Co, n.d., c. 1925.

Ypres League Map of the Salient. London: Geographia for Ypres League, 1925.

Miscellaneous

Harry Ash Tours, traveller's wallet of tickets and itinerary for battlefield tour, 1938.

J. Carrigan, 'My Experiences in France, March 1918', Typescript record.

Queen Victoria Rifles Rest House, Hill 60, business card and leaflet, n.d, c. 1928.

'The Red Cars of Blankenberghe', leaflet listing tour itineraries, n.d., c. 1925.

Southern Railways, book of tickets for battlefields group concessions rate, London to Ypres (via Dover–Ostend), n.d., c. 1928?

Ypres League badges and membership certificates.

PUBLISHED MATERIAL

47th (London) Division War Memorials. High Wood and Martinpuich. London: no imprint, 1925.

51st Highland Division War Memorial Beaumont-Hamel (Somme). Glasgow: Aird and Coghill, 1924?

American Battle Monuments Commission. *A Guide to the American Battle Fields in Europe*. Washington: United States Government Printing Office, 1927.

Anon. 'The Registration and Care of Military Graves During the Present War'. *RUSI Journal* 62, no. 1 (1917): 297–302.

Ashley, Susan L.T. 'Re-colonizing Spaces of Memorializing: The Case of the Chattri Indian Memorial, UK'. *Organization* 23, no. 1 (2016): 29–46.

Aux Champs de Gloire: Le front Belge de l'Yser. Brussels: Ministry of Railways, Marine, Post and Telegraph Services, c. 1920.

Bairnsfather, Bruce. *From Mud to Mufti: With Old Bill on All Fronts*. London: Grant Richards, 1919.

Baker, Herbert. *Architect and Personalities*. London: Country Life, 1944.

Barker, Michael, and Paul Atterbury. *The North of France: A Guide to the Art, Architecture, Landscape and Atmosphere of Artois, Picardy and Flanders*. London: Heyford, 1990.

Barnes, Julian. *Cross Channel*. London: Jonathan Cape, 1996.

Barr, Niall. *The Lion and the Poppy: British Veterans, Politics and Society, 1921–1939*. New York: Praeger, 2005.

Barthes, Roland. *Camera Lucida: Reflections on Photography.* London: Cape, 1982.

Becker, Annette. 'From Death to Memory: the National Ossuaries in France after the Great War'. *History & Memory: Studies in Representations of the Past* 5, no. 2 (1993): 32–49.

Beckles Willson, Lt. Col. (Henry). *Ypres: The Holy Ground of British Arms.* Bruges: Charles Beyaert, 1920.

– *From Quebec to Piccadilly and Other Places: Some Anglo-Canadian Memories.* London: Jonathan Cape, 1929.

Behrend, Arthur. *As From Kemmel Hill: An Adjutant in France and Flanders, 1917 and 1918.* London: Eyre & Spottiswoode, 1963.

Bennett, Jill. 'Art, Affect, and the "Bad Death": Strategies for Communicating the Sense Memory of Loss'. *Signs* 28, no. 1 (2002): 333–51.

– 'The Aesthetics of Sense-Memory: Theorizing Trauma through the Visual Arts'. In *Regimes of Memory*, edited by S. Radstone and K. Hodgkin, 27–39. New York: Routledge, 2003.

Berleant, Arnold. *Living in the Landscape: Towards an Aesthetics of Environment.* Lawrence, KS: University Press of Kansas, 1997.

Bird, W.R. *Thirteen Years After: The Story of the Old Front Revisited.* Toronto: Maclean Publishing, 1930.

Blomfield, Reginald. *Memoirs of an Architect.* London: Macmillan, 1932.

Blunden, Edmund, and Sylvia Norman. *We'll Shift Our Ground, or Two on a Tour.* London: Cobden Sanderson, 1933.

Bond, Brian, ed. *The First World War and British Military History.* Oxford: Oxford University Press, 1991.

Bostyn, Franky, Kristof Blieck, Freddy Declerck, Frans Descamps, Jan van der Fraenen. *Passchendaele 1917: The Story of the Fallen and Tyne Cot Cemetery.* Barnsley: Pen and Sword, 2007.

Bourke, Joanna. *An Intimate History of Killing: Face to Face Killing in Twentieth Century Warfare.* London: Granta, 1999.

Bowman Dodd, Anna. *Up the Seine to the Battlefields.* New York: Harper and Brothers, 1920.

Bracco, Rosa Maria. *Merchants of Hope: British Middlebrow Writers and the First World War, 1919–1939.* Oxford: Berg, 1993.

Braithwaite Buckle, Elizabeth. *A Kingly Grave in France.* London: Longmans, Green and Co., 1919.

Brandon, Laura. *Art or Memorial? The Forgotten History of Canada's War Art.* Calgary: University of Calgary Press, 2006.

Braybon, Gail, ed. *Evidence, History and the Great War.* New York and Oxford: Berghahn, 2003.

Brice, Beatrix. *The Battle Book of Ypres.* London: John Murray, 1927.

Bibliography

Brice, Beatrix (with the assistance of Lieutenant-General Sir William Pulteney). *Ypres – Outpost of the Channel Ports: A Concise Historical Guide to the Salient of Ypres*. London: John Murray, 1929.
British War Cemeteries, Ypres, 1914–1918. Brussels: Thill, 1927.
Brophy, John, and Eric Partridge. *The Long Trail: What the British Soldier Sang and Said in the Great War of 1914–18*. London: Andre Deutsch, 1965.
Brown, Malcolm. *The Imperial War Museum Book of the Somme*. London: Sidgwick and Jackson, 1996.
Buckinx, R. *Ypres et ses environs: Petit guide du touriste* [Ypres and Its Surroundings: A Little Guide for the Tourist]. Ypres: n.d., c. 1936.
Budreau, Lisa M. *Bodies of War: World War I and the Politics of Commemoration in America, 1919–1939*. New York: New York University Press, 2010.
Burrin, Nellie. 'Western Front Pilgrimage – 1920'. *Stand To! The Journal of the Western Front Association*, no. 19 (Spring 1987): 10–12.
Butler, A.S.G. *The Architecture of Sir Edwin Lutyens*. Vol. III. London: Country Life, 1950.
Caddick-Adams, Peter. 'Footsteps across Time: The Evolution, Use and Relevance of Battlefield Visits to the British Armed Forces'. PhD thesis, Cranfield University, 2008.
Campbell von Laurentz, Baroness. *My Motor Milestones*. London: Herbert Jenkins, 1913.
Canadian Battlefield Memorials. Ottawa: F.A. Acland, 1929.
Carden-Coyne, Ana. *Reconstructing the Body: Classicism, Modernism and the First World War*. Oxford: Oxford University Press, 2009.
Carmichael, Jane. *First World War Photographers*. London: Routledge, 1989.
Carrington, Charles. *Soldier from the Wars Returning*. London: Hutchinson, 1965.
Casey, Edward S. 'Boundary, Place and Event in the Spatiality of History'. *Rethinking History* 11, no. 4 (2007): 507–12.
Causley, Charles. *Union Street*. London: Rupert Hart-Davis Ltd, 1957.
Chasseaud, Peter. *Rats Alley: Trench Names of the Western Front, 1914–1918*. Stroud: The History Press, 2017.
Chemin de Fer du Nord. *Souvenir de la Grande Guerre. Visite des Régions Dévastées du Nord de la France: Arras, Albert*. Paris: Chemin de Fer Du Nord/Touring Club de France, 1919.
– *Souvenir de la Grande Guerre. Visite des Régions Dévastées du Nord de France: Couchy-le-Château, Moulin de Laffaux, Chemin des Dames, Anizy-le-Château, Soissons*. Paris: Chemin de Fer Du Nord/Touring Club de France, 1919.
– *Souvenir de la Grande Guerre. Visite des regions dévastées du Nord de la France: Lens, Arras, Albert*. Paris: Touring Club de France, 1919.

Christie, Norm. *For King and Country: The Newfoundlanders in the Great War, 1916–1918*. Ottawa: CEF Books, 2003.

Church Army, Annual Reports, 1919–1930. London: The Church Army.

Claeys, Dries. 'World War I and the Reconstruction of Rural Landscapes in Belgium and France: A Historiographical Essay'. *Agricultural History Review* 65, no. 1 (2017): 108–28.

Clayton, P.B. *Tales of Talbot House*. London: Chatto & Windus, 1919.

Clewell, T. *Modernism and Nostalgia: Bodies, Locations and Aesthetics*. New York: Palgrave Macmillan, 2013.

Clout, Hugh. *After the Ruins: Restoring the Countryside of Northern France after the Great War*. Exeter: University of Exeter Press, 1996.

Cobley, Evelyn. *Representing War: Form and Ideology in First World War Narratives*. Toronto: University of Toronto Press, 1993.

Cohen, Deborah. *War Come Home: Disabled Veterans in Britain and Germany, 1914–1939*. Berkeley, CA: University of California Press, 2001.

Comité d'Action des Régions Dévastées. *Dans les régions dévastées: l'oeuvre de reconstitution et la solidarité française*. Paris: Comité d'Action des Régions Dévastées, 1925.

– *L'encyclopédie des régions dévastées*. Paris: Comité d'Action des Régions Dévastées, 1930.

Condé, Anna-Marie. 'A Marriage of Sculpture and Art: Dioramas at the Memorial'. *Journal of the Australian War Memorial* no. 19 (1991): 56–9.

– '"The Strain of Watching": The Origins of the Pozières Diorama'. *Wartime: Official Magazine of the Australian War Memorial*, no. 7 (1999): 34–6.

Connelly, Mark. *The Great War: Memory and Ritual. Commemoration in the City and East London 1916–1939*. Woodbridge, Suffolk: Boydell and Brewer, 2002.

– 'The Ypres League and the Commemoration of the Ypres Salient, 1914–1939'. *War in History* 16, no. 1 (2009): 51–76.

– *Celluloid War Memorials: The British Instructional Films Company and the Memory of the Great War*. Exeter: Exeter Press, 2016.

Connelly, Mark, and Stefan Goebel. *Ypres*. Oxford: Oxford University Press, 2018.

Cook, Thomas. *Battlefields and Cities of Belgium*. London: Thomas Cook, 1919.

– *How To See Paris and the Battlefields*. Paris: Thomas Cook, 1922.

Cook, Tim. *Vimy: The Battle and the Legend*. Toronto: Allen Lane, 2017.

Coombs, Rose E.B. *Before Endeavours Fade: A Guide to the Battlefields of the First World War*. London: Battle of Britain Prints International, 1976.

Coop, J.O. *A Short Guide to the Battlefields*. Liverpool: Daily Post, 1920.

Copp, Michael, ed. *From Emmanuel to the Somme: The War Writings of A.E. Tomlinson, 1892–1968*. Cambridge: Lutterworth, 1997.

Cowman, Krista. 'Touring behind the Lines: British Soldiers in French Towns and Cities during the Great War'. *Urban History* 41, no. 1 (2014): 105–23.

Crane, David. *Empires of the Dead: How One Man's Vision Led to the Creation of WWI's War Graves.* London: William Collins, 2013.

Craster, J.M. *'Fifteen Rounds a Minute': The Grenadiers at War, August to December 1914. Edited from the Diaries and Letters of Major 'Ma' Jeffreys and Others.* London and Basingstoke: Macmillan, 1976.

Cronier, Emmanuel. *Permissionaries dans la Grande Guerre.* Paris: Belin, 2013.

Cross, Alexander. *The White Cross Touring Atlas of the Western Battlefields.* London: White Cross Insurance Association, n.d., 1920?

Cutcher, Leanne, Karen Dale, Philip Hancock, and Melissa Tyler. 'Spaces and Places of Remembering and Commemoration'. *Organization* 23, no. 1 (2016): 3–9.

Dafoe, John W. *Over the Canadian Battlefields: Notes of a Little Journey in France, in March, 1919.* Toronto: Thomas Allen, 1919.

Daily Mail Handbook to the Battlefields. London: Daily Mail, 1919.

Daly, Selena, Maria Salvante, and Vanda Wilcox, eds. *Landscapes of the First World War.* Basingstoke: Palgrave Macmillan, 2018.

Das, Santanu. *Touch and Intimacy in First World War Literature.* Cambridge: Cambridge University Press, 2005.

Dawson, Graham. *Soldier Heroes, British Adventure, Empire and the Imagining of Masculinities.* London: Routledge, 1994.

de Certeau, Michel. *The Practice of Everyday Life.* Translated by Steven Rendall. Berkeley, CA: University of California Press, 1988.

Dendooven, Dominiek. *Ypres as Holy Ground: Menin Gate and the Last Post.* Translated by Ian Connerty. Koksijde: De Klaproos, 2001.

– 'Asia in Flanders Fields: A Transnational History of Indians and Chinese on the Western Front, 1914–1920'. PhD thesis, University of Kent, 2018.

– *De vergeten soldaten van de Eerste Wereldoorlog.* Berchem: EPO, 2019.

Dennis, Gerald. *A Kitchener Man's Bit: An Account of the Great War, 1914–18.* Solihull: Helion, 2014.

Derez, Mark. 'A Belgian Salient for Reconstruction: People and *Patrie*, Landscape and Memory'. In *Passchendaele in Perspective: The Third Battle of Ypres,* edited by Peter H. Liddle, 437–58. Barnsley: Pen and Sword, 1997.

Duffy, Dennis. 'An Ideal Solution: Sculptural Politics, Canada's Vimy Memorial, and the Rhetoric of Nationalism'. *Mosaic* 43, no. 2 (2010): 167–84.

Dupront, Alphonse. *Du sacré: Croisades et pèlerinages. Images et langages.* Paris: Gallimard, 1987.

Eade, John, and Mario Karić, eds. *Military Pilgrimage and Battlefield Tourism: Commemorating the Dead.* London: Routledge, 2017.

Edensor, Tim. 'Staging Tourism: Tourists as Performers'. *Annals of Tourism Research* 27, no. 2 (2000): 322–44.

Edney, Matthew H. 'Theory and the History of Cartography'. *Imago Mundi* 48 (1996): 185–91.

Elton, Oliver. *C.E. Montague: A Memoir*. London: Chatto & Windus, 1929.

Farrer, Reginald. *The Void of War*. London: Constable, 1918.

Fathi, Romain. '"We Refused To Work Until We Had Better Means for Handling the Bodies": Discipline at the Australian Graves Detachment'. *First World War Studies* 9, no. 1 (2018): 35–56.

Fell, Alison S. *Women as Veterans in Britain and France after the First World War*. Cambridge: Cambridge University Press, 2018.

Fell, Alison S., and Christine E. Hallett, eds. *First World War Nursing: New Perspectives*. London: Routledge, 2013.

Fellows, Richard F. *Reginald Blomfield: An Edwardian Architect*. London: Zwemmer, 1986.

Fleming, Captain Atherton. *How to See the Battlefields*. London: Cassell, 1919.

Fortescue, J.W. *The British Soldiers' Guide to Northern France and Flanders*. London: The Times, 1917.

Foster, Jeremy. 'Creating a *Temenos*, Positing "South Africanism": Material Memory, Landscape Practice and the Circulation of Identity at Delville Wood'. *Cultural Geographies* 11 (2004): 259–90.

Fox, Frank. *The King's Pilgrimage*. London: John Murray, 1922.

Fox, James. *British Art and the First World War*. Cambridge: Cambridge University Press, 2015.

Fox, James, and Sue Elliott. *The Children Who Fought Hitler: A British Outpost in Europe*. London: John Murray, 2010.

Frame, John. *My Life of Globe Trotting*. London: Camelot, 1931.

Freeston, Charles L. *France for the Motorist*. London: Cassell for the Automobile Association, 1927.

French National Railways. *Visit the Battlefields of France*. Paris: SNCF, n.d., c. 1935?

Froula, Christina. 'Mrs Dalloway's Postwar Elegy: Women, War and the Art of Mourning'. *Modernism-Modernity* 9, no. 1 (2002): 125–63.

Fry, Ruth. *A Quaker Adventure: The Story of Nine Years' Relief and Reconstruction*. London: Nisbet and Co., 1926.

Fussell, Paul. *The Great War and Modern Memory*. Oxford: Oxford University Press, 1975.

Gammell, Irene. 'The Memory of St Julien: Configuring Gas Warfare in Mary Riter Hamilton's Battlefield Art'. *Journal of War and Culture Studies* 9, no. 1 (2016): 20–41.

Gardner, Brian (ed). *Up the Line to Death: The War Poets, 1914–1918*. London: Methuen, 1964.

Geurst, Jeroen. *Cemeteries of the Great War by Sir Edwin Lutyens*. Rotterdam: 010 Publishers, 2010.

Gibson, Craig. *Behind the Lines: British Soldiers and French Civilians, 1914–1918*. Cambridge: Cambridge University Press, 2014.

Godden, Tim. 'Memory, Landscape and the Architecture of the Imperial War Graves Commission'. In *Landscapes of the First World War*, edited by Selena Daly, Maria Salvante, and Vanda Wilcox, 193–208. Basingstoke: Palgrave Macmillan, 2018.

– 'Designing Memory: The Junior Architects of the Imperial War Graves Commission and the Creation of Spatial Memory in the British Cemeteries of the Western Front'. PhD thesis, University of Kent, 2020.

Goebel, Stefan. *The Great War and Medieval Memory: War Remembrance and Medievalism in Britain and Germany, 1914–1940*. Cambridge: Cambridge University Press, 2007.

Gordon-Smith, Jeremy. *Photographing The Fallen: A War Graves Photographer on the Western Front, 1915–1919*. Barnsley: Pen and Sword, 2017.

Gough, Paul. 'Memorial Gardens as Dramaturgical Space'. *International Journal of Heritage Studies* 3, no. 4 (1998): 199–214.

– 'Sites in the Imagination: the Beaumont-Hamel Newfoundland Memorial on the Somme'. *Cultural Geographies* 11, no. 2 (2004): 235–58.

– *A Terrible Beauty: British Artists in the First World War*. Bristol: Sansom and Co., 2010.

Graham, Stephen. *The Challenge of the Dead*. London: Cassell, 1921.

Graves, Charles. *Gone Abroad*. London: Nicholson and Watson, 1932.

Grayzell, Sue. *Women's Identities at War: Gender, Motherhood, and Politics in Britain and France during the First World War*. Chapel Hill, NC: University of North Carolina Press, 1999.

Great Eastern Railway. *The World War Battlefields of France and Belgium*. London: Belling for Great Eastern Railway, n.d., 1920?

Greenhalgh, Elizabeth. *The French Army and the First World War*. Cambridge: Cambridge University Press, 2014.

Gregory, Adrian. *The Silence of Memory: Armistice Day, 1919–1946*. Oxford: Berg, 1994.

Guides Bleus. *Verdun, Metz et les Champs de Bataille*. Paris: Hachette, 1934.

'Gun Buster' (Richard Austin). *Battle Dress*. London: Hodder and Stoughton, 1941.

Guoqi, Xu. *Strangers on the Western Front: Chinese Workers in the Great War*. Cambridge: Cambridge University Press, 2011.

Gwynn, Stephen, ed. *The Anvil of War: Letters between F.S. Oliver and His Brother, 1914–1918*. London: Macmillan, 1936.

Hammerton, Sir John, ed. *I Was There*. Vol. II. London: Waverley, 1939.

Hammond, Michael. *The Big Show: British Cinema Culture in the Great War, 1914–1918*. Exeter: University of Exeter Press, 2006.

Hammond, Michael, and Michael Williams, eds. *British Silent Cinema and the Great War*. Basingstoke: Palgrave, 2011.

Hankey, Lord. *The Supreme Command, 1914–1918*. Vols. I and II. London: George Allen and Unwin, 1961.

Harris, Cole. 'Power, Modernity and Historical Geography'. *Annals of the Association of American Geographers* 8, no. 1 (1991): 671–83.

Hart-Davis, Rupert, ed. *Siegfried Sassoon: The War Poems*. London: Faber and Faber, 1983.

Haultain-Gall, Matt. *The Battlefield of Imperishable Memory: Passchendaele and the Anzac Legend*. Melbourne: Monash University Publishing, 2021.

Haven Smith, Corinna, and Caroline R. Hill, *Rising Above the Ruins in France: An Account of Progress Made in the Devastated Regions*. London: G.P. Putnam and Sons, 1920.

Hay, Ian. *9th (Scottish) Division Memorial, Arras, 9 April 1922*. London: John Murray, 1922.

Heckstall Smith, Hugh. *Doubtful Schoolmaster*. London: Peter Davis, 1962.

Helmers, M. 'A Visual Rhetoric of World War I Battlefield Art: C.R.W. Nevinson, Mary Riter Hamilton and Kenneth Burke's Scene'. *The Space Between: Literature and Culture* 5, no. 1 (1999): 78–94.

Hetherington, Andrea. *British Widows of the First World War: The Forgotten Legion*. Barnsley: Pen and Sword, 2018.

Hewitt, Tom. 'Diorama Presentation'. *Journal of the Australian War Memorial* 5 (1984): 29–35.

Hill, A.W. 'Our Soldiers' Graves'. *Journal of the Royal Horticultural Society* 45, October (1919): 1–13.

Hodgkinson, Peter E. 'Clearing the Dead'. *Journal of the Centre for First World War Studies* 1, no. 3 (2007): 1–5.

Hodson, J.L. *Return to the Wood*. London: Victor Gollancz, 1955.

Hoffenberg, Peter H. 'Landscape, Memory and the Australian War Experience, 1915–18'. *Journal of Contemporary History* 36, no. 1 (2001): 111–31.

Holbrook, Carolyn. *Anzac: The Unauthorised Biography*. Sydney: NewSouth, 2014.

Holt, Tonie, and Valmai Holt. *Major and Mrs Holt's Battlefield Guide to The Ypres Salient*. 2003 ed. Barnsley: Pen and Sword, 2003.

Housman, A.E. *A Shropshire Lad*. London: Grant Richards, 1903.

Hucker, Jacqueline. '"Battle And Burial": Recapturing the Cultural Meaning of Canada's National Memorial on Vimy Ridge'. *Public Historian* 31, no. 1 (2009): 89–109.

Hucker, Jacqueline, and Julian Smith. *Vimy. Canada's Memorial to a Generation*. Ottawa: Sandling, 2012.

Hurst, Sidney c. *The Silent Cities: An Illustrated Guide to the War Cemeteries and Memorials to the 'Missing' in France and Flanders, 1914–1918*. London: Methuen, 1929.

Hynes, Samuel. *A War Imagined: The First World War and English Culture*. London: Bodley Head, 1990.

Illustrated Michelin Guides for the Visit to the Battle-fields: Battle-Fields of the Marne 1914. Clermont-Ferrand: Michelin, 1917.

Illustrated Michelin Guides to the Battle-fields (1914–1918): Arras, Lens–Douai and the Battles of Artois. Clermont-Ferrand: Michelin, 1919.

Illustrated Michelin Guides to the Battle-Fields (1914–1918): The Somme Battle-Fields. Vol. I. Clermont-Ferrand: Michelin, 1919.

Illustrated Michelin Guides to the Battle-Fields (1914–1918): Ypres and the Battles of Ypres. Clermont-Ferrand: Michelin, 1919.

Imperial War Graves Commission. *The Graves of the Fallen*. London: HMSO, 1920.

Inglis, Ken. 'Men, Women and War Memorials: Anzac Australia'. *Daedalus* 116, no. 4 (1987): 35–59.

Isherwood, Ian. 'The British Publishing Industry and Commercial Memories of the First World War'. *War in History* 23, no. 3 (2016): 323–40.

'Jackstaff'. *The Dover Patrol, The Straits: Zeebrugge: Ostend Including a Narrative of the Operations in the Spring of 1918*. London: Grant Richards, 1919.

James, Gregory. *The Chinese Labour Corps (1916–1920)*. Hong Kong: Bayview Educational, 2013.

Janowski, Monica, and Tim Ingold. *Imaging Landscapes: Past, Present and Future*. London: Routledge, 2016.

Jaspers, Patrik. 'Huib Hoste and the Reconstruction of Zonnebeke, 1919–1924'. In *Living with History, 1914–1964: Rebuilding Europe after the First and Second World Wars and the Role of Heritage Preservation*, edited by Nicholas Bullock and Luc Verpoest, 219–30. Leuven: Leuven University Press, 2011.

Johnson, N. 'Cast in Stone: Monuments, Geography, and Nationalism'. *Environment and Planning D: Society and Space* 13 (1995): 51–65.

Joiner, S. 'The Evolution of Planting Influences on the Imperial War Graves Commission'. *Garden History* 42, no. 1 (2014): 90–106.

Jolly, Martyn. 'Australian First World War Photography: Frank Hurley and Charles Bean'. *History of Photography* 23, no. 2 (1989): 141–48.

– 'Composite Propaganda Photographs during the First World War'. *History of Photography* 27, no. 3 (2003): 154–65.

Karol, Eitan. *Charles Holden: Architect*. Donington: Shaun Tyas, 2007.

Kavanagh, Gaynor. *Museums and the First World War: A Social History*. London and New York: Leicester University Press, 1994.

Keble, Howard (J. Keble Bell, 2nd Lieut. RAF). *The Glory of Zeebrugge and the Vindictive with the Official Narrative of the Operations at Zeebrugge and Ostend*. London: Chatto and Windus, 1918.

Kennedy, Sir Alexander B.W. *Ypres to Verdun: A Collection of Photographs of the War Areas in France and Flanders*. London: Country Life, 1921.

Kingston, Ralph. 'Mind over Matter? History and the Spatial Turn'. *Cultural and Social History* 7, no. 1 (2010): 111–21.

Kühler-Ross, Elizabeth. *On Death and Dying: What the Dying Have To Teach Doctors, Nurses, Clergy and Their Families*. New York: Scribner, 1969.

Laqueur, Thomas W. *The Work of the Dead*. Princeton: Princeton University Press, 2015.

Lauder, Harry. *A Minstrel in France*. London: Andrew Melrose, 1918.

Lauwers, Delphine. 'Le Saillant d'Ypres entre reconstruction et construction d'un lieu de mémoire'. PhD thesis, Florence, Department of History and Civilization, European University Institute, 2014.

Lawson, Tom. '"The Free-Masonry of Sorrow"? English National Identities and the Memorialization of the Great War in Britain, 1919–1931'. *History and Memory* 20, no. 1 (2008): 89–120.

Leed, Eric J. *No Man's Land: Combat and Identity in World War I*. Cambridge: Cambridge University Press, 1979.

LeMahieu, D.L. *A Culture for Democracy: Mass Communication and the Cultural Mind in Britain between the Wars*. Oxford: Clarendon, 1988.

Lever, Tresham. *Clayton of Toc H*. London: John Murray, 1971.

Lewis, Bryan. 'Adoptive Kinship and the British League of Help: Commemoration of the Great War through the Adoption of French Communities'. PhD thesis, University of Reading, 2006.

Liddle, Peter H, ed. *Passchendaele in Perspective: the Third Battle of Ypres*. London: Leo Cooper, 1997.

Lipstadt, Hélène. 'Learning from Lutyens: Thiepval in the Age of the Antimonument'. *Harvard Design Journal* (1995): 65–70.

Lloyd, David W. *Battlefield Tourism: Pilgrimage and the Commemoration of the Great War in Britain, Australia and Canada, 1919–1939*. Oxford: Berg, 1998.

Longworth, Philip. *The Unending Vigil: A History of the Commonwealth War Graves Commission*. London: Constable, 1967.

Lowe, Lieutenant-Colonel T.A. *The Western Battlefield: A Guide to the British Line. Short Account of the Fighting, the Trenches and Positions*. London: Gale and Polden, 1920.

Lowenthal, David. *The Past is a Foreign Country: Revisited*. Cambridge: Cambridge University Press, 2015.

McKenna, Mark, and Stuart Ward, '"It Was Really Moving, Mate": The Gallipoli Pilgrimage And Sentimental Nationalism in Australia'. *Australian Historical Studies* 38, no. 129 (2007): 141–51.

McKibbin, Ross. *Classes and Cultures: England 1918–1915*. Oxford: Oxford University Press, 1998.

MacDougall, Sarah, ed. *William Rothenstein and His Circle*. London: Ben Uri Gallery and Museum, 2016.

Macfarlane, Robert. *The Old Ways: A Journey on Foot*. London: Penguin, 2013.

Macleod, Jenny. *Reconsidering Gallipoli*. Manchester: Manchester University Press, 2004.

Magrath, C.J. *Ypres – Yper: A Few Notes on Its History before the War – With a Plan of the Town*. London: YMCA, 1918.

Maguire, Anna. 'Looking for Home? New Zealand Soldiers Visiting London during the First World War'. *London Journal* 41, no. 3 (2016): 281–98.

Malvern, Sue. 'War Tourisms: "Englishness", Art and the First World War'. *Oxford Art Journal* 24, no. 1 (2001): 47–66.

– *Modern Art, Britain and the Great War: Witnessing, Testimony and Remembrance*. New Haven and London: Yale University Press/Paul Mellon Centre for Studies in British Art, 2004.

Margry, Peter Jan, ed. *Shrines and Pilgrimage in the Modern World: New Itineraries into the Sacred*. Amsterdam: Amsterdam University Press, 2008.

Masefield, John. *The Old Front Line*. London: William Heinemann, 1917.

Maskell, H.P. *The Soul of Picardy*. London: Ernest Benn, 1930.

Massart, Jean. *Ce qu'il faut voir sur les champs de bataille et dans les villes détruites de Belgique*. Vol. II, *Le Front de Flandre*. Brussels: Touring-Club de Belgique, 1919.

Messenger, Gary S. *British Propaganda and the State in the First World War*. Manchester: Manchester University Press, 1992.

Meyer, Jessica, ed. *British Popular Culture and the First World War*. Leiden: Brill, 2008.

– *Men of War: Masculinity and the First World War in Britain*. Basingstoke: Palgrave Macmillan, 2009.

Middlebrook, Martin. *The First Day on the Somme*. London: Allen Lane, 1971.

Middlebrook, Martin, and Mary Middlebrook. *The Somme Battlefields: A Comprehensive Guide from Crécy to the Two World Wars*. London: Viking, 1991.

Middleton, G.A.T. *Ypres As It Was before the Great War*. Brussels, De Standaard, 1919.

Millington, Chris. *From Victory to Vichy: Veterans in Inter-war France*. Manchester: Manchester University Press, 2012.

Moriarty, Catherine. '"Representations of Patriotism": The Commemorative Representation of the Greatcoat after the First World War'. *Oxford Art Journal* 27, no. 3 (2003): 291–309.

– '"Though in a Picture Only": Portrait Photography and the Commemoration of the First World War'. In *Evidence, History and the Great War*, edited by Gail Braybon, 30–47. New York: Berghahn, 2005.

Morris, Mandy S. 'Gardens for Ever England: Landscape, Identity and the First World War British Cemeteries on the Western Front'. *Ecumene* 4, no. 4 (1997): 410–34.

Mosse, George. *Fallen Soldiers: Reshaping the Memory of Two World Wars*. New York: Oxford University Press, 1990.

Mottram, R.H. *The Spanish Farm Trilogy*. London: Chatto and Windus, 1927.
– *Through the Menin Gate*. London: Chatto and Windus, 1932.
– *A Journey to the Western Front*. London: G. Bell, 1936.
Muirhead, Finlay, ed. *Muirhead's Belgium and the Western Front*. London: Macmillan, 1920.
– *Muirhead's North-Eastern France*. London: Macmillan, 1922.
Murray, W.W., ed. *The Epic of Vimy*. Ottawa: The Legionary, 1936.
Nasson, Bill. 'Delville Wood and South African Great War Commemoration'. *English Historical Review* 119, no. 480 (2004): 57–86.
– *Springboks on the Somme*. Johannesburg: Penguin, 2007.
Newman, Bernard. *Cycling in France (Northern)*. London: Herbert Jenkins, 1936.
Nicholls, Jonathan. *Cheerful Sacrifice. The Battle of Arras, 1917*. London: Leo Cooper, 1990.
Noakes, Lucy. *Dying for the Nation: Grief and Bereavement in Second World War Britain*. Manchester: Manchester University Press, 2020.
Nora, Pierre. 'Between Memory and History: Les Lieux de Mémoire'. *Representations*, no. 26 (1989): 7–24.
O'Keeffe, Eleanor K. 'The Great War and "Military Memory"'. *Journalism Studies* 17, no. 4 (2016): 432–47.
Olson, Kory. 'Maps for A New Kind of Tourist: The First Guides Michelin France (1900–1913)'. *Imago Mundi* 62, no. 2 (2001): 205–20.
Orpen, William. *An Onlooker in France, 1917–1919*. London: Williams and Norgate, 1921.
Osborne, Brian S. 'In the Shadows of Monuments: The British League for the Reconstruction of the Devastated Areas of France'. *International Journal of Heritage Studies* 7, no. 1 (2001): 59–82.
– 'Warscapes, Landscapes, Inscapes, France, War and Canadian National Identity'. In *Place, Culture and Identity: Essays in Historical Geography in Honour of Alan R.H. Baker*, edited by A.R.H. Baker, 311–33. Quebec: Les Presses de L'Universite Laval, 2001.
Oxenham, John. *High Altars: The Battle-fields of France and Flanders as I Saw Them*. London: Methuen, 1918.
Parker, D. 'Maintaining Memories'. *The Horticulturalist* 13, no. 4 (2004): 5–8.
Pegum, John. 'Foreign Fields: Identity and Location in Soldiers' Writings of the First World War'. DPhil thesis, University of Cambridge, 2005.
– 'The Old Front Line: Returning to the Battlefields in the Writings of Ex-servicemen'. In *British Popular Culture and the First World War*, edited by Jessica Meyer, 217–36. Leiden: Brill, 2008.
Pilgrim's Guide to the Ypres Salient. London: Herbert Reiach, 1920.
Porteous, J.D. *Planned to Death: The Annihilation of a Place Called Hodendyke*. Toronto: University of Toronto Press, 1989.

Price, Julius M. 'The Reconstruction of Belgium'. *Fortnightly Review* 113, no. 639 (1920): 476.

Printemps. *Little Guide to Paris and Battlefields of France.* Paris: Le Printemps, c. 1920.

Prior, Robin, and Trevor Wilson. *The Somme.* New Haven and London: Yale University Press, 2005.

Pulteney, Lieutenant-General Sir William, and Beatrix Brice. *The Immortal Salient: An Historical Record and Complete Guide for Pilgrims.* London: John Murray, 1925.

Railways of Northern and Eastern France, *Battlefields of France.* Paris: J. Cussac, c. 1920.

Reader, Ian, and Tony Walter, eds. *Pilgrimage in Popular Culture.* Basingstoke: Macmillan, 1993.

Rency Georges, Paul Prist, and Ligue Patriotique du Tourisme. *Dixmude et Nieuport, Ypres.* Brussels: Touring-Club de Belgique, c. 1919.

Rice, Stanley. *Neuve Chapelle: India's Memorial in France.* London: HMSO, 1927.

Robin, Ron. '"A Foothold in Europe": The Aesthetics and Politics of American War Cemeteries in Western Europe'. *Journal of American Studies* 29, no. 1 (1995): 55–72.

Roper, Michael. *The Secret Battle: Emotional Survival in the First World War.* Manchester: Manchester University Press, 2009.

– 'Nostalgia as an Emotional Experience in the Great War'. *Historical Journal* 54, no. 2 (2011): 421–51.

Rowlands, Michael. 'The Role of Memory in the Transmission of Culture'. *World Archaeology* 25, no. 2 (1993): 141–51.

Rutherford, Andrew, ed. *Rudyard Kipling: War Stories and Poems.* Oxford: Oxford University Press, 1990.

St Barnabas Hostels, France [Reverend Matthew Mullineux]. *How to Reach 'The Hallowed Acres' in France and Belgium.* No imprint, or date, but likely to be 1924.

St Barnabas Pilgrimages: The Menin Gate. London: Eyre and Spottiswoode, 1927.

St Barnabas Pilgrimages: Ypres, Somme, 1923. London: Eyre and Spottiswoode, 1923.

St John, Isabella. *A Journey in War-Time.* London: John Murray, 1919.

Saler, Michael T. *The Avant-Garde in Interwar England: Medieval Modernism and the London Underground.* New York: Oxford University Press, 1999.

Sanders, M.L., and Philip M. Taylor. *British Propaganda during the First World War.* London: Macmillan, 1982.

Saunders, Nicholas J., and Paul Cornish. *Contested Objects: Material Memories of the Great War.* Abingdon: Routledge, 2013.

Scates, Bruce. *Return to Gallipoli: Walking the Battlefields of the Great War.* Cambridge: Cambridge University Press, 2006.

– 'The First Casualty of War: A Reply to Mckenna's and Ward's "Gallipoli Pilgrimage and Sentimental Nationalism"'. *Australian Historical Studies* 38, no. 130 (2007): 312–21.
Schama, Simon. *Landscape and Memory*. London: Penguin, 1995.
Scott, Jill. 'Vimy Ride: Stone with a Story'. *Queen's Quarterly* 114, no. 4 (2007): 506–20.
Scott, Major-General Sir Arthur B., ed. *The History of the 12th (Eastern) Division in the Great War, 1914–1918*. Compiled by P. Middleton Brumwell. London: Nisbett and Co, 1923.
Seton Hutchinson, Graham. *Pilgrimage*. London: Rich and Cowan, 1935.
Shaw, Amy. 'Expanding the Narrative: A First World War with Women, Children, and Grief'. *Canadian Historical Review* 95, no. 3 (2014): 398–406.
Sheftall, David. *Altered Memories of the Great War: Divergent Narratives of Britain, Australia, New Zealand and Canada*. London: I.B. Tauris, 2009.
Shelby, Karen. 'National Identity in First World War Belgian Military Cemeteries'. *First World War Studies* 6, no. 3 (2016): 257–76.
Sherman, Daniel J. *The Construction of Memory in Interwar France*. Chicago: University of Chicago Press, 1999.
Siebrecht, Claudia. 'Imaging the Absent Dead: Rituals of Bereavement and the Place of War Dead in German Women's Art during the First World War'. *German History* 29, no. 2 (2011): 202–23.
Sillars, Stuart. *Art and Survival in First World War Britain*. London: Macmillan, 1987.
Skelton, Tim, and Gerald Gliddon. *Lutyens and the Great War*. London: Frances Lincoln, 2008.
Smyth, Hanna. 'The Material Culture of Remembrance and Identity: South Africa, India, Canada and Australia's Imperial War Graves Commission Sites on the First World War's Western Front'. DPhil thesis, University of Oxford, 2019.
Soja, Edward W. *Thirdspace: Journey to Los Angeles and Other Real-and-Imagined Places*. Oxford: Wiley–Blackwell, 1996.
Souvenir for Visitors to the British Front. London: R. Clay and Sons, 1918.
Spagnoly, Tony, and Ted Smith. *Cameos of the Western Front: A Walk Around Plugstreet*. Revised ed. Barnsley: Pen and Sword, 2015.
'Special Canadian War Memorials Number'. *Colour Magazine* 9, no. 2 (1919): 39–40.
Stamp, Gavin. *The Memorial to the Missing of the Somme*. London: Profile, 2006.
Stephens, John. '"The Ghosts of the Menin Gate": Art, Architecture and Commemoration'. *Journal of Contemporary History* 44, no. 1 (2009): 7–26.
Stevenson, Randall. *Literature and the Great War*. Oxford: Oxford University Press, 2013.
Stewart, Pamela J., and Andrew Strathern, eds. *Landscape, Memory, and History: Anthropological Perspectives*. London: Pluto, 2003.

Stinglhamber, G. *Zeebrugge Museum: Official Guide, 1923*. Bruges: W.L. De Plancke, 1923.

Stone, Dan. 'The Far Right and the Back to the Land Movement'. In *The Culture of Fascism: Visions of the Far Right in Britain*, edited by Julie V. Gottlieb and Tom Linehan, 182–98. London: I.B. Tauris, 2003.

Story of an Epic Pilgrimage: A Souvenir of the Battlefields Pilgrimage. London: British Legion, 1929.

Story, Sommerville. *Present Day Paris and the Battlefields: The Visitor's Handbook with the Chief Excursions to the Battlefields*. New York: D. Appleton and Co., 1920.

– *The Battlefields of France and Belgium*. Vol. II, *Arras, Vimy and Lens*. Paris: La Renaissance du Livre, 1920.

– *The Battlefields of France and Belgium*. Vol. IV, *Verdun, The Forts de Vaux and Douaumont, The Forest of the Argonne etc*. Paris: La Renaissance du Livre, 1920.

– *The Battlefields of France and Belgium*. Vol. V, *Ypres and the Struggle for Calais*. Paris: La Renaissance du Livre, 1920.

– *The Battlefields of France and Belgium*. Vol. VI, *The Somme*. Paris: La Renaissance du Livre, 1920.

Sumartojo, Shanti, and Ben Wellings, eds. *National and Great War Commemoration: Mobilizing the Past in Europe, Australia and New Zealand*. Bern: Peter Lang, 2014.

Swinton, Ernest, ed. *Twenty Years After: The Battlefields of 1914–1918 Then and Now*. Vols. I and II. London: Newnes, 1936–38 [part-work].

Switzer, Catherine. *Ulster, Ireland and the Somme: War Memorials And Battlefield Pilgrimages*. Dublin: The History Press, 2013.

Switzer, Catherine, and Brian Graham. '"Ulster's Love in Letter'd Gold": The Battle of the Somme and the Ulster Memorial Tower, 1918–1935'. *Journal of Historical Geography* 36, no. 2 (2010): 183–93.

Tarlow, Sarah. 'An Archaeology of Remembering: Death, Bereavement and the First World War'. *Cambridge Archaeological Journal* 7, no. 1 (1997): 105–21.

Taylor, H.A. *Good-Bye to the Battlefields*. London: Stanley Paul and Co., 1930.

Taylor, William M. 'War Remains: Contributions of the Imperial War Graves Commission and the Australian War Records Section to Material and National Cultures of Conflict and Commemoration'. *National Identities* 317, no. 2 (2015): 217–40.

Territorial Army Association (Essex). *Territorial Cadets Visit to France*. Chelmsford: Meggy, Thompson and Creasey, 1923.

Thurlow, Colonel E.G.L. *The Pill-Boxes of Flanders*. London: Nicholson and Watson for the British Legion, 1933.

Tippett, Maria. *Art at the Service of War: Canada, Art and the Great War*. 2013 ed. Toronto: University of Toronto Press, 2013.

Toc H. *Over There: A Little Guide for Pilgrims*. London: Toc H, 1935.

Todman, Dan. *The Great War: Myth and Memory*. London: Hambledon, 2007.
Townroe, B.S. *A Wayfarer in Alsace*. London: Methuen, 1926.
– *A Pilgrim in Picardy*. London: Chapman and Hall, 1927.
Trott, Vincent. *Publishers, Readers and the Great War: Literature and Memory since 1918*. London: Bloomsbury, 2017.
Tuan, Yi-Fu. *Space and Place: The Perspective of Experience*. Minnesota: University of Minnesota Press, 1977.
Tucker, John F. *Johnny Get Your Gun: A Personal Narrative of the Somme, Ypres and Arras*. London: William Kimber, 1978.
Ugolini, Laura. *Civvies: Middle-Class Men on the English Home Front, 1914–1918*. Manchester: Manchester University Press, 2013.
Ulster's Tribute to Her Fallen Sons. Belfast: Ulster Division Battlefield Memorial Committee, 1921.
van Emden, Richard. *The Quick and the Dead: Fallen Soldiers and Their Families in the Great War*. London: Bloomsbury, 2011.
– *Missing: The Need for Closure after the Great War*. Barnsley: Pen and Sword, 2019.
Vance, Jonathan F. *Death So Noble: Memory, Meaning, and the First World War*. Vancouver: University of British Columbia Press, 1997.
– *Maple Leaf Empire: Canada, Britain and Two World Wars*. Toronto: Oxford University Press, 2012.
Wagons-Lits, Thomas Cook, R. van de Kerckhove and Son. *Motor Tours in Belgium: Bruges, Knocke Blankenberghe, Heyst*. Bruges: R. van de Kerckhove and Son, c. 1930.
Ward, Lock, and Co. *Handbook to Belgium and the Battlefields*. London: Ward, Lock and Co., 1921.
Ward, Mrs Humphrey. *Fields of Victory*. London: Hutchinson, 1919.
Ware, Fabian. *The Immortal Heritage*. Cambridge: Cambridge University Press, 1937.
Watson, Janet S.K. *Fighting Different Wars: Experience, Memory and the First World War*. Cambridge: Cambridge University Press, 2004.
Wearn, James Alexander, Andrew Philip Budden, Sarah Catherine Veniard, and David Richardson. 'The Flora Landscape of the Somme Battlefield: A Botanical Perspective on a Post-conflict Landscape'. *First World War Studies* 8, no. 1 (2017): 63–77.
Webber, Nick, and Paul Long. 'The Last Post: British Press Representations of Veterans of the Great War'. *Media, War and Conflict* 7, no. 3 (2014): 273–90.
Wellington, Jennifer. 'Narrative History, Image as Memory: Exhibiting the Great War in Australia, 1917–41'. In *Curating Empire: Museums and the British Imperial Experience*, edited by Sarah Longair, and John McAleer, 104–21. Manchester: Manchester University Press, 2012.

– *Exhibiting War: The Great War, Museums, and Memory in Britain, Canada and Australia*. Cambridge: Cambridge University Press, 2017.
West, M.S. *Three Days' Journey in the Devastated Areas of Belgium and France in May, 1919*. Monmouth: S.A. Bucknell, 1996.
Western Front: Drawings by Muirhead Bone. London: Country Life, 1917.
White, Geoffrey M. 'Travelling War: Memory Practices in Motion'. *History and Memory* 27, no. 2 (2015): 5–19.
Wilkinson, Alan. *The Church of England and the First World War*. London: SPCK, 1978.
Williamson, Henry. *The Wet Flanders Plain*. London: Faber and Faber, 1929.
Wilson, R.J. 'Memory and Trauma: Narrating the Western Front 1914–1918'. *Rethinking History* 13, no. 2 (2009): 251–67.
Winter, Caroline. 'Commemoration of the Great War on the Somme: Exploring Personal Connections'. *Journal of Tourism and Cultural Change* 10, no. 3 (2012): 248–63.
Winter, Jay. *Sites of Memory, Sites of Mourning*. Cambridge: Cambridge University Press, 1995.
– 'Shell Shock and the Cultural History of the Great War'. *Journal of Contemporary History* 35, no. 1 (2000): 7–11.
– *Remembering War: The Great War Between Memory and History in the Twentieth Century*. New Haven and London: Yale University Press, 2006.
Winter, Jay, and Emmanuel Sivan, eds. *War and Remembrance in the Twentieth Century*. Cambridge: Cambridge University Press, 1999.
With the Leyland Tigers to Wipers, 1929: Impressions of the Great Leyland Works Trip. No imprint.
Woolley, W.E. 'National Congress of Belgian Architects'. *The Journal of the Society of Architects* 13, no. 4 (1920): 145–46.
Worthy, Scott. 'Communities of Remembrance: Making Auckland's War Memorial Museum'. *Journal of Contemporary History* 39, no. 4 (2004): 599–618.
Wright, Patrick. *Tank. The Progress of a Monstrous War Machine*. London: Faber and Faber, 2000.
Wylly, H.C., ed. *Sherwood Foresters, Notts and Derby Regiment, Regimental Annual, 1926*. London: George Allen and Unwin, 1927.
Yilmaz, Ahenk. 'Memorialization on War-Broken Ground: Gallipoli War Cemeteries and Memorials Designed by Sir John James Burnet'. *Journal of the Society of Architectural Historians* 73, no. 3 (2014): 328–46.
Ypres Centre de Tourisme. *Guide d'Ypres*. La Panne: Editions Publica, 1935.
Zenere, Silvio. 'The Military Territorialization of the Asiago Plateau during the First World War'. MA thesis, University of Padua, 2014.
Zieman, Benjamin. *Contested Commemorations: German War Veterans and Weimar Political Culture*. Cambridge: Cambridge University Press, 2012.

Ziino, Bart. *A Distant Grief: Australians, War Graves and the Great War.* Cambridge: Cambridge University Press, 2007.

WEBSITES

The Great War, 1914–1918: A Guide to the Western Front WW1, Battlefields and History of the First World War, WW1 Monuments and Memorials on the Western Front Battlefields: Demarcation Stones/Bornes du Front (http://www.greatwar.co.uk/Article/demarcation-stone.htm, accessed 16 July 2019).

Memorial Dormans. Le rempact contre l'oubli, sur internet. Les Bournes Vauthier (http://Memorialdormans.free.fr/BornesVauthierHistoire.htm, accessed 16 July 2019).

Ribbon of Stone: Demarcation Stones, list of demarcation stones across France and Belgium (https://sites.google.com/site/wraros/demarcationstonesww1, accessed 16 July 2019).

Index

accommodation/hotels, 29, 34, 238, 262, 317: American demands, 225, 226; Amiens, 47, 66, 72; Arras, 293–4, 302; Bailleul, 366; Béthune, 368; class issues and nature of accommodation, 65; IWGC and accommodation issue, 50, 62; lack of and plans to increase the number, 40, 41, 225, 226, 235; Lille, 65; local people as hoteliers, 380; Poperinghe, 352, 353, 354; problems locating and distribution across former battlefields, 42, 43, 45, 57, 70, 100; quality and facilities, 56, 65–9, 73; sharing rooms to save costs, 185; veterans and hotels, 171–4, 187, 302; war graves visitation services and accommodation, 54–60; Ypres, 66, 69, 164, 183, 271

Albert (Somme), 93, 182, 186, 320, 322, 325, 326: Albert-Bapaume road, 23, 132, 186, 250, 318, 320, 332, 343, 345; Albert-Doullens road, 68; devastation, 133, 239, 320; DGRE offices, 33, 189; expatriate community, 72; gateway to Somme battlefields, 61, 82, 86, 131, 298; proposed site for memorial, 340, 343, 344; rebuilding and restoration, 27, 161, 166; station, 122, 123, 333; war graves hostels, 54, 57

alcohol, 127, 238, 262: veterans and drink, 144, 147, 171, 173, 270, 373

Americans, 34, 79, 224–6, 235–6, 260: Black Americans, 125

Amiens (Somme), 58, 113, 126, 203, 210, 243, 325: hotels, 68, 69, 72; hub for battlefield visiting, 47, 66, 74, 109, 110, 121, 183–4, 205, 242, 325; possible location for memorial, 341, 342; return to pre-war standards, 130; station, 61, 80, 131, 133; war graves hostels, 54, 61

Armentières (as 'behind the lines' town), 55, 308, 361–4: 'Mademoiselle of Armentières', 361–2; possible site of memorial, 369

Arras: battlefields, 168, 247, 290–2, 298–300; comparison with Lens, 300–2; comparison with Pompeii, 114; comparison with Ypres, 294–8; destruction of, 27, 224, 290–2, 294–5; history of, 290, 294–8; hotels, 68, 69, 226, 293–4; Hull and Oppy Wood, 303–4; Newcastle (British League of Help adoption), 302–3; pilgrimages to, 208; silence of, 120; wartime city, 292–4

attitudes towards Belgian and French people, 131, 161–70, 378: businesses at Hill 60, 256; disillusioned by, 364; disputes with, 270–1; respect for, 264; revival of local customs, 264,

301, 304; sympathy towards, 296; welcomed by, 220–1, 367, 370–3; women, 170, 275

Baker, Sir Herbert, 14, 139–40, 248
Bailleul (as 'behind the lines' town), 365–7
Bairnsfather, Bruce, 34, 172, 354, 357, 375
Bapaume (Somme), 31, 75, 298, 326: Albert-Bapaume road, 23, 132, 184, 250, 318, 319, 332, 343, 345; devastation, 133, 134, 225; veterans, 165
bars, 161, 262, 351, 355, 368: 'Ginger' of 'La Poupée' bar, 355; Suzette of Armentières, 365; Tina of Bailleul, 365; Zoë of Skindles bar, 354, 355, 365
Beaumont-Hamel (Somme), 33, 95, 184, 217, 218, 307, 340: Newfoundland Park, 251, 332
Belgian and French people of the battlefields, 145, 220–1, 380. *See also* destruction and devastation; tourists; veterans; Ypres
Benson, H.A.: Arras, 295, 299, 301; IWGC, 196; Last Post ceremony, 282; pilgrims, 57, 178–9, 191, 204, 205; Salvation Army and war graves visitation services, 55, 58; veterans, 149; visitor numbers: 182
bereaved, 8, 190–5, 200–11: Victorian bereavement rituals, 203
Béthune (as 'behind the lines' town), 367–369
Belgian and French customs officials and processes, 81–4
Bird, Will R.: Armentières, 363; Arras, 317; attitude of locals, 270; Béthune, 368; devastation, 131, 132; Newfoundland Park, 252; pillboxes, 245; Ploegsteert, 141–2; Poperinghe, 354–5; Thiepval, 347; Vimy, 309; Ypres, 105, 270, 283, 285, 287
Blomfield, Sir Reginald, 140, 267, 275, 276, 277, 282, 288, 306, 346

Blunden, Edmund, 332, 350, 368
Brice, Beatrix, 195, 253, 256, 356.
See also Ypres League
British League of Help, 220, 302, 321, 338, 370–2: Newcastle and Arras, 302–3; Sheffield and Serre, 304
British Legion and branches, 49, 56, 69, 147, 181, 209, 276, 356
British Legion 1928 pilgrimage, 180, 183, 316, 322, 346: T.F. Lister account, 153; Mrs E.A. Smith account, 88, 147, 208, 209, 212, 311; veteran comments, 165, 246, 339; Yorkshire contingent, 157

Cable, Boyd, 279
cafes, 220, 352, 380: Albert, 161; Armentières, 363–4; Arras, 293–4, 317; Béthune, 368; Hill 60, 254, 259; Poperinghe, 354, 355; Thiepval, 337, 347, 349; tourists, 234; Ypres, 164, 174, 267, 268–9; veterans and, 161, 170, 172, 173, 174
Campbell, Baroness, 73, 94, 127: accommodation and hotels, 67, 68; customs and frontier posts, 81–2; driving in the devastated region, 69, 76–7, 78; food supplies, 238; petrol supplies, 74
Cambrai (Nord), 120, 121, 225, 244: battlefields, 298; devastation, 112, 113, 125; memorial, 185; reconstruction, 124; roads to, 126, 307, 369; war graves visitation hostels, 54, 55, 57
cemeteries and memorials, 6–13, 43, 55, 71, 116: bereaved, 39, 42, 50, 59–60, 190–7, 200–5, 210; cemetery visitor books and visitor numbers, 181–2; DGRE concentration of bodies into a war cemetery, 41, 117, 119; French cemeteries, 200, 304, 316; German cemeteries, 158, 159, 199, 200; guiding pilgrims to cemeteries, 54–5, 57, 59; IWGC design, planning and building programme, standard design features, 14–16, 48, 95,

Index

195–8, 379–80; Kingston-upon-Hull memorial at Oppy Wood, 303; local reverence for IWGC cemeteries, 220–1; regimental memorials; Queen Victoria Rifles, 253; 1 (Australian) Tunnelling Company, 253, 256; relation to memorials to the missing, 369–70; Kipling, Rudyard, *The Gardener*, 376; wartime visits to, 28, 31–3; Talbot House as supplier of information, 360. *See also* divisional memorials; Directorate of War Graves Registration and Enquiries; IWGC cemeteries; Imperial (now Commonwealth) War Graves Commission (IWGC); pilgrims and pilgrimages; veterans; war graves visitation services

Chamberlain, Neville, 113, 114, 115, 118, 125, 127, 129, 189, 233

Chinese Labour Corps, 24, 119, 259

Christianity (see also consolation), 183: antithesis of Prussianism, 218; redemption through suffering, 150, 206, 211; Toc H movement, 82, 87, 357, 359; war graves visitation services, 53, 61, 64, 183

Church Army, 49, 53, 163, 183, 377: advice and support for pilgrims, 60; Christian mission of consolation, 53, 206; costs of pilgrimages, 59; engagement with IWGC, 48; funding for pilgrimages, 50; gratitude of pilgrims, 60, 187; hostels, 56, 57; infrastructure in Belgium and France, 48, 55; policy on pilgrimages, 52; women guides/members, 29, 63; women pilgrims, 207

class, social status and wealth, 56, 85, 94, 239: accommodation, 65; cost of battlefield tours beyond working class incomes, 45; first-class travel options, 46, 47, 188; London Rifle Brigade, 141; nature of pilgrims, 59; pilgrimages aimed at poorer/less affluent, 49, 184, 204; Poperinghe (entertainments for all classes of customer), 355; war graves passes, 51, 52; war graves visitation services, 61, 62

Coop, Reverend J.O.: accommodation, 66; advice to travellers, 74, 75; Armentières, 361, 363; Arras, 294; Bailleul, 366; Béthune, 368; camera and film, 239; Chinese Labour Corps, 124; customs and frontiers, 72, 84; devastation, 137; directions and maps, 77, 95; food and drink, 238, 239; Lens, 300; preparations for wet weather, 75, 93; road conditions, 85; souvenir collecting, 260; 'tank cemetery', 243; Vimy Ridge, 308, 309; Woods of the Somme, 137, 138

Cowlishaw, W.H., 247

Dafoe, J.W.: administrative arrangements for visiting, 45; admiration for locals, 127; Arras, 294, 296, 299; attitude to Germans, 125; devastation, 113, 114, 192; Lens, 300, 301; restoration and reconstruction, 110

Delville Wood, 139–40

desolation and 'the abomination of desolation', 23, 41, 161, 231, 309: 'abomination of desolation', 26–8, 110, 113; Bailleul, 366; Butte de Warlencourt, 134; deserted landscape, 113, 133; Lens, 300; lesson to be learned from it, 215; reconstruction, 245; ruined buildings, 112; scale of desolation, 299–300; scrubland, 112; Somme and sense of desolation, 131, 136; Thiepval, 320; Ypres, 264

devastation, 26–7, 107–42: monotony of, 114–15; curiosities of the devastated region, 115–19; scale of devastated region, 132–5, 300–1

Directorate of War Graves Registration and Enquiries, Graves Registration Units, 39, 95, 117, 188, 191, 192

disillusionment, 7–8, 83, 150, 216, 217, 286, 346
divisional memorials: 9 Division memorial, 105, 366–7; 12 Division memorial, 307; 18 Division memorial, 348; 34 Divisional memorial, 247; 37 Division memorial, 307; 46 Division memorial, 247; 1 (Australian) Division memorial, 247; 3 (Australian) Division memorial, 248
Doig, D.S., 26, 29

Edis, Olive (photographer), 69, 73, 121, 128, 299; accommodation and subsistence costs, 68; devastation, 111, 113, 299; DGRE/GRU, 191, 192; driving in the devastated region, 69, 74, 75, 76, 77; encountering the people of the devastated region, 124, 125, 127; nature, 128; Vimy Ridge, 308; walking, 93
Ewart, Wilfrid: ability to imagine the past, 167; Arras, 308; Bailleul, 365, 366; Béthune, 368; devastation, 115, 116, 124, 129, 132, 135; road conditions, 75, 94; Vimy, 115, 308; walking, 94, 134

facilities (lack of) in immediate post-war years, 41, 43, 235
food, 160: Arras, 69; advice on packing and purchasing food, 238–40; comfort of familiar food for pilgrims, 55, 147; quality of, 68, 69; shortages of, 41, 43; standard in Ypres, 56, 218; visitors put pressure on food supplies, 238
Fleming, Captain Atherton, 89, 227: advice to motorists, 74, 77; advice to visitors, 88; advice to walkers, 77, 84, 85, 87; Arras, 294; Bailleul, 365; Béthune, 368; cameras and photography, 239; Chinese Labour Corps, 123–4; destroyed tanks, 244; directions and navigation, 285; High Wood, 138; Vimy, 309
French concerns over visiting, 43, 235

Germans, attitude towards: attitude of wartime visitors, 27–8; attitudes of post-war visitors and pilgrims, 79, 114, 116, 121, 125–6, 131; reconciliation with, 126, 158–160; requisitioning of materials, 66, 68; Prisoners of War, 111, 119, 125–6; veteran attitudes towards, 114, 158–60, 221, 294, 301; Ypres, 236
Gibbs, Sir Philip: Arras, 299; directions and maps, 94; devastated region, 119, 132, 133; pilgrims and pilgrimages, 215; Thiepval, 350; tourism, 225, 265; unexploded ammunition, 118; Ypres, 265
Gillick, Ernest, 370
Goldsmith, G.H., 288
Graham, Stephen: accommodation, 66, 67; Albert, 133; Arras, 298, 300; Bailleul, 366; customs and frontier posts, 83; devastation, 112, 113, 132, 133; DGRU, 191, 192; Lens, 300; restorative qualities of nature, 128, 129; trees and woods, 136; unexploded ammunition, 118; Vimy, 309; Ypres, 66, 67, 268
guides and pilgrimage leaders: Barker, Eleanor, 63; Cawston, Lieutenant-Colonel E.P., 48, 259, 275; Chanter, Captain H.H., 283; Cockerell, Captain Malcolm, 72, 333; de Trafford, Captain G.F., 170, 172, 187; Gregson, Captain, 47; Griffiths, Captain, 47; Higgins, Catherine, 63–4; Montague, Major C.E., 22, 30, 34; Mullineux, Reverend Matthew, 57, 58, 59, 64, 65, 186, 377; Oswald, Captain Stuart, 72, 168, 186, 329; Parminter, Captain P.D., 36, 72, 183, 184, 272; Poyntz, Captain R.S.P., 46; Sanderson, W.S., 236, 237; Vyner, Philip, 72, 208

Index

guidebooks: *Good-bye to the Battlefields*, 90, 139, 145; *Immortal Salient*, 105, 245, 356; Michelin guides, 35, 94, 226, 227, 320; *Muirhead's Belgium and the Western Front*, 73, 94, 103, 227, 243; *Pilgrim's Guide to the Ypres Salient*, 44, 66, 67, 73, 77, 243, 352; *Silent Cities*, 102; *Ward, Lock and Co. Handbook to Belgium and the Battlefields*, 243; *Ypres: Holy Ground of British Arms*, 267

Hay, Ian, 281

Imperial (now Commonwealth) War Graves Commission (IWGC), 5, 8, 15, 16: Arras as possible site for memorial, 304–6; attitude to visitors, 39, 43, 62, 65, 230; building plan, 190; controversies, 283; costs of visiting and free pilgrimages, 47–50; design elements, 201, 207, 317, 369–70; directions and signage, 95–8, 100–4; funding of war graves visits, 39, 49–51, 62; gardeners, 142, 158, 199, 204, 245, 276, 349, 376–7; horticulture, 196–7, 198, 201, 210, 317, 376; Kipling, Rudyard, 377; local visitors, 220; loneliness of gardeners, 198–9; monitoring visitor numbers, 181; personnel, Charles Smith (gardener), 337, 338, 349; personnel in Ypres, 275–6; personnel, war veterans), 197–199, 275, 349, 377, 379–80; Ploegsteert, 141; recovery of bodies, 117, 349; significance of Étaples cemetery, 109; Thiepval, 323, 339–49; Ulster Tower, 336–7; visitor appreciation of work of IWGC, 164, 195–9, 210; visitor numbers, 181–2; war graves passes, 53; war graves visitation services, 97–8
itineraries, 88: zarial tours, 225; avoiding tourist routes, 175; 'behind the lines' tours, 363, 365; crowded nature of pilgrimage and tourist itineraries, 88, 183–5, 211, 380; Notre Dame de Lorette, 316; Paris as part of the excursion, 234; recommended and well-selected routes, 102, 227, 298; Thiepval, 332; wartime predictions of tourist itineraries, 34; war graves visitation services, 59

IWGC cemeteries: Adinkerke, 100; Anglo-French cemetery, Thiepval memorial, 347–8; Artillery Wood, 374; Birr Crossroads, 87; Blighty Valley, 103; Bray Road, 100; Connaught, 325; Coxyde, 186; Delville Wood, 140; Derry House No. 2, 247; Duhallow, 204; Dozinghem, 97; Dury, 100; Essex Farm, 164, 284, 374; Étaples, 109, 193–4, 374; Fauborg D'Amiens, 306–7; Gordon Dump, 197; Heath British, 85–6; Hunter's, 218; Klein Vierstraat, 169; Landrecies, 97; Le Trou, 201; LRB, 141; Maple Copse, 201; Mill Road, 323, 332, 349, 350; Motor Car Corner, 35; Mud Corner, 201; Noeux-les-Mines, 205; Polygon Wood, 96; Poperinghe Old, 354; Roeux, 97; St Pierre, 210; Sanctuary Wood, 202, 211; Terlincthun British, 96, 210; Tyne Cot, 139, 158, 196, 211, 214, 216, 247–8, 285; Vendresse, 157; Villers-Bretonneux, 97; Vis-en-Artois, 369–70; Vlamertinghe, 156, 175; Westoutre, 156; Windmill, 307; Ypres Ramparts, 288; Ypres Reservoir, 211; Ypres Town, 188; 51 Division, 101

Jones, David, 135, 378

Kerr-Lawson, James, 'The Cloth Hall, Ypres', 265
Kipling, Rudyard: advice to visitors, 195, 230–1; assistance for visitors, 48; *The Gardener*, 376, 377; scale of the IWGC's task 196

Lavery, Sir John ('The Cemetery, Étaples, 1919), 193–5
Lens (Pas-de-Calais), 21, 82, 131, 170, 291, 298, 299, 309: devastation, 293, 300–2, 308; reconstruction, 165
Longstaff, Will, 'Menin Gate at Midnight', 281–2
Lowe, Lieutenant-Colonel T.A., 47: advice to visitors, 238; 'behind the lines' towns, 352, 365–6; cameras and photography, 239; devastated regions, 115; importance of weatherproof clothing, 87, 93, 94; maps, 94; nature, 128; Poperinghe, 354, 355; ruined tanks, 243; souvenirs, 260; Vimy Ridge, 308; walking, 85, 87
Lutyens, Sir Edwin: 15, 276, 305–6, 314, 339, 341–7, 369, 370

maps, 4, 14, 31, 75, 78, 86, 94–5, 145: commemorative, 186; graves/GRU maps, 33, 41, 95; IWGC/Sir Fabian Ware, 100, 103; maps for pilgrims and tourists, 58, 94–5; pre-war maps, 190; sketch maps, 95, 153, 349; trench maps, 21, 106, 163, 252, 255; Ypres League, 80
Masefield, John, 26, 27, 30, 31, 318, 320, 332
Moore, Mary Macleod, 59, 62, 195
Mottram, R.H.: Armentières, 361, 362; Arras, 296, 297, 301; Bailleul, 366; Béthune, 368; Thiepval: 323, 331, 332; Vimy trenches, 315; walking, 143; Woods of the Somme, 138; Ypres, 296

nature, flowers and the battlefields: devastation of the natural world, 107, 109–10, 122, 135–7; flowers in hostels, 53; flowers placed on graves, 122, 203, 206, 208, 210, 220, 221, 278; IWGC horticulture, 196–7, 199, 201, 317, 326, 336; Newfoundland Park, 252; regenerative force, 27, 127–30, 138, 142, 147, 225, 251, 298, 310; shell cases as flower vases, 261; Thiepval, 326; wildflowers, 27, 109–10, 128–9, 137, 202–3, 218, 225, 255, 288. *See also* wreaths
Newfoundland Park, Beaumont-Hamel, 11, 184, 205, 212, 230, 251, 252, 332
Newfoundland memorial, Monchy-le-Preux, 307

Oliver, F.S., 22–8

Péronne (Somme), 21, 76, 113, 298: devastation, 27, 131; dangers of, 41, 42; war graves visitation services hostels, 54, 55
photography, 4–6, 125, 252, 292: Arras, 292, 300; Hill 60 as subject, 254; pilgrims, 60, 186; ruined tanks as subject, 242, 245; tourist, 116, 220, 236, 239–40, 244, 250, 251, 257, 348; visitors recommended to make a photographic record, 239; war graves, 72, 186, 198, 209, 284, 377; war graves pass, 53; war photographers, 135, 224, 244, 251, 375
pilgrims: consolation, 53, 60, 61–2, 63, 128, 180, 195, 205–6, 222, 377; exploitation of, 57–8, 219, 224; hostels for, 54–6; Charles Jones, 89, 90, 112, 115, 209, 295
pilgrimages and pilgrimage organisers: costs and funding war graves pilgrimages, 40, 45, 48–54, 59, 65, 185, 186; Morpeth annual pilgrimage, 236, 237; pilgrimages for young people and children, 214–18; Sheffield annual pilgrimage, 213, 220, 256; spiritual value of pilgrimage, 205–11. *See also* Beckles Willson, Henry; British Legion and branches; British Legion 1928 pilgrimage; Church Army; Salvation Army; St. Barnabas; YMCA; YMCA personnel; Ypres League

Index 451

Poperinghe (West Flanders), 29, 121, 167, 185: 'behind the lines' town, 351–61, 365, 368; hostels and huts, 54; Poperinghe-Ypres road, 26, 151, 166, 167, 169; strategically placed for battlefield tours, 67; Talbot House/ Toc H, 145, 162, 185, 211; veterans and, 165, 166, 167, 169, 175, 351–61. *See also* Talbot House

Pozières (Somme), 86: Australian memorial, 184, 326; pillboxes, 247; proposed site for memorial, 340, 341, 342, 343; ruined tanks, 244; Tank Corps memorial, 184; windmill, 248

railways, trains and stations, 71, 97, 119, 143, 246, 320: across the devastated region, 111; Albert, 86, 122, 123, 161, 333; Amiens, 121, 131, 133; Arras, 292, 299, 306, 317; Château Thierry, 78, 79; costs of rail travel, 50, 58; connections across the battlefields, 80, 84, 109, 145; cross-Channel railway services, 43–4, 237; customs officials, 83, 84; Étaples cemetery, 109, 194; Hill 60, 253; Poperinghe, 353, 355; pressure on the Belgian and French railway systems, 39, 41; Railway Transport Officers (Army), 120; reconstruction and repair, 226, 234; relaxation of travel restrictions on French railways, 40; Somme, 121, 145, 346; war graves visitation services collection points, 60, 72; veterans, 83, 147–9, 170, 171, 174, 175, 181–2, 187; Vimy, 88; war graves visitation services, 58–9, 60, 61, 186, 187; Ypres, 67, 158, 171, 188, 219, 236, 264

reconstruction, 108–14, 130, 132, 135: Arras, 294–6, 305, 308; Béthune, 368, 369; maturing of, 249; Somme region, 133; Thiepval, 322, 332, 338; veterans and, 95, 138, 160, 164, 165, 168, 332, 368; Ypres, 236, 251, 265–7

road conditions, 67–8, 73–80, 85–6, 117, 136, 295, 332

Salvation Army, 53, 183, 377: Christian mission of consolation, 53, 206; costs of pilgrimages, 58; engagement with IWGC, 48; funding for pilgrimages, 50; gratitude of pilgrims, 59; hostels, 55, 57; infrastructure in Belgium and France, 48; women guides/ members, 29, 63–4; women pilgrims, 204

Sassoon, Siegfried, 144, 279, 373

scale of battlefield visiting, 180–3: cemetery visitor book signatures, 181–2

Seton Hutchison, Lieutenant-Colonel Graham: Armentières, 362; Arras, 295, 297, 298; veterans relationship with the landscape, 91, 149–51; walking, 151; Woods of the Somme, 138; Ypres, 149–51, 273, 285

signage, 95–106: demarcation stones, 104–5

Singer Sergeant, Sir John, 'A Street in Arras', 292, 293, 297, 306

souvenirs and souvenir trade, 79, 93, 271, 351, 368, 380: collecting in wartime, 19; criticisms of the trade, 253–61; Hill 60, 250, 256, 259; Newfoundland Park, 252; trade in, 224; tourists and, 235–6, 229–30, 235, 236; Ypres, 267, 269

St. Barnabas, 60, 183, 204, 264, 281, 377: Christian mission of consolation, 206; costs of pilgrimages, 58, 59, 186, 278; fears that pilgrims are exploited, 58; funding and fund-raising, 57, 186, 278; gratitude of pilgrims, 59, 63; growth and expansion, 54; Menin Gate pilgrimage, 1927, 186; organisation and structure, 56; pilgrims, 204; women guides/ members, 61, 63

Story, Sommerville, 45, 65, 76, 84, 228, 300, 308

Taylor, Captain H.A./Raymond Bridgeway: Armentières, 361; Arras, 293, 296, 297, 298, 299; cemeteries and memorials around Arras, 105, 303, 307, 316; Béthune, 367, 368; British expatriates in Albert, 72; Delville Wood memorial, 140; Guynemer memorial, 200; IWGC staff, 199; landscape characteristics, 90–1; Lone Tree Crater, 251; Menin Gate memorial, 287; mystical draw of the battlefields for the veteran, 145; Passchendaele, 285; pillboxes, 245, 247, 248; Poperinghe, 354, 356; Raymond Bridgeway pseudonym, 90; Thiepval, 323, 323, 350; Thiepval, Ulster Tower, 336–7; memorial to the missing at Thiepval, 348; Tyne Cot cemetery, 248; Vimy, 177; Vimy trenches, 312, 313; walking, 90–1; Woods of the Somme, 139; Ypres, 287

Thiepval: 'Heroine of the Ruins', 338–9; scale of devastation, 320–5, 329, 332; Northern Ireland (and Ulster), 323–6, 329–32; re-emergence of a community at, 349–50; William McMaster, 333–7; Thiepval memorial to the missing, 339–49

Thomas, Sir William Beach, 245

Toc H, 82, 243: Clayton, Reverend Philip 'Tubby', 82, 88, 162, 211, 212, 249, 250, 357, 358, 359; Talbot, Gilbert, 357, 211; Talbot, Reverend (later Bishop) Neville, 162, 202, 279, 357; Talbot House, Poperinghe, 145, 162, 211, 356, 357, 359, 360, 361; Toc H members, 359; Toc H pilgrimages, 87, 88, 162, 184, 185, 212, 255, 357, 358, 360

tourism and tourist sites: behaviour of tourists, 228–38, 236–7, 259–61, 267–70; cameras, 233, 236, 239, 242, 245, 254; comparison with pilgrims, 238–9; craters: 250–1; gun emplacements, 241–2; Hill 60, 98, 99, 163, 175, 236, 252, 253–60, 273; pillboxes, 21, 166, 245–49, 254, 255, 309, 358; preserved trenches, 251–53, 310–13, 329–30; restaurants, 66, 173, 293, 317, 352, 368; tanks, 242–45; Ulster Tower, 326–9. *See also* Americans; bars; cafes

Townroe, B.S.: 91, 198, 242, 294, 296, 297, 298

transport: car, 72–8; cross-Channel ferries, 37–8, 43, 242; cycling, 32, 80–1, 308; rumours of tours by aeroplane, 225; walking, 84–93. *See also* railways, trains and stations

travel agents and agencies: Battlefields Bureau, 273; Thomas Cook, 38, 40, 45, 46, 214, 234, 245, 354; Fields of Honour Society, 64; Frame, John, 45, 60, 66, 109, 113, 126, 209; Franco-British Travel Bureau, 46; Imperial Travel Bureau, 46; Wipers Auto Services, 56, 72, 272

travel restrictions, 40, 95, 188, 229, 235, 242, 354

Truelove, J.R., 248, 369–70

veterans: attitudes towards tourists, 234–5, 255, 261, 297; attitude towards reconstruction, 6–7, 102, 364, 379; and the bereaved, 311–2; and drinking, 144, 171, 173, 272, 378; escape from disillusionment, 150; encounters with Germans, 159–60; humour, 170–6, 187, 363; interaction with Belgian and French women, 170, 173, 209; nostalgia/ stepping back in time, 92–3, 283, 312–3, 327, 328, 351, 355, 361, 365, 378; reasons for visiting, 144–51; remembering the dead, 155–9, 378; St Julien memorial, 286; tourists, 234–5, 255, 261, 297; trauma, 6–7, 148–50, 160–2, 170, 286; trees and woods prompt memory, 137–8; understanding the fighting, 151–5, 242, 245–6, 248, 285, 327; Vimy Ridge, 312–6; walking, 86–91;

Index 453

women veterans, 371–2. *See also* journalists and writers; attitudes towards Belgian and French people
veterans (individuals): Allinson, W.A., 87, 164, 233; Baumgartner, W.J., 80, 155, 173, 176, 254, 363; Fair, Charles, 126, 151, 156, 269; Francis, W.A., 69, 295, 307; 'H.D.W.G.', 157; Keith, A.W., 56, 69, 86, 259–60; Lineton, F.J., and West, J., 92, 149, 150, 156, 158, 159, 175; Michell, W.A., 137, 140, 295, 322, 329, 343; Planck, C.D., 152; Salvesen, Charles, 147, 174, 363; Sargeant, Rex, 145, 148, 169, 187, 283; Skinner, Alfred, 170, 171, 198; 'Yarl', 171–2
veteran regimental associations, old comrades associations and groups: 1/4 East Yorkshire, 181; 85 Field Ambulance, 170, 199; 1/5 Gloucester, 81, 242; Great Western Railway OCA, 182; Leyland Motor Works Association, 173, 200, 219, 256; 5 London Field Ambulance, 69; 2 London Regiment, 69, 173, 246, 257, 295, 363; 7 London Regiment, 152; 19 London/St Pancras Rifles, 151; London Rifle Brigade, 141; Manchester Regiment, 81, 148, 173; Metropolitan Railway OCA, 181; National Federation of Discharged and Demobilised Sailors and Soldiers, 49, 52, 54; Northumberland Hussars, 69, 135; Queen Mary's Army Auxiliary Corps OCA, 372–3; 236 Siege Battery Association, 147, 174, 363; 5 Sussex, 371
Von Berg, Wilfred, 141
Vimy Ridge and memorial, 307–16

war graves passes, 50, 59, 181
war graves visitation services: *see* pilgrimage organizations
Ware, Sir Fabian: architects (Blomfield and Lutyens), 305–6;

building programme, 195; costs of visiting and funding for the poor, 48, 50; Étaples cemetery, 109, 194; sensitivity to French concerns at visitor numbers, 43; signage, 96, 100, 101, 102, 104; Thiepval, 341, 342, 343, 344, 345, 348; visitor behaviour, 230; war graves visitation services, 57, 62, 186; Ypres battlefields, 266, 273; Hill 60, 257–9
wartime visiting: Chamberlain, Austen, 22–3; Lauder, Harry, 31–2; Hankey, Maurice, 20–1; informing the home front, 30–2; moral value of visiting, 27–9; predicting post-war tourism, 33–5; soldiers as battlefield visitors, 32–3; Trade Unions, 22, 23, 27; VIPs, 19–22; visitor experiences, 23–7
Williamson, Henry, 84: Americans, 236; customs and frontier posts, 83, 84; derogatory about local people, 83, 84; derogatory about local people in Arras, 301, 302, 309; derogatory about local people in Ypres, 256, 260, 261, 269–70; disturbed by frogs, 322; nature, 128, 138; Passchendaele, 285; Ploegsteert, 141; Poperinghe, 355, 356; Talbot House in Poperinghe, 357, 358; St Julien memorial, 285, 286; Thiepval (Ulster Tower), 327; walking, 87; Ypres, 260–1, 354; Ypres (commercialisation of), 269–70; Ypres (Hill 60), 256; Ypres (ramparts), 287, 288
Willson, Henry Beckles: pilgrimages, 180, 205, 206; signboards, 98–9, 103; Ypres and commercialisation, 267, 268. *See also* Ypres League
women: Belgian and French women married to IWGC staff, 273–4; bereaved, 61–2; exposed to danger, 41–2; free pilgrimages for, 278; guides, 63, 72; 'innocents abroad', 206–7, 278–9; interaction with veterans, 148; pilgrims, 55, 61,

63, 204, 208–10, 312, 363, 377; Ulster Women's Unionist Council, 335; unseemly behaviour, 229–30; vulnerable to exploitation, 377; wealthy women as pioneer visitors, 73–4, 239; working class, 62–3; younger women, 218–9

women (individuals): Bax-Ironside, Helen, 64–5; Braithwaite Buckle, Elizabeth, 68, 109, 112, 188, 202; Burrin, Nellie, 61, 85, 203, 234, 250; Myers, Efga, 109, 114, 129; Norman, Lady, 73, 74, 75, 76, 77, 93, 94, 121; St. John, Isabella, 19, 25, 28, 31; Smith, Mrs E.A., 88, 147, 148, 208, 209, 212, 311, 313; West, Marjory/West sisters, 65, 66, 120–1, 191, 192, 298, 300. *See also* Campbell, Baroness; Edis, Olive (photographer)

wreaths, 208, 211, 218, 230, 377: Étaples cemetery, 194; Hunter's Cemetery, 218; inscriptions, 204; laying services, 72; Menin Gate, 278, 279; veterans, 144. *See also* horticulture and flowers

YMCA, 109, 113, 124, 127, 183, 294, 299, 377: Christian mission of consolation, 53, 206; costs of pilgrimages, 65; criticisms of, 65; funding and fund-raising, 50, 123; gratitude of pilgrims, 60, 61, 210; infrastructure in Belgium and France, 45, 48, 52, 54, 55, 122; hostels and huts, 54, 57, 70, 86, 122, 123, 226; IWGC/Ware, Sir Fabian, 48; pilgrims, 60; pilgrimages and tours, 136; *Red Triangle*, 130, 192; women as guides/members: 29, 61, 66, 79

YMCA personnel, 122: Eastwood, John Hastings, 75, 115, 116, 128, 294; Knight, Alice, 78, 79, 86, 109, 112, 116, 121, 122, 123, 124, 125, 193, 239; 'Rover', 57, 112, 113, 124, 127, 133, 134; Yapp, Sir Arthur, 54, 113, 134, 192, 301

young people and the battlefields (criticism of), 257

Ypres: battlefields, 283–7, 243–6, 248–50, 253–9, 263–5; British Tavern, 261, 269, 272; commercialisation, 233, 266–9; Cloth Hall, 20, 218, 226, 233, 234, 264, 265, 268, 271, 272; devastation, 26, 108, 118, 129, 131, 135, 264–6, 291; Eton Memorial School, Ypres, 274, 276; Hell Fire Corner, 23, 92, 102, 103, 105, 213, 284; holy ground: 149–50, 266–7; hotels, 66–7, 69, 174, 183; Menin Gate, 276–283; Leo Murphy, 272–3; local people, 164, 269–71, 283; ramparts, 25, 128, 219, 277, 284, 287, 288, 289; reconstruction, 166, 225, 265–6, 284; St. George's Church, Ypres, 275–6; St Julien (Canadian) memorial, 163, 166, 246, 286; state of roads, 75–7, 80. *See also* Arras, destruction of; Lens, devastation

Ypres League, 47, 155, 195, 249, 278: Arras representative, 72, 208; comradeship, 150; costs, 187, 198, 250; guidebooks, 105; funding for pilgrimages, 185; hotels, 69; map, 80; meaning of pilgrimage, 179; pilgrimages, 82, 87, 92, 180, 183, 203, 205, 250, 252, 254, 255, 275; pilgrimage itineraries, 184, 185, 332; pilgrims, 186, 198; scale of pilgrimage activity, 181; signboards, 98, 99, 102, 103, 104, 105, 148, 155, 175; veterans, 159, 170, 171, 175, 187, 246; Ypres representatives and projects, 56, 274. *See also* Willson, Henry Beckles